Fluid Mechanics (Vol. 1)

Shiv Kumar

Fluid Mechanics (Vol. 1)

Basic Concepts and Principles

Fourth Edition

Ane Books Pvt. Ltd.

Shiv Kumar
New Delhi, India

ISBN 978-3-030-99764-9   ISBN 978-3-030-99762-5   (eBook)
https://doi.org/10.1007/978-3-030-99762-5

Jointly published with ANE Books Pvt. Ltd.
In addition to this printed edition, there is a local printed edition of this work available via Ane Books in South
Asia (India, Pakistan, Sri Lanka, Bangladesh, Nepal and Bhutan) and Africa (all countries in the African
subcontinent).
ISBN of the Co-Publisher's edition: 978-9-384-72699-7

This Springer imprint is published by the registered company Springer Nature Switzerland AG
The registered company address is: Gewerbestrasse 11, 6330 Cham, Switzerland

Dedicated
to

My Parents

My Wife Dr. Kusum and My Son Tanishq

Preface

This book has been written for the introductory course on Fluid Mechanics at the undergraduate level. This book fulfills the curriculum needs of UG students of Mechanical Engineering, Mechanical and Automation Engineering, Chemical Engineering, Electrical Engineering, Civil Engineering, Production Engineering, Automobile Engineering, aeronautical Engineering, Manufacturing Engineering, Tool Engineering and Mechatronics Engineering etc.Fluid Mechanics is dividing into two volumes. Fluid Mechanics Volume-I includes seven chapters: 1.Properties of Fluids, 2. Pressure and its Measurement, 3. Hydrostatic Forces on Surface, 4. Buoyancy and Floatation. 5. Kinematics of Fluid Motion, 6. Dynamics of Fluid Flow, 7. Dimensional and Model Analysis. Fluid Mechanics deals with the innovative use of the laws of Fluid Mechanics in solving the relevant technological problems. This introductory textbook aims to provide undergraduate engineering students with the knowledge (basics principles and fluid mechanics laws) they need to understand and analyze the fluid mechanics problems they are likely to encounter in practice.

The book is developed in the context of the author's simpler methodology to present even complex things. The most positive factor about the book is that it is concise, and everything is described from an elementary and tangible perspective.

The book presents the concepts in a very logical format with complete word descriptions. The subject matter is illustrated with a lot of examples. A great deal of attention is given to select the numerical problems and solving them. The theory and numerical problems at the end of each chapter also aim to enhance the creative capabilities of students. Ultimately as an introductory text for the undergraduate students, this book provides the background necessary for solving the complex problems in thermodynamics.

Writing this book made me think about a lot more than the material it covers. The methods I used in this book are primarily those that worked best for my students. The suggestions from the teachers and students for the further improvement of the text are welcome and will be implemented in the next edition. The readers are requested to bring out the error to the notice, which will be gratefully acknowledged.

Shiv Kumar

Acknowledgements

First of all, I would like to express my deep gratitude to God for giving me the strength and health for comleting this book. I am very thankful to my colleagues in the mechanical engineering department for their highly appreciable help and my students for their valuable suggestions.

I am also thankful to my publishers Shri Sunil Saxena and Shri Jai Raj Kapoor of Ane Books Pvt. Ltd. and the editorial group for their help and assistance.

A special thanks goes to my wife Dr. Kusum Lata for her help, support and strength to complete the book.

Shiv Kumar

Contents

7. Dimensional and Model Analysis

Properties of Fluids

1.1 INTRODUCTION

Matter exists basically in three states—solid, liquid, and gas. Fluid is the common name given to liquid and gas. The differences among solid, liquid and gas are given below:

S. No.	Solid	Liquid	Gas
1.	A given mass of solid has a definite shape and volume. **Fig. 1.1:** Definite shape and volume.	A given mass of liquid has a definite volume but the shape or size changes according to the shape of the container **Fig. 1.2:** Shape changes but volume remains same.	A given mass of gas has no fixed volume and shape. It expands continuously to fill the container in which it is placed. **Fig. 1.3:** Both shape and volume change
2.	The molecules of a solid are bonded together with high degree of force (cohesion) which give them a rigid and compact form.	The molecules of a liquid are bonded together with low degree of force (cohesion); therefore it can easily acquire the shape of the container in which it is placed.	The intermolecular attraction is practically absent, consequently the gas molecules have a greater freedom of movement even inside the container in which the gas is kept.
3.	It is incompressible in nature.	It is incompressible in nature.	It is compressible in nature.
4.	Silver, iron, stone *etc.* are solids.	Water, mercury, petrol, vegetable oil, *etc.* are liquids.	Air, carbon dioxide, carbon monoxide *etc.* are gases.

© The Author(s) 2023
S. Kumar, *Fluid Mechanics (Vol. 1)*,
https://doi.org/10.1007/978-3-030-99762-5_1

1.2 SOLID AND FLUID

1.2.1 Solid

A solid resists force which tends to deform flows upto elastic limit, it regains its shape and size when load is removed.

1.2.2 Fluid

A fluid is a substance that deforms (flows) continuously when an external shear force is applied on it. The continuous deformation of the fluid does not stop until the force is removed. In other words, fluid is a substance which offers no resistance to shear deformation. This continuous deformation of a substance is known as flow.

Fluids may be classified into two categories:

(i) Liquids   (ii) Gases.

The differences between liquids and gases are given below:

S. No.	Liquids	Gases
1.	Liquids are incompressible in nature.	Gases are compressible in nature.
2.	Liquids can have a free surface.	Gases do not have a free surface.
3.	A given mass of liquid occupies a definite volume of the container.	They fill the container fully regardless of their mass.
4.	Liquids have a high density. For example: Density of water at STP is 1000 kg/m^3 (approximately).	Gases have a very low density. For example: Density of air at STP is 1.22 kg/m^3.

1.3 STANDARD TEMPERATURE AND PRESSURE (STP) AND NORMAL TEMPERATURE AND PRESSURE (NTP)

(i) **Standard Temperature and Pressure (STP):** It refers to the conditions of standard atmospheric pressure of 760 mm of mercury (1.01325 bar) and a temperature of 15°C (or 288 K), i.e., the values of temperatures and pressure at STP are 15°C and 760 mm of Hg respectively.

(ii) **Normal Temperature and Pressure (NTP):** It refers to the conditions of atmospheric pressure of 760 mm of mercury (1.01325 bar) and a temperature of 0°C (or 273 K), i.e., the values of temperature and pressure at NTP are 0°C and 760 mm of Hg.

1.4 MECHANICS OF FLUIDS

Mechanics of fluids or Fluid Mechanics is the branch of science that deals with the behaviour of fluid at rest or in motion. This subject gives a detailed study of the kinds of energies present in fluids (i.e., pressure energy, kinetic energy, potential energy etc.), how one form of energy is transferred to another form by means of nozzle or diffuser and nature of fluid flow (i.e., flow either laminar or turbulent) etc.

Fluid Mechanics may be classified into three categories:

(*i*) **Fluid statics:** The study of fluids at rest is called fluid statics.

(*ii*) **Fluid kinematics:** The study of fluids in motion, without considering the pressure forces or causes of motion.

(*iii*) **Fluid dynamics:** The study of fluids in motion, considering the pressure forces or causes of motion.

1.5 PROPERTIES OF FLUIDS

In order to study the behaviour of a fluid, some important properties of fluids are listed below:

1. Density	2. Specific volume
3. Specific weight	4. Specific gravity
5. Adhesion	6. Cohesion
7. Viscosity	8. Vapour pressure
9. Surface tension	10. Capillarity, and
11. Compressibility.	

1.6 DENSITY

The density of a fluid is defined as the ratio of the mass of a fluid to its volume. It is denoted by ρ (rho).

Mathematically,

$$\text{Density: } \rho = \frac{\text{Mass of fluid} : M}{\text{Volume of fluid} : V}$$

$$\rho = \frac{M}{V}$$

SI unit of density: ρ is kg/m^3

The density (ρ) is also known as specific mass or mass density.

Must Remember		
Density of water	:	$\rho = 1000$ kg/m^3 at temperature 4°C
Density of mercury	:	$\rho = 13600$ kg/m^3 at NTP
Density of air	:	$\rho = 1.29$ kg/m^3 at NTP

1.7 SPECIFIC VOLUME

It is defined as volume per unit mass of a fluid. It is denoted by v.

Mathematically,

$$\text{Specific volumec: } v = \frac{\text{Volume of fluid} : V}{\text{Mass of fluid} : M}$$

$$v = \frac{V}{M} = \frac{1}{M/V}$$

$$v = \frac{1}{\rho}$$

It is the reciprocal of density.

SI unit of specific volume: $v = \dfrac{1}{\rho} = \dfrac{1}{kg/m^3} = \dfrac{m^3}{kg}$

It is commonly applied in gases.

1.8 SPECIFIC WEIGHT

It is defined as weight per unit volume of a fluid. It is denoted by w.

Mathematically,

$$\text{Specific weight: } w = \frac{\text{Weight of fluid}: W}{\text{Volume of fluid}: V}$$

$$w = \frac{W}{V}$$

$$= \frac{Mg}{V} \qquad\qquad \because W = Mg$$

$$w = \rho g \qquad\qquad \because \frac{M}{V} = \rho$$

It is also defined as the product of the density of a fluid (ρ) and acceleration due to gravity (g).

SI unit of specific weight:

$$w = \rho g$$

$$= \frac{kg}{m^3} \times \frac{m}{s^2}$$

$$= \frac{N}{m^3} \qquad\qquad \because \frac{kg\ m}{s^2} = N$$

Specific weight is also known as weight density.

1.9 SPECIFIC GRAVITY

It is defined as the ratio of specific weight of a given fluid to the specific weight of a standard fluid. It is denoted by S.

We know, fluids are classified in two groups:

(*i*) Liquids   (*ii*) Gases

1.9.1 Specific Gravity for Liquids

It is defined as the ratio of the specific weight of a given liquid to the specific weight of water. It is denoted by S_l.

$$S_l = \frac{w}{w_{water}} = \frac{\rho g}{\rho_{water} \cdot g} = \frac{\rho}{\rho_{water}}$$

Specific gravity for liquids is also defined as the ratio of the density of a given liquid to the density of the water. It is also called relative density. Note that the specific gravity is a ratio, it has no dimension *i.e.*, dimensionless. A hydrometer is used to measure the specific gravity of liquid

Density of water at 4°C temperature is 1000 kg/m³

i.e., ρ_{water} = 1000 kg/m³ at 4°C

∴   The density of given liquid: $\rho = S_l \times \rho_{water}$ = 1000 S_l kg/m³

1.9.2 Specific Gravity for Gases

It is defined as the ratio of the specific weight of a given gas to the specific weight of air at NTP. It is denoted by S_g.

Mathematically,

Specific gravity for gases: $S_g = \dfrac{\text{Specific weight of given gas}}{\text{Specific weight of air at NTP}}$

$$= \frac{w}{w_{air}} = \frac{\rho g}{\rho_{air} \cdot g} = \frac{\rho}{\rho_{air}}$$

Specific gravity for gases is also defined as the ratio of the density of a given gas to the density of air at NTP.

Density of air at NTP is 1.29 kg/m³.

i.e., ρ_{air} = 1.29 kg/m³.

∴   The density of given gas: $\rho = S_g \times \rho_{air}$

$$\rho = 1.29\ S_g \text{ kg/m}^3$$

Must Remember

Specific gravity of water	:	$S = 1$
Specific gravity of mercury	:	$S = 13.6$
Specific gravity of air	:	$S = 1$

1.10 ADHESION

The property of a liquid which enables the molecules of a liquid to adhere (stick) the molecules of a solid boundary surface with which it comes in contact is called adhesion. Due to adhesion liquid wets the surface of the container which carries the liquid. For instance, when a dry rod or finger is dipped in water or milk, it becomes wet. It shows that water or milk has strong property of adhesion. But when a dry rod or finger is dipped in mercury, it remains dry, which means mercury does not show adhesion towards rod or finger.

1.11 COHESION

The property of a liquid by which the molecules of the same liquid attract each other is called cohesion. In the other words, it is defined as the intermolecular attraction between the molecules of same liquid. For instance, mercury has high cohesive property.

1.12 VISCOSITY

Viscosity is defined as the property of a fluid which offers resistance to fluid flow under the influence of a shear force. As long as the shear force exists in the fluid it undergoes a continuous deformation. The rate of deformation is dependent upon the magnitude of the shear force. In other words, viscosity is the property which controls the flow of liquid on a horizontal surface. For instance. Water and alcohol are less viscous liquids so they spread easily and quickly on flat horizontal surface. Honey and glycerine are more viscous liquids so they do not spread on flat horizontal surface easily so viscosity controls the flow of liquid.

The nature of viscosity of a fluid can be understood most easily by the following experiment: Consider the flow between two long parallel plates, one fixed and the other moving at constant velocity U. The distance between the plates is h as shown in Fig. 1.4. To maintain velocity U of upper plate, a tangential or shearing force in the direction of motion must act on the upper plate.

Now if we consider two layers of fluid ab and cd, the distance between two layers is dy, layer ab moving with velocity u and upper layer cd moving with velocity $u + du$, the viscosity and relative velocity causes a shear stress acting between the fluid layers. The upper layer (cd) causes a shear stress on the adjacent lower layer (ab) while the lower layer causes a shear stress on the adjacent upper layer. This shear stress is directly proportional to the rate of change of velocity with respect to y (velocity gradient). It is denoted by τ (tau).

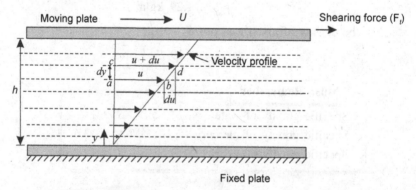

Fig. 1.4: Velocity distribution of a viscous fluid between two parallel plates, one is fixed and other is moving.

Mathematically,

Shear stress: $\tau \propto$ Velocity gradient $\left(\dfrac{du}{dy}\right)$

or $\qquad \tau \propto \dfrac{du}{dy}$

$$\tau = \mu \frac{du}{dy} \qquad \qquad ... \ (i)$$

where μ (mu) is the constant of proportionality and is known as the coefficient of viscosity or dynamic viscosity (because it involves dimension of force). It is also known as the 'absolute viscosity' or simply the 'viscosity' of the fluid.

Equation (i) is known as **Newton's law of viscosity**. This law states that the shear stress (τ) on a fluid element layer is directly proportional to the velocity gradient. The constant of proportionality is called viscosity (μ).

From Eq. (i), we have

$$\text{Viscosity: } \mu = \frac{\tau}{du/dy}$$

($\frac{du}{dy}$ is also known as **rate of shear strain or angular deformation**)

Viscosity may be defined as the shear stress (τ) required to produce unit velocity gradient.

Units of viscosity:

In SI unit: $\quad \mu = \dfrac{\tau}{du/dy} = \dfrac{N/m^2}{\dfrac{m/s}{m}} = Ns/m^2 = kg/ms \quad (\because \ 1 \ N = 1 \ kgm/s^2)$

In MKS unit: $kgf.s/m^2$

In CGS unit: poise or $\dfrac{dyne–s}{cm^2}$

$$1 \text{ poise} = \frac{dyne–s}{cm^2} = \frac{1}{10} \text{ Pa.s (pascal second)} \ (\because \ 1 \text{ pascal, Pa} = \frac{N}{m^2})$$

$$1 \text{ centipoise (cP)} = 10^{-2} \text{ poise} = 10^{-3} \ Ns/m^2 = 10^{-3} \text{ Pa.s}$$

Viscosity of water and air at 20°C and at atmospheric pressure are:

$$\mu = 1 \ cP = 10^{-3} \ Ns/m^2 \quad \text{for water}$$
$$= 0.0181 \ cP = 0.0181 \times 10^{-3} \ Ns/m^2 \quad \text{for air}$$

(Water is nearly 55 times more viscous than air)

1.12.1 Kinematic Viscosity

Kinematic viscosity is defined as the ratio of dynamic viscosity (μ) to density (ρ) of fluid. It is denoted by v (called nu).

$$\text{Kinematic viscosity: } v = \frac{\text{Dynamic viscosity } (\mu)}{\text{Density of fluid } (\rho)}$$

$$v = \frac{\mu}{\rho}$$

In SI unit: $\quad v = \dfrac{\mu}{\rho} = \dfrac{kg/ms}{kg/m^3} = \dfrac{m^2}{s}$

In CGS unit: cm^2/s or stoke

$$1 \text{ stoke} = 1 \text{ cm}^2/s \qquad\qquad (\because 1 \text{ stoke} = 1 \text{ cm}^2/s)$$
$$1 \text{ stoke} = 10^{-4} \text{ m}^2/s$$

$$1 \text{ centistoke or } cSt = \frac{1}{100} \text{ stoke} = 10^{-2} \text{ stoke} = 10^{-6} \text{ m}^2/s$$

It is known as kinematic viscosity because it can be defined dimensionally by only length and time dimensions mass or force dimensions being not involved.

Table 1.1: Values of Dynamic Viscosity and Kinematic Viscosity for Some Common Fluids at 20°C and 1.01325 bar.

Fluids	Dynamic Viscosity Ns/m^2	Kinematic Viscosity m^2/s
Liquid		
Water	1×10^{-3}	1×10^{-6}
Seawater	1.07×10^{-3}	1.04×10^{-6}
Petrol	2.92×10^{-4}	4.29×10^{-7}
Kerosene	1.92×10^{-3}	2.39×10^{-4}
Glycerin	1.49×10^{-3}	1.18×10^{-3}
Mercury	1.56×10^{-3}	1.15×10^{-7}
Castor Oil	9.80×10^{-1}	1.02×10^{-3}
Gases		
Air	1.80×10^{-5}	1.49×10^{-5}
Carbon Dioxide	1.48×10^{-5}	0.80×10^{-5}
Hydrogen	0.90×10^{-5}	10.71×10^{-5}
Nitrogen	1.76×10^{-5}	1.52×10^{-5}
Oxygen	2×10^{-5}	1.50×10^{-5}
Water Vapor	1.01×10^{-5}	1.352×10^{-5}

1.12.2 Effect of Temperature on Viscosity

The viscosity of a fluid is due to two causes:

(*i*) Due to intermolecular cohesion and

(*ii*) Due to transfer of molecular momentum.

In case of gases the interspace between the molecules is large and so the intermolecular cohesion is negligible. But in the case of liquids the molecules are very close to each other and therefore more cohesion exists. Hence in liquids, the viscosity is mainly due to intermolecular cohesion, while in gases, viscosity is mainly due to molecular momentum exchange. The intermolecular cohesive forces decrease with the increase in temperature and hence the viscosity of liquid decreases. The following formula shows the dependence of the viscosity of a liquid on temperature.

$$\mu = \mu_0 \left(\frac{1}{1 + at + bt^2} \right)$$

where μ = viscosity of the liquid at $t°C$ in poise

μ_0 = viscosity of the liquid at 0°C in poise a, b are constant
for the liquid

For water, $\mu_0 = 1.79 \times 10^{-2}$ poise

$a = 0.03368$

$b = 0.000221$

In the case of gases, we know the intermolecular cohesion being negligible, the viscosity depends mainly on transfer of molecular momentum in a direction normal to the flow. As the temp. increases, molecular momentum transfer increases and hence viscosity increases.

Holman gave the following expression for the viscosity of a gas:

$$\mu = \mu_0 + at^2 - bt^2$$

where μ = viscosity of gas at $t°C$ in poise

μ_0 = viscosity of the gas at 0°C in poise a, b are constant for
the gas.

For air, $\mu_0 = 1.72 \times 10^{-4}$ poise

$a = 4.716 \times 10^{-7}$ poise

$b = 5.84 \times 10^{-11}$.

1.12.4 Effect of Pressure on Viscosity

For most liquids, viscosity increases with increasing pressure (except water) because the amount of free volume in the internal structure decreases due to compression. Consequently, the molecules can move less freely and the internal friction force increase.

According to Barus, the relationship between the viscosity and pressure is given as follows:

$$\mu = \mu_0 e^{\alpha p}$$

where μ = Dynamic viscosity at the pressure 'p' (Pa.s)

μ_0 = Dynamic viscosity at the atmospheric pressure (Pa.s)

α = Viscosity-pressure constant coefficient (mm^2/N)

p = Pressure (N/mm^2)

where α is multiplied by pressure, p, with units N/mm^2, in the Barus equation, the units cancel giving the dimensionless exponent of the natural number, e. Thus, viscosity increases exponentially with pressure.

The viscosities of water and ammonia at pressure values ranging from 1MPa to 20 MPa is shown in Table 1.2. It can be seen that the viscosity of water falls with increasing pressure, while the viscosity of liquid ammonia rises.

Table 1.2: Viscosities of Water and Ammonia are varied
with increasing pressure

S.No.	Pressure (MPa)	Dynamics Viscosity (µPa.s.)	
		Water	Ammonia
1.	1	889.87	131.68
2.	2	889.64	132.50
3.	3	889.42	133.30
4.	4	889.20	134.11
5.	5	888.99	134.90
6.	6	888.78	136.69
7.	7	888.58	136.48
8.	8	888.38	137.26
9.	9	888.19	138.03
10.	10	888.00	138.80
11.	11	887.82	139.57
12.	12	887.64	140.33
13.	13	887.47	141.08
14.	14	887.31	141.84
15.	15	887.14	142.58
16.	16	886.99	143.32
17.	17	886.84	144.06
18.	18	886.69	144.80
19.	19	886.55	145.53
20.	20	886.41	146.26

1.12.4 Types of Fluids

The fluids may be classified into following types:

(i) Ideal fluid (ii) Real fluid
(iii) Newtonian fluid (iv) Non-Newtonian fluid
(v) Ideal plastic fluid (vi) Thyxotropic fluid.

(i) **Ideal Fluid:** A fluid which is inviscid (no viscosity), no surface tension and incompressible, is known as an ideal fluid or perfect fluid. In nature, no fluid is having such properties thus it is an imaginary fluid. But there are a few liquids which can be considered ideal for all practical purposes like water which has low viscosity, low surface tension and great resistance to compression. Thus water can be considered an ideal fluid for all practical purposes without incurring much appreciable error in arriving at the result.

(*ii*) **Real Fluid:** A fluid which possesses properties such as viscosity, surface tension and compressibility, is known as real fluid or practical fluid. All fluids available in nature are real fluids.

(*iii*) **Newtonian Fluid:** A real fluid which obeys the Newton's law of viscosity [shear stress (τ) is directly proportional to velocity gradient $\left(\dfrac{du}{dy}\right)$], is known as Newtonian fluid. For example; water, air, glycerine, kerosene *etc.* Thus, in Newtonian fluids:

Shear stress: $\tau = \mu \dfrac{du}{dy}$

where $\dfrac{du}{dy}$ = velocity gradient or shear strain.

μ = viscosity of fluid.

(*iv*) **Non-Newtonian Fluid:** A real fluid which does not obey the Newton's law of viscosity, is known as non-Newtonian fluid. Blood, grease and sugar solutions are some common non-Newtonian fluids. In non-Newtonian fluid:

Shear stress: $\tau \neq \mu \dfrac{du}{dy}$.

A general relationship between shear and stress and velocity gradient for non-Newtonian fluid may be written as

$$\mu = A\left(\frac{du}{dy}\right)^{n} + B$$

where A and B are constant which depend upon the type of fluid and condition imposed on the flow.

For Newtonian fluid, power index; $n = 1$, $B = 0$ and the constant A varying only with the type of fluid.

(*v*) **Dilatant Fluids:** Dilatant Fluids exhibit an increase in viscosity with increasing rate of shear deformation.

In these fluids, $n > 1$, $B = 0$

Examples: Butter, Sugar solution, wet beach sand, starch in water etc.

(vi) **Bingham Plastics:** They resist a small shear stress but flow very easily under large shear stress.

In these fluids, $n = 1$, $B \neq 0$

Examples: Seewage sludge, drilling muds, gel, toothpaste etc.

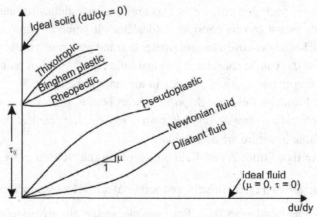

Fig. 1.5: Types of fluids.

(vii) **Pseudoplastic Fluids:** Pseudo plastic fluids exhibit a viscosity which decreases with increasing velocity gradient.

In these fluids, $n < 1$, $B = 0$

Examples: Paper pulp, Polymer solutions, blood, milk etc.

(viii) **Thixotropic fluids:** Thixotropic fluids exhibit a decrease in viscosity with time. Common example is honey, if you keep stirring solid honey will become liquid.

In these fluids, $n < 1$, $B \neq 0$

Examples: Crude Oils, Yoghurt, Honey, Paints, Synovial Fluid (present in the knee joints).

(ix) **Rheopectic Fluids:** The viscosity increases with time for which shearing forces are applied.

In these fluids, $n > 1$, $B \neq 0$

Examples: Gypsum suspension, bentonite clay suspension etc.

1.13 VAPOUR PRESSURE

All liquids possess a tendency to evaporise when exposed to air or gaseous atmosphere. The evaporisation takes place due to liquid molecules escaping from the liquid free surface. The rate of evaporisation depends upon:

(*i*) **Nature of Liquid:** Low or high volatile, petrol is highly volatile and diesel is low volatile liquid.

(*ii*) **Temperature of Liquid:** Evaporisation increases with rise in temperature.

(*iii*) Condition of the atmosphere adjoining it.

Consider a liquid in a sealed container and maintained at a constant temperature within the container (Fig. 1.6). The temperature of both the liquid and its surrounding atmosphere is same. By nature of liquid, the molecules of liquid start evaporizing and leaving the liquid surface. After a certain time, the equilibrium condition will be reached, *i.e.,* the rate at which the molecules are leaving the liquid surface will be the same as the rate of return of molecules *i.e.,* rate of evaporisation and condensation is same. In this condition the air above the liquid is saturated with liquid vapour molecules. **The partial pressure of water vapour of saturated air exerted on the liquid surface is called vapour pressure.** It is denoted by p_v. Vapour pressure increases with increase in temperature (Table 1.3).

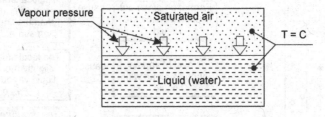

Fig. 1.6: Vapour Pressure.

Table 1.3: Vapour Pressure of Water

Temperature °C	Vapour Pressure (p_v) m of water
0	0.63
10	0.125
15	0.165
20	0.239
30	0.437
40	0.762
50	1.275
60	2.075
80	4.960
100	10.790

1.13.1 Evaporisation and Boiling

The phenomenon of vaporisation and boiling are differentiated as follows:

(*i*) **Evaporisation:** It is defined as rate of liquid leaving from its free surface when the vapour pressure of liquid is less than surrounding pressure (pressure above a liquid surface) at given temperature.

Condition for evaporisation:

$$p_v < p \quad \text{at } T = C$$

(*ii*) **Boiling:** Boiling is high rate of evaporisation, when the vapour pressure of liquid is greater than or equal to the surrounding pressure (pressure above a liquid surface) at given temperature.

Condition for boiling:

$$p_v \geq p \quad \text{at } T = C$$

1.13.2 Cavitation

The mechanism of cavitation is defined as the following steps:

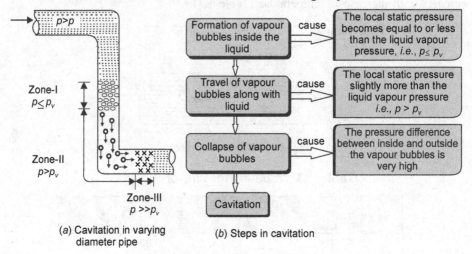

(a) Cavitation in varying   (b) Steps in cavitation
    diameter pipe

Fig. 1.7: Mechanism of cavitation.

 (*a*) Formation of vapour bubbles (Zone-I)

 (*b*) Travel of vapour bubbles (Zone-II)

 (*c*) Collapse of vapour bubbles (Zone-III)

 In order to easily understand the phenomenon of cavitation we consider a liquid passing through the varying diameter pipe.

 (*a*) **Formation of Vapour Bubbles (Zone-I):** If the local static pressure at any point in a conduit becomes less than or equal to the vapour pressure of liquid at same operating temperature, boiling starts and vapour bubbles form inside the liquid. This initiates the cavitation. The formation of vapour bubbles inside the liquid is depicted as Zone-I in Fig. 1.7 (*a*).

 (*b*) **Travel of Vapour Bubbles (Zone-II):** After the formation of vapour bubbles, these bubbles travel along with the liquid into regions of higher pressure. The path covered by vapour bubbles is shown as Zone-II in Fig. 1.7(*a*).

 (*c*) **Collapse of Vapour Bubbles (Zone-III):** When the vapour bubbles reach the region of high pressure, the pressure outside the bubbles is greater than the pressure inside the bubbles. Therefore, bubbles collapse and create vacuum, the surrounding higher pressure rush to the region of vacuum with very high velocity, which develop very high pressure wave (100 times of atmospheric pressure). This high pressure wave will act as a hammering action on the metal. This hammering action is known as **Cavitation.**

1.14 SURFACE TENSION

At surface of contact between a gas and a liquid or between two immiscible liquids, tensile force (*i.e.*, tension) acts on the surface due to unbalanced molecular forces acting downwards. This tensile force is known as surface tension. It is denoted by σ (sigma). Surface tension is the property of the liquid surface.

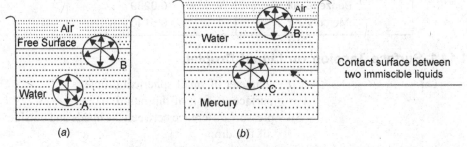

Fig. 1.8: Surface tension

Surface tension is expressed as the force acting normal per unit length of the free surface or contact surface between two immiscible liquids.

Mathematically,

$$\text{Surface tension: } \sigma = \frac{\text{Unbalanced molecular force act normal to the free surface}}{\text{Length of the free surface}}$$

SI unit of σ is N/m

Surface tension is also defined as surface energy per unit area.

$$\text{Surface tension: } \sigma = \frac{\text{Surface energy}}{\text{Surface area}} = \frac{Nm}{m^2} = N/m$$

Due to surface tension, free surface of liquid behaves like a thin membrane. This enables the free surface to support very small load placed on it. For instance, a small needle gently placed on the liquid surface will not sink due to surface tension.

To understand more about surface tension, let us consider equilibrium of force within a liquid. We consider a molecule *A* inside the liquid as shown in Fig. 1.8 (*a*). This molecule *A* is attracted by cohesive forces on all sides by the molecules of same liquid. Thus resultant force acting on molecule *A* is zero. On the other hand, a molecule at the surface of liquid at point *B* does not have any liquid molecule above it and therefore, there is a net downward force on the molecule *B* due to the attraction on molecules below it. (*i.e.*, cohesive force). This downward force will act normal to the free surface and create tensile force (*i.e.*, tension) on the surface. This downward unbalanced force rises to surface tension. Similar reason in case of the contact surface between two immisable liquids as shown in Fig. 1.8 (*b*).

Effect of Temperature: Surface tension depends directly upon the intermolecular cohesion since this cohesion decreases with rise in temperature, thus surface tension decreases with temperature rise.

The surface tension of some liquids at 20°C and at atmospheric pressure are given in the Table 1.4:

Table 1.4

Liquid in Contact with Air	Surface Tension (σ): N/m
Water–air	0.0728
Kerosene–air	0.0277
Glycerine–air	0.0633
Benzene–air	0.0289
Mercury–air	0.5140

1.14.1 Surface Tension on Liquid Droplet

Let d = diameter of small spherical droplet.

σ = surface tension of the liquid.

Δp = pressure difference between the inside and outside of the drop

Let the droplet in Fig. 1.9 (a) is cut into two halves. The surface tension and excess pressure acting on left half portion is shown in Fig. 1.9 (b) and Fig. 1.9 (c)

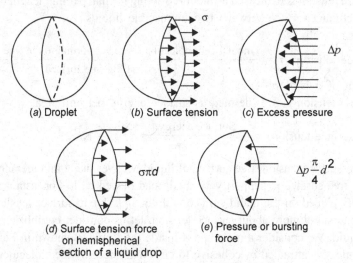

(a) Droplet   (b) Surface tension   (c) Excess pressure

(d) Surface tension force
on hemispherical
section of a liquid drop

(e) Pressure or bursting
force

Fig. 1.9: Surface tension on liquid droplet.

Considering the equilibrium of one half of the droplet, from Figs. 1.9 (d) and 1.9 (e);

Bursting force = Surface tension force

$$\Delta p. \frac{\pi}{4} d^2 = \text{surface tension} \times \text{circumference}$$

$$= \sigma.\pi d$$

$$\Delta p = \frac{\sigma.\pi d}{\frac{\pi}{4} d^2} = \frac{4\sigma}{d}$$

$$\Delta p = \frac{4\sigma}{d} \qquad \qquad \ldots (i)$$

Above Eq. (i), shows that with the decrease of diameter of the droplet, pressure intensity inside the droplet increases.

1.14.2 Surface Tension on a Hollow Bubble

A hollow bubble like a soap bubble in air has two surfaces in contact with air, one inside and other the outside, as shown in Fig. 1.10. The surfaces that act on the hemispherical section are same as that of the droplet, but two surfaces of bubble are subjected to surface tension. At equilibrium condition in this case, we have

$$\Delta p \frac{\pi}{4} d^2 = 2 \cdot \sigma \pi d$$

$$\therefore \qquad \Delta p = \frac{8\sigma}{d} \qquad \qquad \ldots (ii)$$

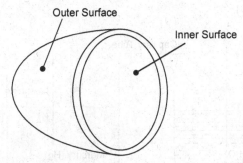

Outer Surface

Inner Surface

Fig. 1.10: Two surfaces of a soap bubble or hollow bubble.

The above Eq. (ii), shows that a soap solution has a high value of surface tension (σ), which causes a soap bubble to be large in diameter for small pressure of blowing.

1.14.3 Surface Tension on a Liquid Jet

Consider a liquid jet of diameter d and length l as shown in Fig. 1.11.

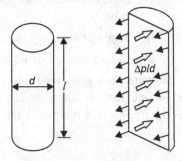

d l

$\Delta p l d$

Fig. 1.11: Force on liquid jet.

Let Δp = pressure difference between the inside and outside
                                         of the liquid jet.

                                    σ = surface tension of the liquid.

Consider the equilibrium of the semi jet,

Force due to pressure, Δp = Force due to surface tension

$$\Delta p.\ ld = \sigma \times 2l$$

$$\Delta p = \frac{2\sigma}{d}$$

where d is diameter of jet.

1.15 CAPILLARITY

The phenomenon of rise and fall of liquid in a capillary tube is known as capillarity.
The rise of water in capillary tube is called capillary rise in water and the fall of
mercury in capillary tube is called capillary fall or depression in mercury as shown in
Figs. 1.12 (*a*) and (*b*).

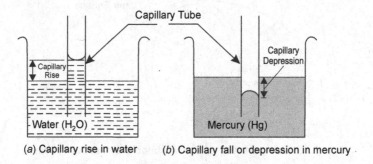

(a) Capillary rise in water   (b) Capillary fall or depression in mercury

Fig. 1.12: Capillarity.

The capillary action (*i.e.*, liquid rise or fall in capillary tube) is due to the cohesion
and adhesion (*i.e.*, surface tension).

Capillary rise in water (Fig. 1.13):

Let θ = angle of contact between liquid surface and glass
                                         tube.

                                    h = height of capillary rise.

                                    σ = surface tension of liquid.

We may assume that the water in a glass tube will continue to rise until the
vertical component of the surface tension force ($\sigma \cos \theta$) which acts over the
circumference of the tube at the free surface is equal to the weight of water column.

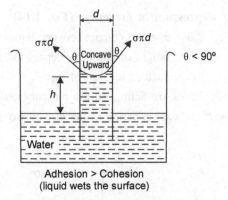

Fig. 1.13: Capillary rise.

For equilibrium condition,

Upward surface tension force= Weight of the water column in the glass tube

$$\pi d.\ \sigma\ \cos\ \theta = mg$$

$$\pi d.\ \sigma\ \cos\ \theta = \rho V.\ g \qquad\qquad (\because \text{Mass: } m = \rho V)$$

$$\pi d.\ \sigma\ \cos\ \theta = \rho \frac{\pi}{4} d^2.h.g$$

$$h = \frac{4\sigma\cos\theta}{\rho g.d}$$

$$h = \frac{4\sigma\cos\theta}{wd} \qquad\qquad\qquad ... \ (i)$$

where w is specific weight of the liquid which is equal to ρg.

For pure water and clean glass tube surface, take the value of $\theta \approx 0°$

Then, $$h = \frac{4\sigma}{wd}$$

But in actual, the water is neither pure nor the glass surface clean.

For water in contact with glass surface and air, Gibson has obtained the value $\theta = 25.53°$ and $\sigma = 0.0735$ N/m.

Substituting the value of θ, σ and $w\ (= \rho g)$ in Eq. (i), we get

$$h = \frac{4 \times 0.0735 \times \cos 25.53°}{1000 \times 9.81 \times d} \qquad\qquad \text{for water}$$

$$h = \frac{2.7043 \times 10^{-5}}{d}\ \text{m} \qquad\qquad (ii)$$

In above Eq. (ii), units of the both h and d in metre (m)

$$h = \frac{27.043}{d}\ \text{mm} \qquad\qquad (iii)$$

From above Eq. (iii), we get the value of h in mm. If substituting the value of d in mm.

Capillary fall or depression in mercury: (Fig. 1.14)

Let $\qquad\qquad\theta$ = angle of contact between liquid surface and glass tube

h = height of capillary depression in glass tube

σ = surface tension of liquid.

In equilibrium, two forces are acting on the mercury inside the tube.

Ist one is due to surface tension acting in the downward direction and is equal to $\pi d.\sigma\cos\theta$.

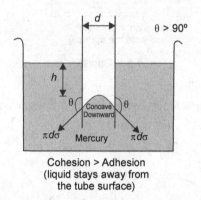

Cohesion > Adhesion
(liquid stays away from
the tube surface)

Fig. 1.14: Capillary fall in mercury.

2nd force is due to hydrostatic force acting upward and is equal to intensity of pressure at a depth h × area

$$= p\frac{\pi}{4}d^2 \ = \ wh.\frac{\pi}{4}d^2 \qquad\qquad (\because p = \rho gh = wh)$$

Equating the two forces, we get

$$\pi d\sigma \, \cos\theta = \ wh.\frac{\pi}{4}d^2$$

$\therefore\qquad\qquad$ $$h = \frac{4\sigma\cos\theta}{wd} \qquad\qquad\qquad (iv)$$

The expression is same as capillary rise in water.

For mercury in contact with glass surface and air, Gibson has obtained the value of θ = 128.37° and σ = 0.53 N/m

Substituting the value of θ, σ and $w(= \rho g)$ in Eq. (iv), we get

$$h = \frac{4\times0.53\times\cos 128.37°}{13600\times9.81\times d}$$

$$(\because \rho_{Hg} = 13600 \text{ kg/m}^3, \ g = 9.81 \text{ m/s}^2, \ \cos 128.37° = -0.6207)$$

$$h = -\frac{9.86\times10^{-6}}{d}\text{m} \qquad\qquad\qquad (v)$$

In above Eq. (v), units of the both h and d in metre (m), –ve sign shows mercury fall in glass tube.

and $\qquad\qquad\qquad h = -\dfrac{9.86}{d}$ $\qquad\qquad\qquad\qquad\qquad$ (vi)

From Eq. (vi), we get the value of d in mm.

Table 1.5 gives the value of h for glass tube of different diameter for water and mercury as calculation from the above Eq. (iii) and (vi) respectively:

Table 1.5: Capillary Rise of Water and Fall of Mercury in Glass Tube

Diameter of Glass Tube, d in mm	Capillary Rise of Water, h in mm	Capillary Fall of Mercury, h in mm
2	13.52	4.93
5	5.40	1.97
10	2.70	0.98
15	1.80	0.65
20	1.35	0.49
25	1.08	0.39

1.16 COMPRESSIBILITY

The compressibility is the measure of change of volume or density of fluid with respect to change in applied pressure. It is measured by the following two parameters:

(i) Bulk modulus of elasticity (K)

(ii) Coefficient of compressibility (β)

(i) **Bulk modulus of elasticity (K):** It is defined as the ratio of the change in pressure to corresponding volumetric strain.

Consider a cylinder with a piston containing gas as shown in Fig. 1.15.

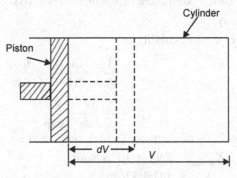

Fig. 1.15: Cylinder containing a gas.

Let $\qquad\qquad\qquad$ V = volume of a gas enclosed in the cylinder

$\qquad\qquad\qquad\qquad$ p = pressure of the gas when volume is V

Let the piston be pushed so as to increase the pressure p + dp and the corresponding decrease in the volume of the gas be dV.

$\therefore \qquad$ Volumetric strain $= \dfrac{\text{Change in volume}}{\text{Original volume}} = \dfrac{dV}{V}$

$\therefore$ Bulk modulus of elasticity:

$$K = \dfrac{\text{Increase in pressue}}{\text{Volumetric strain}} = \dfrac{dp}{-\dfrac{dV}{V}} \qquad \dots (i)$$

$$\boldsymbol{K = -\dfrac{V\,dp}{dV}}$$

The $-ve$ sign indicates that there will be decrease in volume due to increase in pressure.

We know,

$$\text{Mass: } m = \text{density} \times \text{volume}$$
$$m = \rho V$$

Taking $\log_e$ both sides, we get

$$\log_e m = \log_e \rho + \log_e V$$

Since the mass of a certain volume is constant, differentiating

$$0 = \dfrac{d\rho}{\rho} + \dfrac{dV}{V}$$

or

$$-\dfrac{dV}{V} = \dfrac{d\rho}{\rho}$$

Substituting the value of $-\dfrac{dV}{V}$ in Eq. (i), we get

$$K = \dfrac{dp}{\dfrac{d\rho}{\rho}}$$

$$\boldsymbol{K = \dfrac{\rho\,dp}{d\rho}}$$

(ii) **Coefficient of compressibility (β):** The reciprocal of the bulk modulus of elasticity is called coefficient of compressibility (β).

Mathematically,

Coefficient of compressibility:

$$\beta = \dfrac{1}{K} = \dfrac{-dV/V}{dp} = \dfrac{d\rho/\rho}{dp}$$

$$\boldsymbol{\beta = -\dfrac{dV}{V dp} = \dfrac{d\rho}{\rho dp}}$$

The bulk modulus of elasticity (K) for water and air at NTP are:

$$K_{\text{water}} = 2.94 \times 10^6 \text{ kN/m}^2$$
$$K_{\text{air}} = 101.3 \text{ kN/m}^2$$

It means that air is 20,000 times more compressible than water.

Relation between pressure (p) and bulk modulus of elasticity (K)

The relation between p and K for a gas in two different processes of compression are as:

(*i*) Isothermal process.

(*ii*) Adiabatic process.

(*i*) **Isothermal Process:** In this process, the specific volume of a given mass of a perfect gas varies inversely with the absolute pressure when the temperature is kept constant.

Mathematically,

$$v \propto \frac{1}{\rho} \quad \text{at } T = C$$

$$pv = \text{constant} = C$$

Taking $\log_e$ both sides, we get

$$\log_e p + \log_e v = \log_e C$$

Differentiating the above equation, we get

$$\frac{dp}{p} + \frac{dv}{v} = 0$$

or

$$\frac{dp}{p} = -\frac{dv}{v}$$

or

$$p = \frac{-dp}{dv/v}$$

or

$$\frac{-dp}{dv/v} = p$$

Substituting the value of $\dfrac{-dp}{dv/v}$ in Eq. (*i*), we get

$$K = p$$

(*ii*) **Adiabatic Process:** If the process is such that no heat can be added to or withdrawn from the system, it is said to be an adiabatic process. In this process, there is no generation of heat within the system even due to friction. For adiabatic process the relationship between pressure and density is given by

$$\frac{p}{\rho^\gamma} = C$$

where γ = the ratio of specific heats at constant pressure and constant volume $\left(i.e., \dfrac{c_p}{c_v} \right)$. It is also called adiabatic index.

$$pv^\gamma = C \qquad\qquad \because \frac{1}{\rho} = v$$

Taking $\log_e$ both sides, we get

$$\log_e p + \gamma \log_e v = \log_e C$$

Differentiating the above equation, we get

$$\frac{dp}{p} + \gamma \frac{dv}{v} = 0$$

$$\frac{dp}{p} = -\gamma \frac{dv}{v}$$

or

$$\frac{-dp}{dv/v} = \gamma p$$

Substituting the value of $\dfrac{-dp}{dv/v}$ in Eq. (i), we get

$$K = \gamma p$$

The value of adiabatic index: $\gamma = 1.4$ for air

Hence, $K = p$ for isothermal process

and $K = \gamma p$ for adiabatic process

Problem 1.1:   A certain mass of a liquid has a volume of 5 m³ and a weight of 39240 N. Find the specific weight, specific mass and specific gravity of the liquid.

Solution: Given data:

Volume of liquid: $V = 5$ m³

Weight of liquid: $W = 39240$ N

(i)   Specific weight: $w = \dfrac{\text{Weight of liquid} : W}{\text{Volume of liquid} : V}$

$$= \frac{39240}{5} \text{ N/m}^3 = \textbf{7848 N/m}^3$$

(ii)   Specific mass: $\rho = \dfrac{\text{Mass of liquid} : M}{\text{Volume of liquid} : V}$

$$= \frac{W/g}{V} \qquad (\because W = Mg \text{ or } M = W/g)$$

$$= \frac{W}{gV} = \frac{39240}{9.81 \times 5} = \textbf{800 kg/m}^3$$

(iii) Specific gravity of the liquid:

$$S_l = \frac{\text{Density of the liquid}}{\text{Density of water at } 4°\text{C}}$$

$$= \frac{\rho}{\rho_{\text{water}}} = \frac{800}{1000} = \textbf{0.8}$$

Problem 1.2:   One litre of liquid ghee at 40°C temperature has mass 900 gm. Calculate the mass density, specific weight, specific volume, and specific gravity of liquid ghee.

Solution: Given data:

Volume of liquid ghee: $V = 1$ litre $= 1 \times 10^{-3}$ m³

Mass of liquid ghee: $M = 900$ gm $= 0.9$ kg

Now,

(i) Mass density: $\rho = \dfrac{M}{V} = \dfrac{0.9}{1 \times 10^{-3}} =$ **900 kg/m³**

(ii) Specific weight: $w = \rho g = 900 \times 9.81$ N/m³ $=$ **8829 N/m³**

(iii) Specific volume: $v = \dfrac{1}{\rho} = \dfrac{1}{900} =$ **1.11 × 10⁻³ m³/kg**

(iv) Specific gravity: $S_l = \dfrac{\rho}{\rho_{water}} = \dfrac{900}{1000} =$ **0.9**

Problem 1.3: A liquid has a specific gravity of 0.85. Find its mass density and specific weight. Also find the weight per litre of the liquid.

Solution: Given data:

Specific gravity of liquid: $S_l = 0.85$

(i) Mass density: ρ

We know, specific gravity of liquid:

$$S_l = \frac{\rho}{\rho_{water}}$$

$$S_l = \frac{\rho}{100}$$

or $\rho = 1000 \times S_l = 1000 \times 0.85 =$ **850 kg/m³**

(ii) Specific weight: $w = \rho g = 850 \times 9.81$ N/m³ $=$ **8338.5 N/m³**

(iii) Weight per litre of the liquid:

$$= \frac{8338.5}{1000} \frac{N}{litre} \quad 1m^3 = 10^3 \text{ litre})$$

$$= \textbf{8.338 N/litre}$$

Problem 1.4: A large airship of volume 90000 m³ contains helium under standard atmospheric conditions. Determine the density and total weight of the helium.

Solution: Given data:

Volume: $V = 90000$ m³

At standard atmospheric condition,

$$p = 101.325 \text{ kPa}$$
$$T = 15°C = (15 + 273) \text{ K} = 288 \text{ K}$$

Molecular weight of helium: $M = 4$

Gas constant: $R = \dfrac{\text{Universal gas constant}}{\text{Molecular weight}}$

$$= \frac{\bar{R}}{M} = \frac{8.314}{4} = 2.078 \text{ kJ/kgK}$$

Equation of state, $pV = mRT$

$$101.325 \times 90000 = m \times 2.079 \times 288$$

or $m = 15230.42$ kg

Density of Helium: $\rho = \dfrac{\text{Mass}:m}{\text{Volume}:V}$

$$= \dfrac{15230.42}{90000} = \textbf{0.1692 kg/m}^3$$

Weight of helium: $W = mg$

$$= 15230.42 \times 9.81$$

$$= 149410.42 \text{ N} = \textbf{149.41 kN}$$

Problem 1.5:   Define mass density, specific weight, specific volume and specific gravity. 3.2 m³ of a certain oil has weight 27.5 kN. Calculate its specific weight, mass density, specific volume and specific gravity with respect to water. If kinematic viscosity of the oil is 7×10^{-3} stoke, what would be its dynamic viscosity in centipoise?

Solution: Refer article 1.5: Properties of fluids.

Given data:

Volume of oil: $V = 3.2 \text{ m}^3$

Weight of oil: $W = 27.5 \text{ kN} = 27.5 \times 10^3 \text{ N}$

Kinematic viscosity of the oil: $v = 7 \times 10^{-3}$ stoke $= 7 \times 10^{-3}$ cm²/s

$$= 7 \times 10^{-7} \text{ m}^2/\text{s}$$

$$(1 \text{ stoke} = 1 \text{ cm}^2/\text{s} = 10^{-4} \text{ m}^2/\text{s})$$

(*i*) Specific weight of oil: $w = \dfrac{\text{Weight of oil}:W}{\text{Volume of oil}:V}$

$$= \dfrac{27.5 \times 10^3}{3.2} \text{ N/m}^3 = 8593.75 \text{ N/m}^3$$

$$= \textbf{8.593 kN/m}^3$$

(*ii*) Mass density: $\rho = \dfrac{M}{V} = \dfrac{W}{gV}$ $(W = Mg \text{ or } M = W/g)$

$$= \dfrac{27.5 \times 10^3}{9.81 \times 3.2} = \textbf{876.02 kg/m}^3$$

(*iii*) Specific volume: $v = \dfrac{1}{\rho} = \dfrac{1}{876.02} = \textbf{1.1415} \times \textbf{10}^{-3} \textbf{ m}^3/\textbf{kg}$

(iv) Specific gravity of oil: $S = \dfrac{\rho}{\rho_{water}} = \dfrac{876.02}{1000} = \mathbf{0.876}$

(v) Dynamic viscosity: μ = density × kinematic viscosity

$= \rho v = 876.02 \times 7 \times 10^{-7}$

$= 6132.14 \times 10^{-7}$ kg/m.s or Ns/m²

$= 6132.14 \times 10^{-7} \times 10$ poise ($\because$ 1 Ns/m² = 10 poise)

$= 6132.14 \times 10^{-6}$ poise

$= 6132.14 \times 10^{-6} \times 100$ centipoise

($\because$ 1 poise = 100 cP)

$= \mathbf{0.613\ cP}$

Problem 1.6: The shear stress at a point in an oil is 0.30 Pa and the velocity gradient at that point is 0.25 per second. Find the viscosity of the oil, if the density of the oil is 970 kg/m³. Find also the kinematic viscosity of the oil.

Solution: Given data:

Shear stress: $\tau = 0.3$ Pa $= 0.3$ N/m²   ($\because$ 1 Pa = 1 N/m²)

Velocity gradient: $\dfrac{du}{dy} = 0.25$ s^{-1}

Density of oil: $\rho = 970$ kg/m³

(i) Viscosity of oil: $\mu = ?$

We know that the shear stress:

$$\tau = \mu \dfrac{du}{dy}$$

or $\mu = \dfrac{\tau}{du/dy} =$ Ns/m² $= \mathbf{1.2\ Ns/m^2\ or\ Pa.s}$

(ii) Kinematic viscosity: $v = \dfrac{\mu}{\rho} = \dfrac{1.2}{970} = 1.237 \times 10^{-3}$ m²/s

$= 12.37 \times 10^{-4}$ m²/s $= \mathbf{12.37\ cm^2/s\ or\ stoke}$

Problem 1.7: Two horizontal plates are placed 11.5 mm apart, the space between them being filled with oil of viscosity 14 poise. Calculate the shear stress in the oil if the upper plate moved with a velocity of 3.5 m/s.

Solution: Given data:

Distance between plates: $dy = 11.5$ mm $= 0.0115$ m

Viscosity of the oil: $\mu = 14$ poise $= \dfrac{14}{10}$ Ns/m²

($\because$ 1 poise $= \dfrac{1}{10}$ Ns/m²)

$$= 1.4 \text{ Ns/m}^2$$

Relative velocity between plates:

$$du = \text{velocity of moving plate} - \text{velocity of fixed plate}$$

$$= 3.5 - 0 = 3.5 \text{ m/s}$$

$\therefore$ Velocity gradient: $\dfrac{du}{dy} = \dfrac{3.5}{0.0115} = 304.34 \text{ s}^{-1}$

We know that the shear stress: $\tau = \mu\dfrac{du}{dy} = 1.4 \times 304.34 = \mathbf{426.07 \text{ N/m}^2}$

Problem 1.8:   A painter is painting a wall 3 m × 4 m with a brush 10 cm wide and 1.25 cm thick. The thickness of one coat of paint is 0.5 mm and the viscosity of the paint is 3 Ns/m². Calculate the total work required for painting one side of the wall if he moves the brush with a velocity of 10 cm/s.

Solution: Given data:

Area of wall : $A = 3 \times 4 \text{ m}^2 = 12 \text{ m}^2$

Width of brush : $b = 10 \text{ cm} = 0.1 \text{ m}$

Thickness of brush : $t = 1.25 \text{ cm} = 0.0125 \text{ m}$

$\therefore$ Area of brush : $a = b \times t = 0.1 \times 0.0125 = 0.00125 \text{ m}^2$

Thickness of one coat : $y = 0.5 \text{ mm} = 0.5 \times 10^{-3} \text{ m}$

Viscosity of the paint : $\mu = 3 \text{ Ns/m}^3$

Velocity of brush : $V = 10 \text{ cm/s} = 0.10 \text{ m/s}$

Let us assume a linear variation in velocity, then the velocity gradient:

$$\frac{du}{dy} = \frac{0.10}{0.5 \times 10^{-3}} = 200 \text{ s}^{-1}$$

Shear stress on the brush tip: $\tau = \mu\dfrac{du}{dy} = 3 \times 200 = 600 \text{ N/m}^2$

Force required to move the brush :

$$F = \text{Shear stress} \times \text{Area} = \tau a$$
$$= 600 \times 0.00125 = 0.75 \text{ N}$$

Power required $= \text{Force} \times \text{Velocity}$
$$= 0.75 \times 0.10 = 0.075 \text{ W}$$

Total time required to paint one side of wall

$$= \frac{A}{bV} = \frac{12}{0.1 \times 0.10} = 1200 \text{ s}$$

Total work required for painting one side of the wall

$$= \text{Power} \times \text{Time}$$
$$= 0.075 \times 1200$$
$$= \mathbf{90 \text{ J}}$$

Problem 1.9: A three cylinder Maruti car has pistons of 75 mm and cylinders of 75.1 mm. Find the percentage decrease in force required to drive the piston when the lubricant warms from 25°C to 100°C.

Temperature: °C	Dynamic Viscosity μ: Ns/m^2
25	2
100	0.4

Solution: Given data:

Number of cylinder: $\quad\quad n = 3$

Diameter of piston: $\quad\quad d = 75$ mm $= 0.075$ m

Dynamic viscosity: $\quad\quad \mu_1 = 2$ Ns/m^2 at 25°C

$\quad\quad\quad\quad\quad\quad\quad\quad\quad \mu_2 = 0.4$ Ns/m^2 at 100°C

According to Newton's law of viscosity;

Shear stress: $\quad\quad\quad\quad\quad \tau = \mu\dfrac{du}{dy}$

also $\quad\quad\quad\quad\quad\quad\quad \tau = \dfrac{F}{A}$

$\therefore \quad\quad\quad\quad\quad\quad\quad \dfrac{F}{A} = \mu\dfrac{du}{dy}$ or $F = A.\ \mu\dfrac{du}{dy}$

$\quad\quad\quad\quad\quad\quad\quad\quad F \propto \mu$

where, A, $\dfrac{du}{dy}$ are constant.

$$\frac{F}{\mu} = \text{constant}$$

$$\frac{F_1}{\mu_1} = \frac{F_2}{\mu_2}$$

$$\frac{F_2}{F_1} = \frac{\mu_2}{\mu_1} = \frac{0.4}{2} = 0.2$$

or $\quad\quad\quad\quad\quad\quad -\dfrac{F_2}{F_1} = -0.2$

Adding one on both sides, we get

$$1 - \frac{F_2}{F_1} = 1 - 0.2 = 0.8; \quad \frac{F_1 - F_2}{F_1} = 0.8$$

$\therefore$ Percentage decrease in force: $\dfrac{(F_1 - F_2)}{F_1} \times 100 = \textbf{80\%}$

Problem 1.10: The clearance between a 70 mm diameter shaft and its journal bearing is 0.7 mm. The speed is 120 rpm. Find the shear stress induced in the lubricant oil. Take $\mu = 0.10$ Ns/m².

Solution: Given data:

Diameter of shaft: $\quad\quad\quad\quad d = 70$ mm $= 0.07$ m

Speed of shaft: $\quad\quad\quad\quad\quad N = 120$ rpm

Thickness of oil film or

Clearance: $\quad\quad\quad\quad\quad\quad\quad dy = 0.7$ mm $= 0.0007$ m

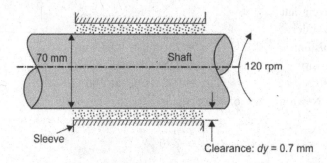

Fig. 1.16: Schematic for Problem 1.10

Now, peripheral or tangential velocity of shaft:

$$u = \frac{\pi dN}{60} = \frac{3.14 \times .07 \times 120}{60} \text{m/s} = 0.4396 \text{ m/s.}$$

Change in velocity: $\quad du = $ velocity of lubricant oil on shaft –

velocity of lubricant oil on sleeve

$= u - 0 = 0.4396$ m/s

Velocity gradient: $\quad\quad \dfrac{du}{dy} = \dfrac{0.4396 \text{ m/s}}{0.0007 \text{ m}} = 628 \text{ s}^{-1}$

Shear stress induced in the lubricant oil:

$$\tau = \mu \frac{du}{dy} = 0.1 \times 628 = \textbf{62.8 N/m}^2$$

Problem 1.11: The dynamic viscosity of an oil, used for lubrication between a shaft and sleeve, is 6 poise. The shaft is of diameter 0.5 m and rotates at 210 rpm. Calculate the power lost in the bearing for a sleeve length of 90 mm. The thickness of the oil film is 1.4 mm.

Solution: Given data:

Diameter of shaft: $\quad\quad d = 0.5$ m

Speed of shaft: $\quad\quad\quad N = 210$ rpm

Thickness of oil film: $\quad dy = 1.4$ mm $= 0.0014$ m

Dynamic viscosity: $\quad\quad \mu = 6$ poise $= \dfrac{6}{10}$ Ns/m² $= 0.6$ Ns/m²

Sleeve length: l = 90 mm = 0.09 m

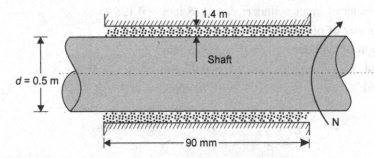

Fig. 1.17: Schematic for Problem 1.11

We know that the peripheral or tangential velocity of shaft:

$$u = \frac{\pi dN}{60} = \frac{3.14 \times 0.5 \times 210}{60} = 5.495 \text{ m/s}$$

Change in velocity: du = velocity of oil on shaft – velocity of oil on sleeve

$$= u - 0 = 5.495 \text{ m/s}$$

∴ Velocity gradient: $\dfrac{du}{dy} = \dfrac{5.495}{0.0014} = 3925 \text{ s}^{-1}$

Oil exerted shear stress on the shaft:

$$\tau = \mu\frac{du}{dy} = 0.60 \times 3925 = 2355 \text{ N/m}^2$$

∴ Shear force on the shaft: F = shear stress × circumference area of the shaft

$$= 2355 \; 5 \; \pi dl$$
$$= 2355 \times 3.14 \times 0.5 \times 0.09 = 332.761 \text{ N}$$

Torque on the shaft: T = shear force × $\dfrac{d}{2}$

$$= F.\frac{d}{2} = 332.761 \times \frac{0.5}{2} = 83.19 \text{ Nm}$$

Power lost in bearing: $P = \dfrac{2\pi NT}{60}$ watt or W

where T = torque in Nm
N = speed in rpm

∴ $P = \dfrac{2 \times 3.14 \times 210 \times 83.19}{60}$ W

$$= \textbf{1828.52 W} = \textbf{1.828 kW}$$

Problem 1.12:   A 125 mm diameter vertical cylinder rotates concentrically inside a fixed cylinder of diameter 130 mm. Both cylinders are 325 mm long. Find the dynamic viscosity of the liquid that fills the space between the cylinders, if a torque of 0.92 Nm is required to maintain a speed of 70 rpm.

Solution: Given data:

Diameter of inner cylinder: $d_1 = 125$ mm $= 0.125$ m

Diameter of outer cylinder: $d_2 = 130$ mm $= 0.13$ m

Length of cylinder: $\quad\quad\quad l = 325$ mm

Torque: $\quad\quad\quad\quad\quad\quad T = 0.92$ Nm

Speed: $\quad\quad\quad\quad\quad\quad N = 70$ rpm

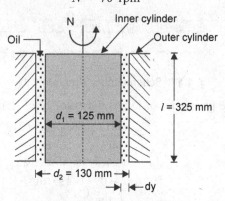

Fig. 1.18 : Schematic for Problem 1.12

Space between the cylinder: $dy = \dfrac{d_2 - d_1}{2} = \dfrac{130 - 125}{2} = 2.5$ mm $= 0.0025$ m

Now, tangential velocity at the periphery of the inner cylinder:

$$u = \frac{\pi d_1 N}{60} = \frac{3.14 \times 0.125 \times 70}{60} = 0.4579 \text{ m/s}$$

Change in velocity: $\quad\quad du = u - 0 = 0.4579$ m/s

$\therefore$ Velocity gradient: $\quad\quad \dfrac{du}{dy} = \dfrac{0.4579}{0.0025} \dfrac{\text{m/s}}{\text{m}}$

$$\frac{du}{dy} = 183.16 \text{ s}^{-1}$$

Shear stress: $\quad\quad\quad\quad \tau = \mu \dfrac{du}{dy} = \mu \times 183.16$

Shear force on the inner cylinder: $F = \tau \times$ circumference area of the inner cylinder

$\quad\quad\quad\quad\quad\quad\quad\quad\quad = \mu \times 183.16 \times \pi d_1 l$

$\quad\quad\quad\quad\quad\quad\quad\quad\quad = \mu \times 183.17 \times 3.14 \times .125 \times 0.325$

$\quad\quad\quad\quad\quad\quad\quad\quad F = 23.36 \ \mu$ N

Here dynamic viscosity, μ in $\dfrac{\text{Ns}}{\text{m}^2}$

Torque on the inner cylinder: $\quad T = F . \dfrac{d_1}{2}$

$$0.92 = 23.36 \ \mu \times \frac{0.125}{2}$$

or Dynamic viscosity: $\quad\quad\quad \mu = 0.630 \text{ Ns/m}^2 = \textbf{6.30 poise}$

Problem 1.13: Calculate the dynamic viscosity of an oil, which is used for lubrication between square plate of size 700 mm × 700 mm and inclined plane with angle of inclination 25° as shown in Fig. 1.19. The weight of the square plate is 250 N and it slides down the inclined plane with a uniform velocity of 0.35 m/s. The thickness of oil film is 1.4 mm.

Solution: Given data:

Weight of plate: $\qquad W = 250$ N

Area of plate: $\qquad A = 700 \times 700$ mm² = 490000 mm² = 0.49 m²

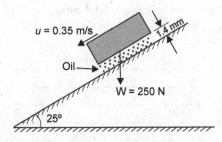

Fig. 1.19 : Schematic for Problem 1.13

Velocity of plate: $\qquad u = 0.35$ m/s

Angle of plane: $\qquad \theta = 25°$

Thickness of oil film: $\qquad dy = 1.4$ mm = 0.0014 m

Now, the component of weight (W), along the plane

$$W \sin \theta = 250 \times \sin 25° = 105.65 \text{ N}$$

Thus the shear force (F) on the bottom surface of the plate:

$$F = W \sin \theta = 105.65 \text{ N}$$

and $\quad$ Shear stress: $\tau = \dfrac{\text{Shear force}}{\text{Area}} = \dfrac{F}{A} = \dfrac{105.65}{0.49} = 215.61 \text{ N/m}^2$

According to Newton's law of viscosity,

$$\text{Shear stress: } \tau = \mu \frac{du}{dy}$$

where $\qquad du = $ change of velocity

$$= u - 0 = 0.35 - 0 = 0.35 \text{ m/s}$$

∴ $\qquad 215.61 \text{ N/m}^2 = \mu \times \dfrac{0.35 \text{ m/s}}{0.0014 \text{ m}}$

or $\qquad$ Dynamic viscosity: $\mu = 0.8624$ Ns/m² (∵ 1 Ns/m² = 10 poise)

$$= 0.8624 \times 10 \text{ poise} = \textbf{8.624 poise}$$

Problem 1.14: Castor oil at 20°C has relative density and kinematic viscosity of 0.96 and 1.03×10^{-3} m²/s respectively. Its velocity at a distance of 50 mm was found to be 1 m/s. Determine the velocity gradient and shear stress at the boundary and at points 25 mm and 50 mm from the boundary assuming a straight line velocity variation.

Solution: Given data:

Relative density: $S = 0.96$

∴ Density : $\rho = 1000\,S$ kg/m³ $= 1000 \times 0.96 = 960$ kg/m³

Kinematic viscosity : $\nu = 1.03 \times 10^{-3}$ m²/s

∴ Dynamic viscosity : $\mu = \rho\nu = 960 \times 1.03 \times 10^{-3} = 0.988$ Ns/m²

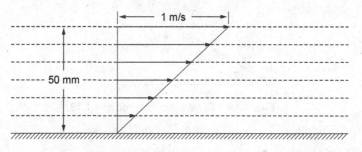

Fig. 1.20 : Schematic for Problem 1.14

Velocity gradient at point 50 mm,

$$\left(\frac{du}{dy}\right)_{y=0.050\,m} = \frac{1}{0.050} = \mathbf{20\ s^{-1}}$$

Shear stress : $(\tau)_{y=0.050\ m} = \mu\left(\frac{du}{dy}\right)_{y=0.050\,m}$

$$= 0.988 \times 20 = \mathbf{19.76\ N/m^2}$$

Velocity gradient at point 25 mm,

$$\left(\frac{du}{dy}\right)_{y=0.025\,m} = \mu\left(\frac{du}{dy}\right)_{y=0.050\,m} \qquad \therefore \text{Straight line velocity variation.}$$

$$= 20\ s^{-1}$$

∴ Shear stress: $(\tau)_{y=0.025m} = \mu\left(\frac{du}{dy}\right)_{y=0.025\,m}$

$$= 0.988 \times 20 = \mathbf{19.76\ N/m^2}$$

Velocity gradient at the boundary *i.e.,* $y = 0$

$$\left(\frac{du}{dy}\right)_{y=0} = 0$$

$\therefore$ Shear stress : $(\tau)_{y=0} = \mu \left(\dfrac{du}{dy} \right)_{y=0}$

$$= 0.988 \times 0 = 0$$

Problem 1.15:   Calculate the velocity gradient at distance of 0, 150, 200 mm from the boundary if the velocity profile is a parabola with the vertex 200 mm from the boundary, where the velocity is 1.4 m/s. Also calculate the shear stresses at these points if the fluid has a viscosity of 8 poise.

Solution: Given data:

Let the equation of the parabolic velocity profile is

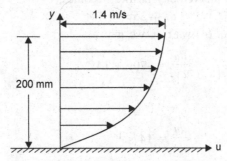

Fig. 1.21 : Schematic for Problem 1.15

$$u = ay^2 + by + c \qquad \qquad \dots (1)$$

where *a*, *b* and *c* are constants to be determined from the following boundary conditions.

Boundary conditions, (*i*) $u = 0$ at $y = 0$

Boundary conditions (*ii*) $u = 1.4$ m/s   at $y = 0.2$ m

Boundary conditions (*iii*) $\dfrac{du}{dy} = 0$ at the vertex, *i.e.*, at $y = 0.2$ m

By using boundary condition (*i*), Ist we find out the values of constants *a*, *b* and *c*.

Substituting the values of $u = 0$ and $y = 0$ in Eq. (1), we get

$$0 = 0 + 0 + c$$

or $c = 0$

By using boundary condition (*ii*), substituting the values of $u = 1.4$ m/s and $y = 0.2$; $c = 0$ in Eq. (1), we get

$$1.4 = a(0.2)^2 + b(0.2) + 0$$

$$1.4 = 0.04a + 0.2b \qquad \qquad \dots (2)$$

And differentiating equation (1) w.r.t. *y*, we get

$$\frac{du}{dy} = 2ay + b \qquad \qquad \dots (3)$$

By using boundary condition (*iii*), substituting the values of $\dfrac{du}{dy} = 0$ and $y = 0.2$ m in Eq. (3), we get

$$0 = 2 \times 2a + b$$
$$0 = 0.4a + b$$

or
$$b = -0.4a$$

Substituting the value of '*b*' in Eq. (2), we get

$$1.4 = 0.04a + 0.2(-0.4)a$$
$$1.4 = 0.04a - 0.08a$$

or
$$a = -35 \quad \text{and} \quad b = -0.4(-35) = 14$$

Now, Eq. (1) for the velocity profile becomes,

$$u = -35y^2 + 14y \qquad \qquad \text{... (4)}$$

Differentiating Eq. (4) w.r.t. *y*, we get

Velocity gradient: $\qquad \dfrac{du}{dy} = -70y + 14 \qquad \qquad$ ... (5)

Case-I

At $\qquad\qquad\qquad\qquad y = 0$

Velocity gradient: $\qquad \dfrac{du}{dy} = 14 \ \text{s}^{-1} \qquad\qquad$ [by using Eq. (5)]

Shear stress: $\qquad\qquad \tau = \mu \dfrac{du}{dy} \qquad$ [$\mu = 8$ poise (given) $= 0.8$ Ns/m^2]

$$= 0.8 \times 14 \ \frac{\text{Ns}}{\text{m}^2}\text{s}^{-1} = \textbf{11.2 N/m}^2$$

Case-II

At $\qquad\qquad\qquad\qquad y = 0.15$ m

Velocity gradient: $\qquad \dfrac{du}{dy} = -70(0.15) + 14 = 3.5 \ \text{s}^{-1}$

Shear stress: $\qquad\qquad \tau = \mu \dfrac{du}{dy} = 0.8 \times 3.5 \ \dfrac{\text{Ns}}{\text{m}^2}.\text{s}^{-1} = \textbf{2.8 N/m}^2$

Case-III

At $\qquad\qquad\qquad\qquad y = 0.2$ m

Velocity gradient: $\qquad \dfrac{du}{dy} = -70(0.2) + 14 = 0$

Shear stress: $\qquad\qquad \tau = \mu \dfrac{du}{dy} = 0.8 \times 0 = \textbf{0}$

Problem 1.16: A plate of metal having dimensions 1.2 m × 1.2 m × 1.2 mm is to be lifted up with a velocity of 0.14 m/s through an infinitely extending gap 26 mm wide containing an oil of specific gravity 0.90 and viscosity 21.6 poise. Find the force required assuming the plate to remain midway in the gap. Assuming the weight of the plate to be 35 N.

Solution: Given data:

Width of gap   = 26 mm = 0.026 m

Dynamic viscosity: μ = 21.6 poise = 2.16 Ns/m^2

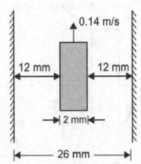

Fig. 1.22 : Schematic for Problem 1.16

Specific gravity of oil: S = 0.9

$\therefore$ Density of oil: $\rho = S \times \rho_{water}$ = 0.9 × 1000 = 900 kg/m^3

Volume of plate: V = 1.2 × 1.2 × 0.002 m^3 = 2.88 × 10^{-3} m^3

Thickness of plate: t = 2 mm = 0.002 m

Velocity of plate: u = 0.14 m/s

Weight of plate: W = 35 N

When plate is in the middle of the gap, distance of the plate from vertical surface:

dy = 12 mm = 0.012 m   as shown in Fig. 1.22.

Now the shear force on the left side of the metallic plate:

F_1 = shear stress × area

$$= \mu \frac{du}{dy} \times A = \frac{2.16 \times 0.14}{0.012} \times 1.2 \times 1.2 = 36.288 \text{ N}$$

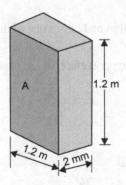

Fig. 1.23 : Dimensions of Plate for Problem 1.16

Due to similarly, the shear force on the right side of the metallic plate:

$$F_2 = F_1$$

$\therefore$ Total shear force = $F_1 + F_2 = 2F_1 = 2 \times 36.288 = 72.576$ N

In this case, the weight of plate acts vertically downward and upward thrust is also to be taken into account.

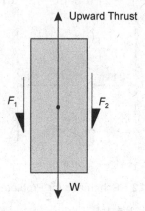

Fig. 1.24: Number of forces exerted on metallic plate.

Upward thrust = weight of oil displaced by the plate

$$= \rho g V$$
$$= 900 \times 9.81 \times 2.88 \times 10^{-3} = 25.427 \text{ N}$$

Total force required to lift the plate at given velocity

= total shear force ($\downarrow$) + weight of plate ($\downarrow$)

– upward thrust ($\uparrow$)

$$= 27.576 + 35 - 25.427 = \textbf{82.149 N}$$

Problem 1.17:   Through a very narrow gap of height h, a thin plate of large extent is pulled at a velocity V. On one side of the plate is oil of viscosity μ_1 and on the other side oil of viscosity μ_2. Calculate the position of the plate so that (i) the shear force on the two sides of the plate is equal, (ii) the pull required to drag the plate is minimum.

Solution:

Let y be the distance of the thin plate from one of the surface as shown in Fig. 1.25.

Force per unit area on the upper surface of the plate = $\mu_1 \dfrac{du}{dy} = \mu_1 \dfrac{V}{(h-y)}$ ...(i)

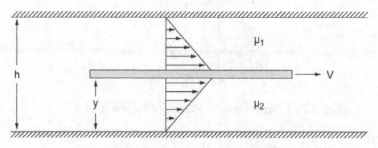

Fig. 1.25 : Schematic for Problem 1.17

Force per unit area on the bottom surface of the plate $= \mu_2 \dfrac{du}{dy} = \mu_2 \dfrac{V}{y}$ ...(ii)

Equating Eqs. (i) and (ii) , we get

$$\mu_1 \frac{V}{h-y} = \mu_2 \frac{V}{y}$$

$$\mu_1 V = \mu_2 V \frac{(h-y)}{y}$$

$$\mu_1 = \mu_2 \frac{(h-y)}{y}$$

$$\mu_1 y = \mu_2 h - \mu_2 y$$

$$y(\mu_1 + \mu_2) = \mu_2 h$$

or
$$y = \frac{\mu_2 h}{\mu_1 + \mu_2}$$

Let F be the pull per unit area required to drag the plate, then

$$F = F_1 + F_2$$

$$= \mu_1 \frac{V}{h-y} + \mu_2 \frac{V}{y}$$

For the force F to be minimum,

$$\frac{dF}{dy} = 0$$

$$\frac{\mu_1 V}{(h-y)^2} - \frac{\mu_2 V}{y^2} = 0$$

or
$$\frac{\mu_1}{(h-y)^2} = \frac{\mu_2}{y^2}$$

or
$$\frac{\mu_1}{\mu_2} = \frac{(h-y)^2}{y^2}$$

$$\frac{\mu_1}{\mu_2} = \frac{h^2 + y^2 - 2hy}{y^2}$$

$$\frac{\mu_1}{\mu_2} = \frac{h^2}{y^2} + 1 - \frac{2h}{y}$$

or
$$\frac{h^2}{y^2} - 2\frac{h}{y} + \left(1 - \frac{\mu_1}{\mu_2}\right) = 0$$

Solving this quadratic equation for $\dfrac{h}{y}$, we get

$$\frac{h}{y} = \frac{2 \pm \sqrt{4 - 4\left(1 - \dfrac{\mu_1}{\mu_2}\right)}}{2} = \frac{2 \pm 2\sqrt{1 - 1 + \dfrac{\mu_1}{\mu_2}}}{2}$$

$$\frac{h}{y} = 1 \pm \sqrt{\frac{\mu_1}{\mu_2}}$$

Since $\dfrac{h}{y}$ cannot be less than unity,

$$\therefore \qquad \frac{h}{y} = 1 + \sqrt{\frac{\mu_1}{\mu_2}}$$

or

$$y = \frac{h}{1 + \sqrt{\dfrac{\mu_1}{\mu_2}}}$$

Problem 1.18: Find the excess pressure inside a water drop of 0.35 mm diameter if the surface tension of water is 0.07 N/m.

Solution: Given data:

   Diameter of water drop: $d = 0.33$ mm $= 0.33 \times 10^{-3}$ m

   Surface tension of water: $\sigma = 0.07$ N/m

   Now, the excess pressure inside a water drop $= \Delta p$

We know $\qquad \Delta p = \dfrac{4\sigma}{d}$

$$= \frac{4 \times 0.07}{0.33 \times 10^{-3}} \frac{\text{N/m}}{\text{m}} = \textbf{848.48 N/m}^2 \textbf{ or Pa}$$

Problem 1.19: A liquid drop 30 mm in radius has an internal pressure of 13 Pa greater than the outside pressure. Find the surface tension of the liquid film.

Solution: Given data:

   Radius at liquid drop: $\qquad r = 30$ mm $= 0.03$ m

   Diameter of liquid drop: $\quad d = 2r = 0.06$ m

   Excess pressure inside the liquid drop:

$$\Delta p = 13 \text{ Pa} = 13 \text{ N/m}^2$$

Now, we know $\qquad \Delta p = \dfrac{4\sigma}{d}$

or Surface tension of liquid: $\sigma = \dfrac{\Delta p.d}{4} = \dfrac{13 \times 0.06}{4} \dfrac{\text{N}}{\text{m}^2}.\text{m} = \textbf{0.195 N/m}$

Problem 1.20: The pressure outside the droplet of water of diameter 0.05 mm is 1.0133 bar (atmospheric pressure). Calculate the pressure within the droplet if surface tension is given as 0.07 N/m of water.

Solution: Given data:

Diameter of droplet: $d = 0.05$ mm $= 0.05 \times 10^{-3}$ m

Pressure outside the droplet: $p_0 = 1.0133$ bar

$= 1.0133 \times 10^5$ N/m^2 (1 bar $= 10^5$ N/m^2)

Surface tension of water: $\sigma = 0.07$ N/m

Now, excess pressure inside the droplet:

$$\Delta p = \frac{4\sigma}{d} = \frac{4 \times 0.07}{0.05 \times 10^{-3}} \frac{\text{N/m}}{\text{m}} = 5600 \text{ N/m}^2$$

$\Delta p = $ pressure inside the droplet (p_i)

– pressure outside the droplet (p_0)

$$\Delta p = p_i - p_0$$

or Pressure inside the droplet: $p_i = \Delta p + p_0 = 5600 + 1.0133 \times 10^5$ N/m^2

$$= 106930 \text{ N/m}^2 = \textbf{1.0693 bar}$$

Problem 1.21: A 5 mm diameter glass tube is immersed in

       (*i*) water and   (*ii*) mercury.

Calculate the capillary effect in millimetres in the glass tube. The values of surface tension for water and mercury are 0.0735 N/m and 0.530 N/m respectively. The angle of contact for water is 25° and that for mercury 128°.

Solution: Given data:

Diameter of glass tube: $d = 5$ mm $= 0.005$ m

Surface tension of water: $\sigma_{\text{water}} = 0.0735$ N/m

Surface tension of mercury: $\sigma_{\text{Hg}} = 0.530$ N/m

Angle of contact in water: $\theta_{\text{water}} = 25°$

(*i*) For water: $h = \dfrac{4\sigma \cos \theta}{wd} = \dfrac{4\sigma \cos \theta}{\rho g d}$

$$= \frac{4 \times 0.0735 \cos 25°}{1000 \times 9.81 \times 0.005} = 5.432 \times 10^{-3} \text{ m}$$

$$= \textbf{5.43 mm (rise of water)}$$

(*ii*) For mercury: $h = \dfrac{4\sigma \cos \theta}{wd} = \dfrac{4\sigma \cos \theta}{\rho g d}$

$$= \frac{4 \times 0.530 \times \cos 128°}{13600 \times 9.81 \times 0.005} = -1.956 \times 10^{-3} \text{ m}$$

$$= -1.956 \text{ mm (depression of mercury)}$$

Problem 1.22: If the pressure of a liquid is increased from 850 N/cm^2 to 1500 N/cm^2, the volume of liquid is decreased by 0.180%. Determine bulk modulus of elasticity of liquid.

Solution: Given data:

Change in pressure: $\quad\quad\quad\quad\quad\quad dp = 1500 - 850 = 650 \text{ N/cm}^2$
$$= 650 \times 10^4 \text{ N/m}^2$$

Volumetric strain: $\quad\quad\quad\quad\quad \dfrac{dV}{V} = -\dfrac{0.18}{100} = -0.0018$

Now,

Bulk modulus of elasticity: $\quad\quad K = \dfrac{\text{Change in pressure}}{\text{Volumetric strain}} = \dfrac{dp}{-\dfrac{dV}{V}}$

$$= \dfrac{650 \times 10^4}{0.0018} \text{ N/m}^2$$
$$= 361111.11 \times 10^4 \text{ N/m}^2$$
$$= \mathbf{3.61 \times 10^9 \text{ N/m}^2}$$

SUMMARY

1. Fluid means a liquid or gas or mixture of both. It can be defined as a substance that deforms continuously under the action of external shearing force.
2. The continuous deformation of the fluid is called **flow**.
3. **Standard Temperature and Pressure (STP):**
$$p = 760 \text{ mm of Hg.}$$
and $\quad\quad\quad T = 15°C$
4. **Normal Temperature and Pressure (NTP):**
$$p = 760 \text{ mm of Hg.}$$
and $\quad\quad\quad T = 0°C$
5. The study of the behaviour of fluids at rest or in motion is called **fluid mechanics.**
   - The study of fluids at rest is called **fluid statics.**
   - The study of fluids in motion, without considering the pressure or causes of motion is called **fluid kinematics.**
   - The study of fluids in motion, while considering the pressure or causes of motion is called **fluid dynamics.**
6. **Density:** $\quad\quad\quad\quad\quad \rho = \dfrac{\text{Mass of fluid} : M}{\text{Volume of fluid} : V}$

$$\rho = \dfrac{M}{V}$$
$$\rho = 1000 \text{ kg/m}^3 \text{ for water at } 4°C$$

Contd...

$$= 13600 \text{ kg/m}^3 \text{ for mercury at NTP}$$

Density is also known as mass density.

7. Specific volume:
$$v = \frac{\text{Volume of fluid} : V}{\text{Mass of fluid} : M}$$

$$v = \frac{V}{M} = \frac{1}{M/V}$$

$$v = \frac{1}{\rho}$$

Specific volume (v) is the reciprocal of the density (ρ)
SI unit of v is m^3/kg.

8. Specific weight:
$$w = \frac{\text{Weight } (W)}{\text{Volume } (V)}$$

$$w = \frac{Mg}{V} = \rho g$$

SI unit of w is N/m^3

9. Specific gravity for liquid:
$$S_l = \frac{\rho}{\rho_{water}}$$

where $\rho_{water} = 1000 \text{ kg/m}^3$
and   Specific gravity for gas:
$$S_g = \frac{\rho}{\rho_{air}}$$

where $\rho_{air} = 1.29 \text{ kg/m}^3 \text{ at NTP}$

10. The intermolecular attraction between the molecules of same liquid is called **cohesion.**

11. The attraction between the molecules of liquid and the molecules of other substance is called **adhesion.**

12. The internal frictional force in a liquid which opposes the relative motion amongst its different layers is called **viscosity.**

13. The shear stress (τ) on a fluid element layer is directly proportional to the velocity gradient, is called **Newton's law of viscosity.**

$$\tau \propto \frac{du}{dy}$$

$$\tau = \mu \frac{du}{dy}$$

where μ = constant of proportionality is called **viscosity** or **dynamic viscosity**

SI unit of μ = Ns/m^2
MKS unit of μ = kg.f s/m^2
CGS unit of μ = poise = dyne-s/cm^2

Contd...

Units conversion of viscosity (μ):

$$1 \text{ poise} = \text{dyne-s/cm}^2 = \frac{1}{10} \text{Ns/m}^2 = \frac{1}{10} \text{Pa.s}$$

$$[\because 1 \text{ pascal, Pa} = \text{N/m}^2;$$
$$1\text{N} = 10^5 \text{ dyne};$$
$$1 \text{ dyne} = 1\text{gm. cm/s}^2; 1\text{N} = \text{kgm/s}^2; 1\text{cP} = 10^{-2} \text{ P}]$$
$$1 \text{ centipoise or cP} = 10^{-2} \text{ poise} = 10^{-3} \text{ Ns/m}^2$$

14. **Kinematic viscosity:** $v = \dfrac{\text{Dynamic viscosity } (\mu)}{\text{Density of fluid } (\rho)}$

$$v = \frac{\mu}{\rho}$$

SI unit of v is m^2/s

CGS unit of v is cm^2/s or stoke

Units conversion of kinematic viscosity, v

$$1 \text{ stoke} = 1 \text{ cm}^2\text{/s} = 10^{-4} \text{ m}^2\text{/s}$$
$$1 \text{ centistoke or cSt} = 10^{-2} \text{ stoke} = 10^{-2} \text{ cm}^2\text{/s} = 10^{-6} \text{ m}^2\text{/s}$$

15. A fluid which is inviscid (no viscosity) having no surface tension and no compressibility is known as an **ideal fluid** or **perfect fluid.**

16. A fluid which possesses properties of viscosity, surface tension and compressibility, is known as **real fluid.**

17. A real fluid which obeys the Newton's law of viscosity is known as **Newtonian fluid.** For example: Water, air, kerosene *etc.* are Newtonian fluids.

   For Newtonian fluids:

$$\tau = \mu \frac{du}{dy}$$

18. A real fluid which does not obey the Newton's law of viscosity, is known as non-Newtonian fluid.

   For **non-Newtonian fluid:**

$$\tau \neq \mu \frac{du}{dy}$$

19. The partial pressure of water vapor of saturated air exerted on the liquid surface is called **vapour pressure.** It is denoted by p_v.

20. The rate of liquid leaving from its free surface in the form of vapors when the vapour pressure (p_v) of a liquid is less than surrounding pressure (p) at given temperature, is called **evaporisation.**

   Condition for evaporisation:

$$p_v < p \quad \text{at } T = C$$

21. The high rate of liquid leaving from its free surface when the vapor pressure (p_v) of a liquid is greater than or equal to the surrounding pressure (p) at given temperature, is known as **boiling.**

Contd...

Condition for boiling:

$$p_v \geq p \quad \text{at } T = C$$

22. The resultant unbalanced force that acts normal to the free surface per unit length, is called **surface tension** (σ).

$$\sigma = \frac{\text{Force}}{\text{Length of free surface}}$$

SI unit of σ is N/m

It is also defined as surface energy per unit area.

$$\sigma = \frac{\text{Surface energy}}{\text{Surface area}} = \frac{\text{N m}}{\text{m}^2} = \text{N/m}$$

23. The relation between the surface tension (σ) and difference of pressure (Δp) between the inside and outside of a liquid drop is given as

$$\Delta p = \frac{4\sigma}{d} \quad \text{where } d = \text{diameter of spherical droplet.}$$

$$\Delta p = \frac{2\sigma}{d} \quad \text{for liquid jet.}$$

$$\Delta p = \frac{8\sigma}{d} \quad \text{for a hollow bubble or soap bubble.}$$

24. The phenomenon of rise or fall of liquid in a capillary tube is known as **capillarity**. The rise in the level of water in capillary tube is called **capillary rise** and the fall in the level of mercury in capillary tube is called **capillary fall** or **depression.**

The capillary rise or fall of a liquid is given by;

$$h = \frac{4\sigma \cos\theta}{wd}$$

- The value of $\theta = 0°$ for pure water and clean glass tube surface.
- In actual condition;

$$\left.\begin{array}{l} \theta = 25.53° \\ \sigma = 0.0735 \text{ N/m} \end{array}\right\} \text{ for water at STP}$$

$$\left.\begin{array}{l} \theta = 128.37° \\ \sigma = 0.53 \text{ N/m} \end{array}\right\} \text{ for mercury at STP}$$

25. **Compressibility** is the measure of change of volume or density of liquid with respect to change in applied pressure.

Bulk modulus of elasticity: $K = \dfrac{dp}{-dv/v} = \dfrac{dp}{d\rho/\rho}$

The $-ve$ sign indicates that there will be decrease in volume due to increase in pressure.

Coefficient of compressibility: $\beta = \dfrac{1}{K}$

$$\beta = \frac{-dv/v}{\Delta p} = \frac{d\rho/\rho}{dp}$$

The relation between pressure (p) and bulk modulus of elasticity (K).
  (i) $K = p$ for isothermal process
  (ii) $K = \gamma p$ for adiabatic process

ASSIGNMENT - 1

1. Distinguish amongst solids, liquids and gases.
2. Define the following fluid properties:
   (i) Specific weight   (ii) Density
   (iii) Specific gravity   (iv) Specific volume.
3. Explain the terms: (i) dynamic viscosity (ii) kinematic viscosity. Give their dimensions.
4. What is the difference between cohesion and adhesion?
5. State the Newton's law of viscosity. Explain the importance of viscosity in fluid motion. What is the effect of temperature on viscosity of water and that of air?
6. Why does dynamic viscosity have the prefix "dynamic"?
7. What is kinematic viscosity? Write its unit.
8. Distinguish between (i) ideal and real fluids (ii) Newtonian and non-Newtonian fluids (iii) Ideal plastic and Thyxotropic fluids.
9. Define surface tension. What are the factors that affect surface tension?
10. A thin blade of steel can be made to float on water. Explain how this is possible.
11. Explain briefly dynamic viscosity and kinematic viscosity.
12. Prove that the relationship between surface tension and pressure inside a droplet of liquid in excess of outside pressure is given by

$$\Delta p = \frac{4\sigma}{d} \quad \text{where } \sigma = \text{surface tension of liquid}$$

$$d = \text{diameter of capillary tube}$$

13. Explain the phenomenon of capillarity. Derive an expression for capillary depression of a mercury.
14. What is meant by vapour pressure? Explain its importance in liquid flow system.
15. What is difference between evaporisation and boiling?
16. Derive an expression for the variation of density of a fluid in terms for its bulk modulus and pressure change.
17. Derive the relation between pressure (p) and bulk modulus of elasticity (K)
    (i) $K = p$ for isothermal process
    (ii) $K = \gamma p$ for adiabatic process
        where γ = adiabatic index

18. Define compressibility. Prove that compressibility for a perfect gas undergoing isothermal compression is $1/p$ and a perfect gas undergoing adiabatic compression is $1/\gamma p$, where p is the initial pressure of the gas and γ is the ratio of specific heats. (*GGSIP University, Delhi Dec. 2001*)

ASSIGNMENT - 2

1. A certain mass of liquid has a volume of 6 m^3 and weight of 40000 N. Find the specific weight, mass density and specific gravity of the liquid: **Ans.** 6666.66 N/m^3 679.57 kg/m^3 0.679

2. One litre liquid ghee at 40°C temperature has mass 980 gm. Calculate the mass density, specific weight, specific volume and specific gravity of liquid ghee. **Ans.** (*i*) 980 kg/m^3 (*ii*) 961.8 N/m^3 (*iii*) 1.020×10^{-3} m^3/kg (*iv*) 0.98.

3. A liquid has a specific gravity of 0.79. Find its mass density and specific weight. Find also the weight per litre of the liquid. **Ans.** (*i*) 790 kg/m^3 (*ii*) 7749.9 N/m^3 (*iii*) 7.749 N/litre

4. Define: mass density, specific weight, specific volume and specific gravity. 4 m^3 of a certain oil weighs 31 kN. Calculate its specific weight, mass density, specific volume and specific gravity with respect to water. If kinematic viscosity of the oil is 7.5×10^{-3} stoke, what would be its dynamic viscosity in centipoise. **Ans.** (*i*) 7.75 kN/m^3 (*ii*) 790 kg/m^3 (*iii*) 1.265×10^{-3} m^3/kg (*iv*) 0.79 (*v*) 0.592 cP

5. The shear stress at a point in an oil is 0.35 Pa and the velocity gradient at the point is 0.28 per second. Find the viscosity of the oil if the density of oil is 920 kg/m^3. Find the kinematic viscosity of the oil. **Ans.** (*i*) 1.25 Pa.s. (*ii*) 13.58 stoke

6. Two horizontal plates are placed 12 mm apart, the space between them being filled with oil of viscosity 15 poise. Calculate the shear stress in the oil if the upper plate moved with a velocity of 4 m/s. **Ans.** 500 N/m^2

7. The clearance between a 75 mm diameter shaft and its journal bearing is 0.8 mm. The speed is 130 rpm. Find the shear stress induced in the lubricant oil. Take $\mu = 0.15$ Ns/m^2 **Ans.** 95.67 N/m^2

8. The dynamic viscosity of an oil, used for lubrication between a shaft and sleeve is 6.5 poise. The shaft is of diameter 0.6 m and rotates at 390 rpm. Calculate the power lost in the bearing for a sleeve length of 100 mm. The thickness of the oil film is 1.5 mm. **Ans.** 12.24 kW

9. A 130 mm diameter vertical cylinder rotates concentrically inside a fixed cylinder of diameter 135 mm. Both cylinders are 350 mm long. Find the dynamic viscosity of the liquid that fills the space between the cylinders if a torque of 0.95 N-m is required to maintain a speed of inner cylinder at 80 rpm. **Ans.** 4.69 poise

10. A square plate 0.6 m × 0.6 m, weighting 245 N slides down an inclined plane with a uniform velocity of 0.3 m/s as shown in Fig. 1.26. The inclined plane is laid at a slope of vertical to 2.4 horizontal and is provided with a 1 mm thick oil film. Find the dynamic viscosity of the oil.   **Ans.** 8.725 poise

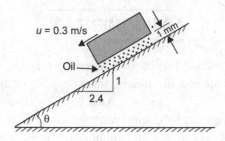

Fig. 1.26: Q-10

11. Calculate the velocity gradient at distance of 0, 100, 150 mm from the boundary if the velocity profile is a parabola with the vertex 150 mm from the boundary, where the velocity is 1 m/s. Also calculate the shear stress at these points if the fluid has a viscosity of 0.804 is Ns/m^2.

       Ans. (*i*) 13.33 s^{-1}, 10.8 N/m^2 (*ii*) 4.45 s^{-1}, 3.575 N/m^2 (*iii*) zero, zero

12. A 60 mm diameter piston moves inside a 60.10 mm diameter cylinder. Find the percentage decrease in force needed to move the piston if the lubricant warms up from 0°C to 120°C. Take viscosity of the lubricant at 0°C and 120°C is equal to 0.0182 N-s/m^2 and 0.0021 N-s/m^2 respectively.   **Ans.** 88.46%

13. A plate of metal having dimensions 1000 mm × 1000 mm × 2 mm is to be lifted up with a velocity of 0.10 m/s through an infinitely extending gap 20 mm wide containing an oil of specific gravity 0.90 and viscosity 2.15 Ns/m^2. Find the force required assuming the plate to remain midway in the gap. Assume the weight of the plate to be 29.5 N.   **Ans.** 59.62 N

14. A space 25 mm wide between two large horizontal plane surfaces is filled with glycerine. What force is required to drag a very thin plate 0.75 m^2 in area between the surfaces at a speed of 0.5 m/s,

   (*i*) if this plate remains equidistant from the two surfaces, and

   (*ii*) if it is at a distance 10 mm from one of the surfaces ?

Take dynamic viscosity, μ = 7.85 poise   **Ans.** (*i*) 47.0 N (*ii*) 49.0 N

15. Find the excess pressure inside a water drop of 0.4 mm in diameter if the surface tension of water is 0.075 N/m.   **Ans.** 750 Pa

16. A liquid drop 35 mm is radius has an internal pressure of 14 Pa greater than the outside pressure. Find the surface tension of the liquid film.

                                                    Ans. 0.245 N/m

17. The pressure outside the droplet of water of diameter 0.08 mm is 1.0133 bar (atmospheric pressure). Calculate the pressure within the droplet if surface tension is given as 0.072 N/m of water.   **Ans.** 1.049 bar

18. Calculate the capillary effect in millimetres in a glass tube of 4 mm diameter, when immersed in:

    (*i*) water and   (*ii*) mercury

    The temperature of the liquid is 20°C and the values of surface tension of water and mercury at 20°C in contact with air are 0.0736 N/m and 0.51 N/m. respectively. The angle of contact for water is zero and that for mercury 130°.

    Ans. (*i*) 7.51 mm (rise of water) (*ii*) 2.46 (fall of mercury)

19. If the pressure of a liquid is increased from 900 N/cm^2 to 1500 N/cm^2, the volume of liquid is decreased by 0.20%. Determine bulk modulus of elasticity and coefficient of compressibility of liquid.

    Ans. (*i*) 3 × 10^9 N/m^2 (*ii*) 3.33 × 10^{-10} m^2/N

20. Given that the bulk modulus of elasticity of water is 2 × 10^6 kN/m^2. Determine what increase in pressure is required to cause reduction in volume by 1% ? What would be the effect on the mass density of water?

    Ans. (*i*) 20,000 kN/m^2 (*ii*) 1%

2

Pressure and Its Measurement

2.1 PRESSURE AND ITS UNITS

Pressure is defined as normal force exerted by a fluid per unit area. It is also known as the intensity of pressure. It is usually more convenient to use pressure rather than force to describe the influences upon fluid behaviour. The counterpart of pressure in solids is normal stress. Since pressure is defined as force per unit area. That is,

$$\text{Pressure:} \quad p = \frac{\text{Force}:F}{\text{Area}:A} = \frac{F}{A}$$

If F is expressed in Newton (N) and A expressed in square metre (m^2), then the unit of p will be N/m^2. The standard unit of pressure is the pascal (Pa) or N/m^2.

Pressure is also used for solids as synonymous to normal stress, which is force acting perpendicular to the surface per unit area. For example, an object sitting on a surface, the force pressing on surface is the weight of the object but in different orientations it might have a different area in contact with the surface and therefore exert a different pressure as shown in Fig. 2.1.

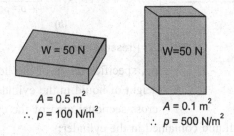

W = 50 N

$A = 0.5 \text{ m}^2$

$\therefore\ p = 100 \text{ N/m}^2$

W = 50 N

$A = 0.1 \text{ m}^2$

$\therefore\ p = 500 \text{ N/m}^2$

Fig. 2.1: Pressure varies with the area of contact with the surface.

© The Author(s) 2023
S. Kumar, *Fluid Mechanics (Vol. 1)*,
https://doi.org/10.1007/978-3-030-99762-5_2

There are many physical situations where pressure is the most important variable. If you are peeling an apple, then pressure is the key variable. If the knife is sharp, then the area of contact is small and you can peel with less force exerted on the knife. If you get an injection, then pressure is the most important variable in getting the needle through your skin. It is better to have a sharp needle than a dull one since the small area of contact implies that less force is required to push the needle through the skin.

Some important pressure with conversion:

$$1 \text{ Pa} = 1 \text{ N/m}^2$$

$$1 \text{ bar} = 10^5 \text{ Pa} = 0.1 \text{ MPa} = 100 \text{ kPa}$$

$$1 \text{ atm} = 101325 \text{ Pa} = 101.325 \text{ kPa} = 1.01325 \text{ bar}$$

$$1 \text{ kgf/cm}^2 = 9.81 \text{N/cm}^2 = 9.81 \times 10^4 \text{ N/m}^2 \text{ or Pa}$$

$$1 \text{ kgf/cm}^2 = 0.981 \text{ bar} = \frac{0.981}{1.01325} \text{ atm} = 0.968 \text{ atm}$$

$$\left[\because 1 \text{ bar} = \frac{1}{1.01325} \text{ atm} \right]$$

Note the pressure units **kgf/cm²**, **bar** and **atm** are almost equivalent to each other.

2.2 PRESSURE HEAD

Consider a vessel containing some liquid. We know that the liquid will exert pressure on all sides as well as bottom of the vessel. Now consider a bottomless cylinder be made to stand in the liquid as shown in Fig. 2.2.

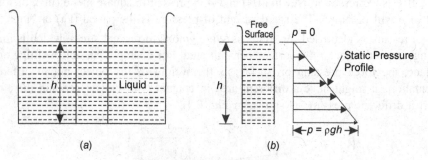

Fig. 2.2: Pressure head.

Let $w = \rho g$, specific weight of the liquid.

h = height of liquid in the cylinder.

A = cross-sectional area of the cylinder.

Then, weight of liquid contained in the cylinder:

= mass of liquid × acceleration due to gravity

$= m \times g = \rho V g$ ($\because$ Mass : $m = \rho V$, Volume : $V = Ah$)

$= \rho A h g$

Therefore, intensity of pressure at the base of the cylinder is given by:

$$p = \frac{\text{Weight of liquid in the cylinder}}{\text{Cross - sectional area of the cylinder base}}$$

$$p = \frac{\rho A h g}{A} = \rho g h \qquad \qquad ...(i)$$

where $\rho g = w = $ constant for a given liquid.

$$p \propto h$$

Thus, it is proved that the intensity of pressure at any point in a liquid is directly proportional to the depth of the point measured from free surface of liquid as shown in Fig. 2.2 (*b*).

From Eq. (*i*), we get

$$h = \frac{p}{\rho g}$$

Pressure head: $\qquad h = \frac{p}{\rho g} = \frac{p}{w}$ $\qquad$ (Specific weight: $w = \rho g$)

Here, p is called intensity of static pressure and h is called pressure head of the liquid.

If p is expressed in N/m^2 and

w is expressed in N/m^3

Then $\qquad\qquad h = \frac{N/m^2}{N/m^3} = m$

2.3 LAWS OF LIQUID PRESSURE, HYDROSTATIC EQUATION AND ITS APPLICATION

2.3.1 Laws of Liquid Pressure

(*i*) The surface of a liquid which is subjected to atmospheric pressure is called free surface of the liquid.

(*ii*) The free surface of a liquid is always horizontal.

(*iii*) Intensity of pressure at any point in a liquid is the same in all directions.

(*iv*) Intensity of pressure at any point in a liquid is directly proportional to the depth of the point from the free surface of liquid.

(*v*) Pressure on a surface submerged in a liquid acts normal to the surface.

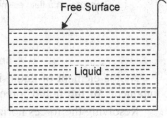

Fig. 2.3. Free Surface

2.3.2 Hydrostatic Equation

$$p = \rho g h \qquad \qquad ...(i)$$

where
- p = intensity of pressure of a static liquid at any point in it.
- ρ = density of liquid.
- g = acceleration due to gravity = 9.81 m/s²

and
- h = depth of the point from the free surface of liquid.

The above Eq. (i) is called **hydrostatic equation.**

2.3.3 Application

Hydrostatic equation is used to find out "total pressure" on immersed surface in a liquid and the position of the "centre of pressure".

2.4 PASCAL'S LAW

Pascal's law states that the intensity of pressure at any point in a fluid at rest is same in all directions.

Proof

Let us consider an arbitrary small fluid element of wedge shape ABC as shown in Fig. 2.4. Let the width of the fluid element is unity and p_x, p_y and p_z are the intensity of pressure acting on the face AC, BC and BA respectively.

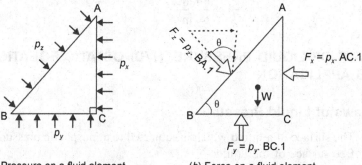

(a) Pressure on a fluid element   (b) Force on a fluid element

Fig. 2.4.

θ = angle of the element of the fluid.

As the element of liquid is at rest, therefore sum of the horizontal and vertical components of the liquid forces must be equal to zero.

Resolving the forces horizontally:

$$F_z . \sin\theta - F_x = 0$$
$$p_z\ BA \sin\theta - p_x AC = 0$$

$$p_z AC - p_x AC = 0$$

or $\qquad\qquad p_z = p_x \quad ...(i)$

$\because\ F_z = p_z . BA$
$\quad\ F_x = p_x AC$

$\angle ABC;\ \sin\theta = \dfrac{AC}{BA}$

or $\ AC = BA.\sin\theta$

Now resolving the forces vertically

$$F_z \cos\theta + W - F_y = 0$$

where $\qquad\qquad$ W = weight of the liquid element acts vertical downward direction

$$F_z = p_z.BA$$

and $\qquad\qquad$ $F_y = p_y.BC$

$$\therefore \quad p_z.BA \cos\theta + W - p_y.BC = 0$$

Since the element is very small and hence neglecting the weight of the liquid,

i.e., $\qquad\qquad$ $W \approx 0$ $\qquad\qquad\qquad\qquad$ $\because \angle ABC,$

$$\therefore \quad p_z.BA\cos\theta - p_y.BC = 0 \qquad\qquad \cos\theta = \frac{BC}{BA}$$

or $\quad p_z.BC - p_y.BC = 0$ $\qquad\qquad$ or $\quad BC = BA \cos\theta$

or $\qquad\qquad p_z = p_y$ $\qquad\qquad\qquad\qquad\qquad\qquad$ ...(ii)

From Eqs. (i) and (ii), we get

$$p_x = p_y = p_z$$

Hence, the intensity of pressure at any point in a fluid at rest is the same in all directions.

2.5 ATMOSPHERIC PRESSURE AND ITS MEASUREMENT

The atmospheric pressure is defined as the normal force exerted by atmospheric air per unit area. It is classified into two categories:

(i) Standard atmospheric pressure.

(ii) Local atmospheric pressure.

(i) **Standard Atmospheric Pressure:** This pressure is measured at sea level and 15°C temperature. The value of the standard atmospheric pressure is fixed:

i.e., $\qquad$ 760 mm of Hg = 101.325 kPa

$\qquad\qquad\qquad\qquad\qquad$ = 1.01325 bar

(ii) **Local Atmospheric Pressure:** This pressure is measured at any condition and place. The value of the local atmospheric pressure is not fixed, varies at given place with change in weather.

2.5.1 Measuring Atmospheric Pressure

The standard instrument used to measure the atmospheric pressure, is called Barometer. It is also known as *Torricelli Barometer*, because, the first measurement of atmospheric pressure began with a simple experiment performed by Evangeslista Torricelli in 1643. In this experiment, Torricelli immersed a tube, sealed at one end, in a container of mercury as shown in Fig. 2.5. Atmospheric pressure then forced mercury up into the tube that was considerably higher than the mercury in a container.

Writing a force balance in the vertical direction gives

$$p_{atm} \, A = W$$

$$p_{atm} \, A = \rho g h A$$

$$\boldsymbol{p_{atm} = \rho g h}$$

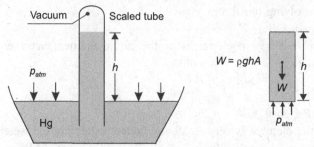

Fig. 2.5: Mercury Barometer (Torricelli's Barometer)

where ρ = density of mercury,

g = acceleration due to gravity, and

h = height of the mercury column above the free surface.

Note that the length and the cross sectional area of the tube have no effect on the height of the fluid column of a barometer as shown in Fig. 2.6.

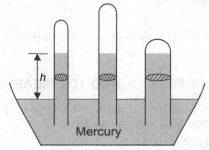

Fig. 2.6: Fluid column of a barometer is independent of the cross-sectional area of the tube.

If mercury is used in barometer, it is called mercury barometer and if water is used in barometer, it is called water barometer. Both types of barometers are used to measure standard atmospheric pressure, as shown in Fig. 2.7.

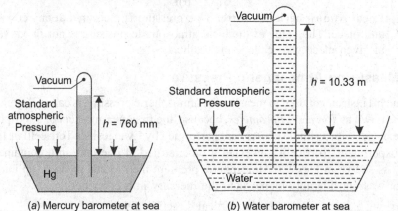

(a) Mercury barometer at sea level and 15°C atmospheric temp.

(b) Water barometer at sea level and 15°C atmospheric temp.

Fig. 2.7: Barometer.

The standard atmospheric pressure provides the force necessary, to push the mercury up (in mercury barometer) the evacuated tube is 760 mm. And to push the water up (in water barometer) the evacuated tube is 10.33 m.

Mathematically,

For Mercury Barometer	**For Water Barometer**
Standard atmospheric pressure	Standard atmospheric pressure
$\qquad$ = density $\times$ g $\times$ height	$\qquad$ = density $\times$ g $\times$ height
$\qquad$ = $\rho g h$	$\qquad$ = $\rho g h$
Substituting the value of	Substituting the value of
$\qquad \rho = 13600$ kg/m^3	$\qquad \rho = 1000$ kg/m^3
$\qquad g = 9.81$ m/s^2	$\qquad g = 9.81$ m/s^2
and $\qquad h = 760$ mm $= 0.76$ m	and $\qquad h = 10.33$ m
in above equation, we get	in above equation, we get
$\qquad$ = $13600 \times 9.81 \times 0.76$	$\qquad$ = $1000 \times 9.81 \times 10.33$
$\qquad$ = 101396.16 N/m^2	$\qquad$ = 101337 N/m^2
$\qquad$ = 101.39 kPa	$\qquad$ = 101.33 kPa
$\qquad$ = 1.013 bar	$\qquad$ = 1.013 bar

So, standard atmospheric pressure

$$= 760 \text{ mm of Hg} = 10.33 \text{ m of water}$$
$$= 1.013 \text{ bar}$$

In practical, water barometer is not used, because of

(*i*) its big size *i.e.*, water rise in evacuated tube is 10.33 m; as compared to mercury is 0.76 m.

(*ii*) the high vapour pressure as compared to mercury; consequently, the water barometer does not give consistent and accurate readings.

2.5.2 Aneroid Barometer

The aneroid barometer consists of a flexible short bellows called aneroid cell, which tightly sealed after removing the air at near zero pressure. The aneroid expands or squeezes due to variation in the atmospheric pressure. This motion causes the link L to move the pointer over the scale S. The deflections indicated by the pointer are relative to complete vacuum. The deflection given by the pointer indicate the absolute pressure.

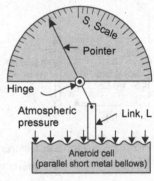

Fig. 2.8: Aneroid Barometer

The aneroid barometer is smaller and compact as compare to mercury barometer and much easier to read. It is the heart of altimeter used in modern aviation (Fig. 2.8).

Problem 2.1: Find the depth of oil of specific gravity of 0.82 which produces an intensity of pressure equal to 2.5 kN/m^2. Also find the pressure head in terms of water and mercury.

Solution: Given data:

Specific gravity of oil: $S = 0.82$

∴ Density of oil: ρ_{oil} = specific gravity of oil × density of water
$$= 0.82 \times 1000 = 820 \text{ kg/m}^3$$

Intensity of pressure: $p = 2.5 \text{ kN/m}^2 = 2500 \text{ N/m}^2$

Now, intensity of pressure:

$$p = \rho_{oil} \, g h_1$$

where h_1 = depth of oil

∴ $$h_1 = \frac{p}{\rho_{oil} g} = \frac{2500}{820 \times 9.81} = \mathbf{0.3107 \text{ m of oil}}$$

Let h_2 and h_3 pressure heads of water and mercury at same intensity of pressure (p).

∴ $$h_2 = \frac{p}{\rho_{water} g} \quad \text{for water}$$

$$= \frac{2500}{1000 \times 9.81} = \mathbf{0.2548 \text{ m of water}}$$

and $$h_3 = \frac{p}{\rho_{Hg} g} \quad \text{for mercury}$$

$$= \frac{2500}{13600 \times 9.81} = \mathbf{0.0187 \text{ m of Hg}}$$

∵ $\rho_{Hg} = 13600 \text{ kg/m}^3$

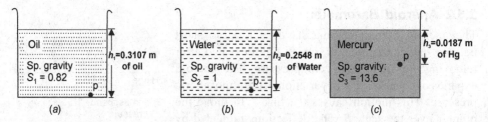

(a)   (b)   (c)

Fig. 2.9: Different values of pressure heads for oil, water and mercury of same intensity of pressure (p).

Problem 2.2:   Convert pressure head of mercury into equivalent pressure head of water.

Solution:

For mercury barometer,

Atmospheric pressure: $p = \rho_{Hg} \, g \, h_{Hg}$

where $\qquad \rho_{Hg}$ = 13600 kg/m³, density of mercury

$\qquad\qquad h_{Hg}$ = pressure head of mercury in metre (m)

$\qquad\qquad p$ = 13600 g h_{Hg} $\qquad\qquad$ ...(i)

For water barometer,

Atmospheric pressure: $\qquad p = \rho_{water}$ g h_{water}

where $\qquad\qquad \rho_{water}$ = 1000 kg/m³, density of water

$\qquad\qquad h_{water}$ = pressure head of water in metre (m)

$\qquad\qquad p$ = 1000 g h_{water} $\qquad\qquad$ ...(ii)

Equating Eq. (i) with Eq. (ii), we get

$\qquad\qquad$ 13600 g h_{Hg} = 1000 g h_{water}

or $\qquad\qquad \boldsymbol{h_{water}} = \boldsymbol{13.6\ h_{Hg}}$

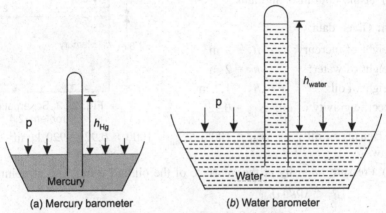

(a) Mercury barometer $\qquad$ (b) Water barometer

Fig. 2.10: Schematic for Problem 2.2

Problem 2.3: An open tank contains water upto a depth of 4 m and above it an oil of specific gravity 0.9 for a depth of 1.5 m. Find the intensity of pressure

$\qquad$ (i) at the interface of the two liquids and

$\qquad$ (ii) at the bottom of the tank.

Solution: Given data:

$\qquad$ Height of water: $\qquad h_1$ = 4 m

$\qquad$ Height of oil: $\qquad\quad h_2$ = 1.5 m

$\qquad$ Specific gravity of oil: $\quad S$ = 0.9

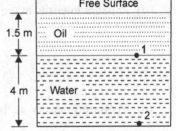

Fig. 2.11: Schematic for Problem 2.3

∴ Density of oil : ρ = 1000 × S kg/m³ = 1000 × 0.9 = 900 kg/m³

Now

$\qquad$ (i) Pressure intensity at the interface of the two liquids, *i.e.,* at point '1':

$$p_1 = \rho g h_2 = 900 \times 9.81 \times 1.5 \frac{kg}{m^3}\frac{m}{s^2}m$$

$$= 13243.5 \text{ N/m}^2 \text{ or Pa} = \mathbf{13.243 \ kPa}$$

(ii) Pressure intensity at the bottom, i.e., at point '2':

$$p_2 = (\rho g h_2)_{\text{oil}} + (\rho g h_1)_{\text{water}}$$

$$= p_1 + (\rho g h_1)_{\text{water}}$$

$$= 13243.5 + 1000 \times 9.81 \times 4 = 52483.5 \text{ N/m}^2 \text{ or Pa} = \mathbf{52.48 \ kPa}$$

Problem 2.4: An open tank contains mercury upto a depth 3 m and above mercury of water upto a depth 2 m and above water an oil of specific gravity 0.92 for a depth of 1.2 m. Find the intensity of pressure

    (i) at the interface of the oil and water.

    (ii) at the interface of the water and mercury

    (iii) at the bottom of the tank.

Solution: Given data:

Height of mercury:	$h_1 = 3$ m
Height of water:	$h_2 = 2$ m
Height of oil:	$h_3 = 1.2$ m
Specific gravity of oil:	$S_{\text{oil}} = 0.92$

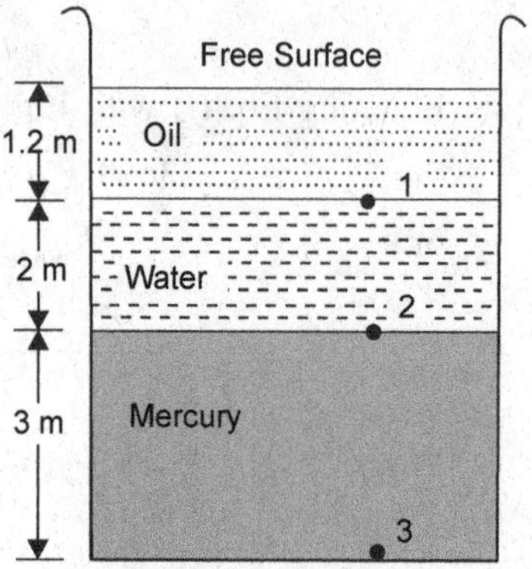

Fig. 2.12: Schematic for Problem 2.4

∴ Density of oil : $\rho_{\text{oil}} = 1000 \times S_{\text{oil}} = 1000 \times 0.92 = 920 \text{ kg/m}^3$

Now,

    (i) Pressure intensity at the interface of the oil and water, i.e., at point '1':

$$p_1 = (\rho g h_3)_{\text{oil}}$$

$$= 920 \times 9.18 \times 1.2 \ \frac{\text{kg}}{\text{m}^3} \frac{\text{m}}{\text{s}^2} \text{m}$$

$$= 10830.24 \text{ N/m}^2 \text{ or Pa} = \mathbf{10.83 \ kPa}$$

    (ii) Pressure intensity at the interface of the water and mercury, i.e., at point '2':

$$p_2 = (\rho g h_3)_{\text{oil}} + (\rho g h_2)_{\text{water}}$$

$$= p_1 + (\rho g h_2)_{\text{water}}$$

$$p_2 = 10830.24 + 1000 \times 9.81 \times 2$$

$$= 30450.24 \text{ N/m}^2 \text{ or Pa} = \mathbf{30.45 \ kPa}$$

    (iii) Pressure intensity at the bottom of the tank, i.e., at point '3':

$$p_3 = (\rho g h_3)_{\text{oil}} + (\rho g h_2)_{\text{water}} + (\rho g h)_{\text{Hg}}$$

$$= p_2 + (\rho g h)_{\text{Hg}}$$

$$= 30450.24 + 13600 \times 9.81 \times 3$$

$$= 430698.24 \text{ N/m}^2 = \mathbf{430.69 \ kPa}$$

Problem 2.5: A closed tank contains 0.5 m of mercury, 2 m of water, 3 m of oil of specific gravity 0.6 and there is air space above the oil. If the gauge pressure at the bottom of the tank is 19.62 N/cm², what is the pressure of air at the top of the tank.

Solution: Give data:

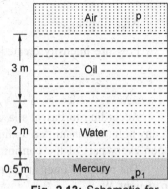

Height of mercury: $h_1 = 0.5$ m

Height of water: $h_2 = 2$ m

Height of oil: $h_3 = 3$ m

Specific gravity of oil: $S_{oil} = 0.6$

∴ Density of oil: $\rho_{oil} = 1000 \, S_{oil}$ kg/m³

     $= 1000 \times 0.6$

     $= 600$ kg/m³

Gauge pressure at the bottom of the tank:

$$p_1 = 19.62 \text{ N/cm}^2$$
$$= 19.62 \times 10^4 \text{ N/m}^2$$

Fig. 2.13: Schematic for Problem 2.5

Let $p = $ Pressure of air at the top of the tank.

Pressure at the bottom of the tank,

$$p_1 = (\rho g h)_{mercury} + (\rho g h)_{water} + (\rho g h)_{oil} + p$$
$$p_1 = \rho_{Hg} \, g h_1 + \rho_{water} \, g h_2 + \rho_{oil} \, g h_3 + p$$

where $\rho_{Hg} = 13600$ kg/m³, density of mercury

     $\rho_{water} = 1000$ kg/m³, density of water

∴ $19.62 \times 10^4 = 13600 \times 9.81 \times 0.5 + 1000 \times 9.81 \times 2 + 600 \times 9.81 \times 3 + p$

     $19.62 \times 10^4 = 6.67 \times 10^4 + 1.96 \times 10^4 + 1.76 \times 10^4 + p$

     $19.62 \times 10^4 = 10.39 \times 10^4 + p$

or $p = 19.62 \times 10^4 - 10.39 \times 10^4$

     $\mathbf{= 9.23 \times 10^4 \text{ N/m}^2 = 9.23 \text{ N/cm}^2}$

Problem 2.6: A hydraulic press has a ram of 400 mm diameter and a plunger of 50 mm diameter. Find the weight lifted by the hydraulic press when the force applied at the plunger is 800 N.

Solution: Given data:

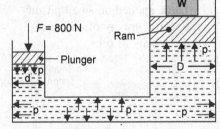

Diameter of ram: $D = 400$ mm

     $= 0.4$ m

Diameter of plunger $d = 50$ mm

     $= 0.05$ m

Force on plunger: $F = 800$ N

Now,

Weight lifted: $W = ?$

Fig. 2.14: Schematic for Problem 2.6

Area of ram: $A = \dfrac{\pi}{4} D^2$

     $= \dfrac{3.14}{4} (0.4)^2 = 0.1256 \text{ m}^2$

Area of plunger: $a = \dfrac{\pi}{4}d^2 = \dfrac{3.14}{4}(0.05)^2 = 1.962 \times 10^{-3}$ m²

Pressure intensity created by 800 N on the water through plunger is given by

$$p = \frac{F}{a} = \frac{800\,\text{N}}{1.962 \times 10^{-3}\,\text{m}^2} = \textbf{407747.19 N/m}^2$$

According to Pascal's law, the intensity of pressure applied to a confined fluid at any point is transmitted undiminished throughout the fluid in all directions and acts upon every part of the confining vessel at right angles to its interior surfaces.

The pressure intensity at the ram due to plunger = pressure intensity at ram due to weight W.

$$407747.19 \ \text{N/m}^2 = \frac{W}{A} \qquad\qquad \left[\because \frac{F}{a} = \frac{W}{A}\right]$$

or   Weight: $W = 407747.19 \times 0.1256$ N/m²·m²

$$= 51213.04 \ \text{N} = \textbf{51.213 kN}$$

Problem 2.7:   The diameter of a small piston and large piston of a hydraulic jack are 40 mm and 200 mm respectively. A force of 500 N is applied on the small piston. Find the load lifted by the large piston when:

(*i*) the pistons are at the same level, and

(*ii*) small piston is 500 mm above the large piston.

The specific gravity of the oil in the jack is 0.92.

Solution: Given data:

Diameter of small piston: $d = 40$ mm $= 0.04$ m

∴ Cross-sectional area of small piston: $a = \dfrac{\pi}{4}d^2 = \dfrac{\pi}{4}(.04)^2 = 1.256 \times 10^{-3}$ m²

Diameter of large piston: $D = 200$ mm $= 0.2$ m

∴ Cross-sectional area of large piston: $A = \dfrac{\pi}{4}D^2 = \dfrac{3.14}{4}(0.2)^2 = 0.0314$ m²

Force applied on small piston: $F = 500$ N

Specific gravity of oil: $S_l = 0.92$

∴ Density of oil: $\rho = 1000 \times 0.92 = 920$ kg/m³

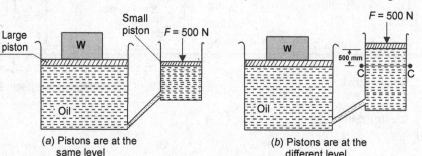

(a) Pistons are at the
same level

(b) Pistons are at the
different level

Fig. 2.15: Schematic for Problem 2.7

Now,   Let the load lifted $= W$

(i) When the pistons are at the same level, as shown in Fig. 2.14 (a),

Pressure intensity on small piston: $p = \dfrac{F}{a}$

$$= \frac{500}{1.256 \times 10^{-3}} \frac{N}{m^2} = 398089.17 \ N/m^2$$

According to Pascal's law, this pressure is transmitted undiminished in all directions.

$\therefore$ $\dfrac{F}{a}$ = pressure intensity on the large piston

$$398089.17 = \frac{W}{A}$$

$$398089.17 = \frac{W}{0.0134}$$

or $W = 5334.39 \ N = \textbf{5.334 kN}$

(ii) When the small piston is 500 mm above the large piston, as shown in Fig. 2.14(b).

Let p = intensity of downward pressure at C-C level

= intensity of pressure due to 500 N

+ intensity of pressure due to 500 mm of oil head

$$= \frac{F}{a} + \rho g h = \frac{500}{1.256 \times 10^{-3}} + 920 \times 9.81 \times 0.5$$

$$= 402601.772 \ N/m^2$$

This intensity of pressure on the large piston:

$$p = \frac{W}{A}$$

$$402601.772 = \frac{W}{0.0134}$$

or $W = 5394.86 \ N = \textbf{5.394 kN}$

2.6 ABSOLUTE, GAUGE AND VACUUM PRESSURE

2.6.1 Absolute Pressure

An absolute zero of pressure will occur when molecular momentum is zero. Such a situation can occur only when there is a perfect vacuum. The pressure measured with reference to absolute zero is called absolute pressure. It is abbreviated as p_{abs}.

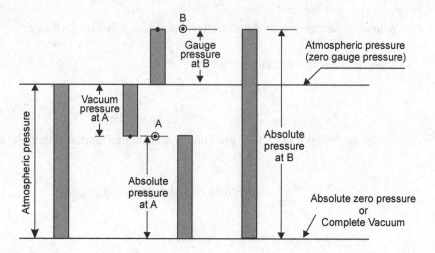

Fig. 2.16: Absolute, gauge, vacuum and atmospheric pressure.

2.6.2 Gauge Pressure

The pressure measured by a gauge (instrument) is relative to the atmospheric pressure is called gauge pressure. The instrument by which gauge pressure is measured, is called pressure gauge in which the atmospheric pressure is marked as zero.

i.e., The pressure gauges always indicate the pressure above the atmospheric pressure. It is abbreviated as p_g.

2.6.3 Vacuum Pressure

The pressure of a fluid to be measured is less than the atmospheric pressure, is called vacuum pressure. It is also known as negative or suction pressure. Vacuum pressure of a liquid is measured by an instrument is called vacuum gauge. It is abbreviated by p_{vac}.

The relations among absolute, atmospheric, gauge and vacuum pressure are shown in Fig. 2.14.

Mathematically,

(*i*) For positive gauge pressure,

Absolute pressure (p_{abs}) = Atmospheric pressure (p_{atm}) + Gauge pressure (p_g)

$$p_{abs} = p_{atm} + p_g$$

(*ii*) For vacuum pressure or negative gauge pressure,

Absolute pressure, (p_{abs}) = Atmospheric pressure (p_{atm}) – Vacuum pressure (p_{vac})

$$p_{abs} = p_{atm} - p_{vac}$$

Problem 2.8: Convert the following reading of pressure to kPa, assuming that the barometer reads 760 mm of Hg.

(*i*) 40 cm of Hg vacuum

(*ii*) 90 cm of Hg gauge

(*iii*) 1.2 m of H_2O gauge.

Solution: Given data:

Barometer reading = 760 mm of Hg = 0.760 m of Hg

$\therefore$ Atmospheric pressure: $p_{atm} = (\rho g h)_{Hg}$

$= 13600 \times 9.81 \times 0.760$ Pa

$= 101396.16$ Pa $= 101.39$ kPa

(*i*) 40 cm Hg vacuum

Vacuum pressure head = 40 cm Hg = 0.40 m Hg

$\therefore$ Vacuum pressure: $p_{vac} = (\rho g h)_{Hg}$

$= 13600 \times 9.81 \times 0.40$ Pa

$= 53366.4$ Pa $= 53.36$ kPa

Absolute pressure: $p_{abs} = p_{atm} - p_{vac}$

$= 101.39 - 53.36 = \textbf{48.03 kPa}$

(*ii*) 90 cm Hg gauge

Gauge pressure head = 90 cm Hg = 0.90 m Hg

$\therefore$ Gauge pressure: $p_g = (\rho g h)_{Hg} = 13600 \times 9.81 \times 0.90$

$= 120074.4$ Pa $= 120.07$ kPa

Absolute pressure: $p_{abs} = p_{atm} + p_g$

$= 101.39 + 120.07 = \textbf{221.46 kPa}$

(*iii*) 1.2 m of H_2O gauge

Gauge pressure head = 1.2 m of H_2O

$\therefore$ Gauge pressure: $p_g = (\rho g h)_{H_2O} = 1000 \times 9.81 \times 1.2$

$= 11772$ Pa $= 11.77$ kPa

Absolute pressure: $p_{abs} = p_{atm} + p_g$

$= 101.39 + 11.77 = \textbf{113.16 kPa}$

Problem 2.9: Determine the vacuum pressure in meter of water when the absolute pressure is 0.622001 bar. Assume atmospheric pressure as 10.33 metre of water.

Solution: Given data:

Absolute pressure: $p_{abs} = 0.622001$ bar $= 62200.1$ N/m^2

Atmospheric pressure: $p_{atm} = 10.33$ m of water

Now,

Absolute pressure: $p_{abs} = \rho g h$

where h = absolute pressure head in metre of water

                               ρ = 1000 kg/m³, density of water

Absolute pressure head in water:

$$h = \frac{p_{abs}}{\rho g} = \frac{62200.1 \text{ N}/\text{m}^2}{1000 \times 9.81 \dfrac{\text{kg}}{\text{m}^3} \cdot \dfrac{\text{m}}{\text{s}^2}} = 6.340 \text{ m}$$

We know,

        Absolute pressure head = atmospheric pressure head – vacuum pressure head

                  6.340 = 10.33 – vacuum pressure head

or   Vacuum pressure head = **3.99 m of water**

Problem 2.10: A point lies 7.5 m below free surface of water. Determine the pressure of the point in: (*i*) kPa (*ii*) mm of Hg.

If atmospheric pressure of 760 mm of Hg, determine the absolute pressure in: (*iii*) m of water (*iv*) bar and (*v*) mm of Hg.

Solution:

    (*i*) The pressure of water at point A:

$$p_A = \rho g h = 1000 \times 9.81 \times 7.5 \ \frac{\text{kg}}{\text{m}^3} \cdot \frac{\text{m}}{\text{s}^2} \cdot \text{m}$$

$$= 73575 \text{ N/m}^2$$

    or p_A = **73.575 kPa**

(*ii*) Let h_{Hg} = pressure head of mercury in metre equivalent to given 7.5 m of water

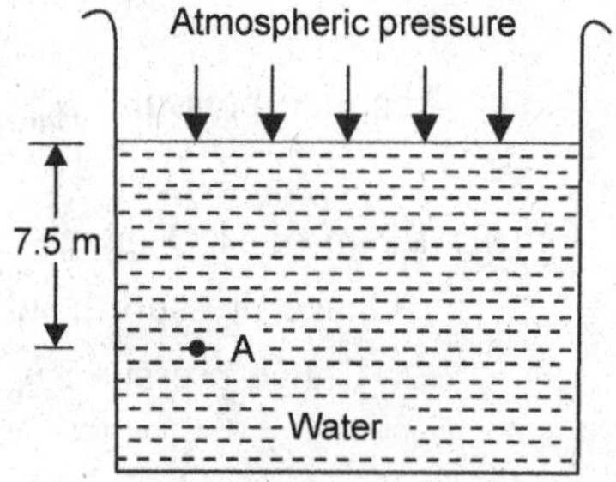

$$(\rho g h)_{Hg} = (\rho g h)_{water}$$

$$\rho_{Hg} \times g \times h_{Hg} = \rho_{water} \times g \times h_{water}$$

$$h_{Hg} = \frac{\rho_{water}}{\rho_{Hg}} \cdot h_{water}$$

Atmospheric pressure

7.5 m

A

Water

Fig. 2.17 Schematic for Problem 2.10

$$= \frac{1000}{13600} \times 7.5 = 0.55147 \text{m} = \textbf{551.47 mm}$$

(*iii*) Absolute pressure at point A:

$$p_{abs} = p_{atm} + (\rho g h)_{water}$$

$$= (\rho g h)_{Hg} + (\rho g h)_{water}$$

$$= 13600 \times 9.81 \times 0.76 + 1000 \times 9.81 \times 7.5$$

$$= 174971.16 \text{ N/m}^2$$

Let h_{water} = absolute pressure head of water in metre,

$$\therefore \qquad p_{abs} = \rho_{water}gh_{water}$$
$$174971.16 = 1000 \times 9.81 \times h_{water}$$
$$\text{or} \qquad h_{water} = \textbf{17.836 m of water}$$

(*iv*) Absolute pressure:

$$p_{abs} = 174971.66 \text{ N/m}^2 = \textbf{1.749 bar} \qquad \because 1 \text{ bar} = 10^5 \text{ N/m}^2$$

(*v*) Let $\quad h_{Hg}$ = absolute pressure head of mercury in metre

$$\therefore \qquad p_{abs} = \rho_{Hg}gh_{Hg}$$
$$174971.16 = 13600 \times 9.81 \times h_{Hg}$$
$$\text{or} \qquad h_{Hg} = 1.31147 \text{ m} = \textbf{1311.47 mm of Hg}$$

OR

$$\text{or} \qquad h_{Hg} = \frac{h_{water}}{13.6} = \frac{17.836}{13.6}$$
$$= 1.31147 \text{ m} = \textbf{1311.47 mm of Hg}$$

2.7 MEASUREMENT OF PRESSURE

The pressure of a fluid is measured by using the following two principles:

Principle-I

The pressure of a fluid is measured at a point in a fluid by balancing the column of same liquid or the column of another liquid. According to the Principle-I, the instruments used to measure the pressure of a liquid, are called manometers:

Manometers are classified as:
(*i*) Simple manometers
(*ii*) Differential manometers.

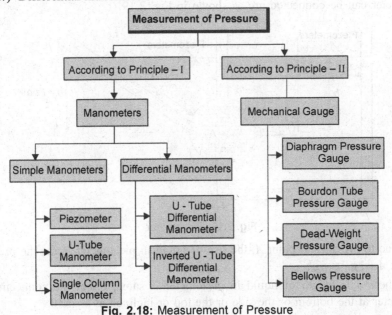

Fig. 2.18: Measurement of Pressure

Principle-II

The pressure of a fluid is measured at a point in a fluid by balancing the force of a spring or dead weight. According to the Principle-II, the instruments used to measure the pressure of a fluid, are called mechanical gauges.

Mechanical gauges are classified as:

(*i*) Diaphragm pressure gauge.

(*ii*) Bourdon tube pressure gauge.

(*iii*) Dead-weight pressure gauge and

(*iv*) Bellows pressure gauge.

2.7.1 Simple Manometers

Simple manometers are used to measure the pressure of the fluid in a pipe or vessel. A simple manometer consists of a glass tube having one of its ends connected to the gauge point where the pressure is to be measured and other end remains open to atmosphere. Some type of simple manometers are as given below:

(*a*) Piezometer

(*b*) U-tube manometer, and

(*c*) Single column manometer (or Micro manometer).

Piezometer

It is the simplest form of manometer which is used for measuring the pressure above the atmospheric pressure *i.e,* gauge pressure. One end of the piezometer is connected to the point where pressure is to be measured and other end is opened to the atmosphere. Piezometers may be inserted in the top or the side or the bottom of the containers, but the liquid will rise the same level. The various manners in which piezometer can be connected are as shown in Fig. 2.19.

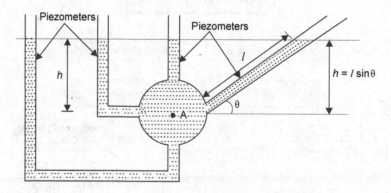

Fig. 2.19: Piezometer

Due to pressure at point A, the rise of liquid in piezometer h, so, the pressure at point A is, $p_A = \rho g h$

where h is height of liquid in piezometer is same, either the connection of piezometer at the bottom or the side or the top or inclined at angle θ.

Limitations of Piezometer

1. Piezometer cannot be used to measure very high pressure of a liquid, because in that case a long glass tube will be required. Long glass tube is inconvenient to be used.

Mathematically, we know

Pressure: $p = \rho g h$

or $p \propto h$

Thus, at same liquid, pressure head: (h) increases with increase in pressure; (p).

2. It cannot be used to measure the gauge pressure of a gas *i.e.*, leakage of gas through piezometer.

3. It cannot measure the vacuum pressure. Instead of column of liquid going up in the tube, the atmospheric air will enter into the pipe through the glass tube. Hence, it can neither measure high pressure nor vacuum pressure of a liquid.

U-tube Manometer

A U-tube manometer consists of a glass tube bent in U-shape. One end of U-tube is connected to the gauge point where pressure of fluid is to be measured and the other end remains open to the atmosphere. The U-tube manometers are used to measure high and low [vacuum] pressure of the liquids or gases. The U-tube generally contains liquid whose:

(*i*) Density is greater than the density of the fluid in pipe or vessel whose pressure to be measured.

(*ii*) Do not mix with the liquid in pipe or vessel whose the pressure is to be measured.

S. No.	Fluid in Pipe or Vessel	Liquid in U-tube
1.	Water, oil	Mercury
2.	Gases at low pressure	Water, oil, alcohol
3.	Gases at high pressure	Mercury

(*i*) **For Gauge Pressure:** Let point A at which pressure is to be measured. The datum line is (1)-(1).

Let h_1 = height of a liquid in vessel or pipe above the datum line.

h_2 = height of a manometric liquid above the datum line.

s_1 = specific gravity of liquid in vessel or pipe

s_2 = specific gravity of liquid in manometer.

As the pressure or pressure head is the same for the horizontal surface. Hence pressure above the horizontal datum line (1)-(1) in the left hand limb and in the right hand limb of U-tube manometer should be same. Refer Fig. 2.20 (*a*).

Gauge pressure of fluid in left hand limb above the datum line = gauge pressure of fluid in right hand limb above the datum line.

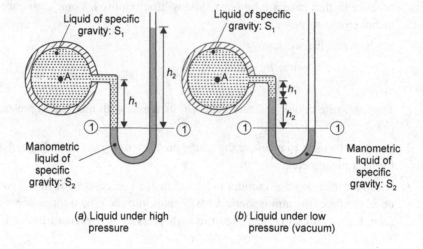

(a) Liquid under high pressure

(b) Liquid under low pressure (vacuum)

Fig. 2.20: U-tube manometer.

OR

Pressure of liquid in pipe at point A + pressure of liquid in left hand limb due to pressure head h_1 = pressure of liquid in right hand limb due to pressure head h_2.

$$p_A + \rho_1 g h_1 = \rho_2 g h_2$$
$$p_A = \rho_2 g h_2 - \rho_1 g h_1$$
$$\rho_1 g h_A = \rho_2 g h_2 - \rho_1 g h_1$$

Pressure head at point A,

$$h_A = \frac{\rho_2}{\rho_1} h_2 - h_1$$

(ii) For Vacuum Pressure: As shown in Fig. 2.20 (b)

Pressure above the horizontal datum line (1)-(1) in the left hand limb and in the right hand limb of U-tube manometer should be same:

Gauge pressure of fluid in left hand limb above the datum line = gauge pressure of fluid in right hand limb above the datum line

$$\therefore \quad p_A + \rho_1 g h_1 + \rho_2 g h_2 = 0$$
$$p_A = -(\rho_1 g h_1 + \rho_2 g h_2)$$
$$\rho_1 g h_A = -(\rho_1 g h_1 + \rho_2 g h_2)$$

Pressure head at point A,

$$h_A = -\left(h_1 + \frac{\rho_2}{\rho_1} h_2\right)$$

Single Column Manometer or Micromanometer

The single column manometer is a modified form of the U-tube manometer in which a reservoir having a large cross-sectional area (nearly 100 times) as compared to the area of the tube is connected to the left limb of the manometer whose end is connected to gauge point where the pressure is to be measure as shown in Fig. 2.21. Due to large cross-sectional area of the reservoir, the change in liquid level in the reservoir, will be very small, which may be neglected. Hence the reading in the right limb only is to be taken. Since there is no need to take any reading corresponding to the liquid surface in the reservoir, left limb and reservoir need not be made transparent. A micromanometer (or single column manometer) is used for measuring low pressure, where accuracy is not much importance. The single column manometer is classified into two types as:

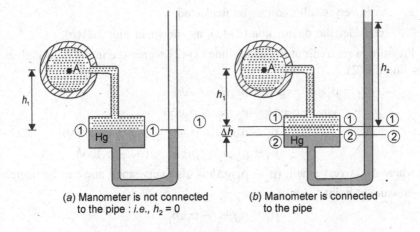

(a) Manometer is not connected to the pipe : i.e., $h_2 = 0$

(b) Manometer is connected to the pipe

Fig. 2.21: Vertical single column manometer.

(i) Vertical single column manometer, and

(ii) Inclined single column manometer.

(i) **Vertical Single Column Manometer:** Let (1)-(1) be the level of mercury in the reservoir and in the right limb of the manometer as shown in Fig. 2.21(a), when it is not connected to the pipe. When manometer is connected to the pipe, due to high pressure at gauge point A, the mercury in the reservoir will be pushed downward (Δh) and mercury rise (h_2) in the right limb as shown in Fig. 2.21 (b).

Let h_2 = rise of mercury (or manometric liquid) in right limb.

h_1 = height of centre of pipe above (1)-(1).

i.e., the value of h_1 is fixed.

Δh = level of mercury (or manometric liquid) fall in reservoir

A = cross-sectional area of the reservoir

a = cross-sectional area of the right limb

ρ_1 = density of liquid in pipe

ρ_2 = density of mercury or manometric liquid

The volume of mercury fall in reservoir is equal to volume rise in right limb.

i.e.,
$$A \cdot \Delta h = a h_2$$

$\therefore$
$$\Delta h = \frac{a}{A} h_2$$

where $A >> a$

ratio $\dfrac{a}{A}$ becomes very small and

$\dfrac{a}{A} h_2$ is also very small and can be neglected.

$\therefore$ Δh is very small and can be neglected.

Now consider the datum line (2)-(2), as shown in Fig. 2.21(*b*).

Pressure in reservoir at the datum line (2)-(2) = pressure in the right limb at the datum line (2)-(2).

$$p_A + \rho_1 g(h_1 + \Delta h) = \rho_2 g(h_2 + \Delta h)$$
$$p_A + \rho_1 g h_1 + \rho_1 g \Delta h = \rho_2 g h_2 + \rho_2 g \Delta h$$
$$p_A = \rho_2 g h_2 - \rho_1 g h_1 + \rho_2 g \Delta h - \rho_1 g \Delta h$$
$$p_A = \rho_2 g h_2 - \rho_1 g h_1 + (\rho_2 - \rho_1) g \Delta h$$

where Δh is very small, $(\rho_2 - \rho_1) g \Delta h$ is also very small and can be neglected.

Pressure at point 'A' in the pipe:

$$p_A = \rho_2 g h_2 - \rho_1 g h_1 \qquad \qquad \ldots(i)$$

Pressure head: $\quad h_A = \dfrac{p_A}{\rho_1 g} = \dfrac{\rho_2}{\rho_1} h_2 - h_1 \qquad \ldots(ii)$

From above Eqs. (*i*) and (*ii*),

$$h_1 = \text{constant}, \rho_1, \rho_2 \text{ are also constant.}$$

So, pressure head or pressure only varies with h_2.

(2) Inclined Single Column Manometer: This is an improvement over the vertical single over the vertical single column manometer. Inclined single column mano-meter is shown in Fig. 2.22. This manometer or more sensitive, due to inclination the distance moved by the mercury or mano-metric liquid in the right limb will be more.

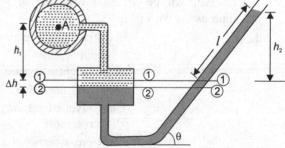

Fig. 2.22: Inclined single column manometer.

Let $\qquad \theta$ = inclination of right limb with horizontal

$\qquad l$ = length of mercury moved in right limb from (1)-(1)

$\qquad h_2$ = vertical rise of mercury in right limb from (1)-(1)

$\qquad = l \sin \theta$.

Now consider the datum line (2)-(2), as shown in Fig. 2.22.

Pressure in reservoir at the datum line (2)-(2) = pressure in the right limb at the datum line (2)-(2).

$$p_A + h_1 \rho_1 g + \rho_1 g \Delta h = \rho_2 g \Delta h \sin\theta + \rho_2 g h_2 \sin\theta$$

$$p_A = \rho_2 g h_2 \sin\theta - \rho_1 g h_1 + (\rho_2 \sin\theta - \rho_1) g \Delta h$$

where Δh is very small,

So, $(\rho_2 \sin\theta - \rho_1) g \Delta h$ can be neglected.

Pressure at point 'A' in the pipe:

$$p_A = \rho_2 g h_2 \sin\theta - \rho_1 g h_1 \qquad \qquad ...(i)$$

or Pressure head: $\qquad h_A = \dfrac{p_A}{\rho_1 g} = \dfrac{\rho_2}{\rho_1} h_2 \sin\theta - h_1 \qquad ...(ii)$

Problem 2.11: The right limb of a simple U-tube manometer containing mercury is open to the atmosphere while the left limb is connected to a pipe in which a liquid of specific gravity 0.95 is flowing. The centre of the pipe is 0.3 m below the level of mercury in the right limb. Find the pressure of liquid in the pipe if the difference between the level of mercury in the two limbs is 0.5 m.

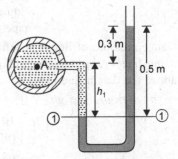

Fig. 2.23: U-tube mercury manometer.

Solution: Given data:

Specific gravity of liquid: $S = 0.95$

$\therefore$ Density of liquid: $\quad \rho_1 = S \times \rho_{\text{water}}$

$\qquad = 0.95 \times 1000$

$\qquad = 950 \text{ kg/m}^3$

Let (1)-(1) is the horizontal datum line; the pressure in the left hand limb and in the right hand limb on the datum line (1)-(1) should be same.

$\therefore$ Gauge pressure of liquid in left hand limb above the datum line = gauge pressure of liquid in right hand limb above the datum line (1)-(1).

OR

Pressure of liquid in pipe at point A + pressure of liquid in left hand limb due to pressure head h_1 = pressure of liquid in right hand limb due to pressure head h_2.

$$p_A + \rho_1 g h_1 = \rho_2 g h_2$$

where $\qquad h_1 = 0.5 - 0.3 = 0.2 \text{ m}$

and $h_2 = 0.5$ m

Density of mercury: $\rho_2 = 13600$ kg/m³

$\therefore \quad p_A + 950 \times 9.81 \times 0.2 = 13600 \times 9.81 \times 0.5$

or $p_A = 64844.1$ N/m² or Pa = **64.844 kPa**

Problem 2.12: A simple U-tube manometer containing mercury is connected to a pipe in which a liquid of specific gravity 0.9 and having vacuum pressure is flowing. The other end of the manometer is open to atmosphere. If the difference of mercury level in the two limbs is 50 cm and the height of liquid in the left limb from the centre of pipe is 20 cm below, find the vacuum pressure in pipe.

Solution: Given data:

Specific gravity of liquid flowing through pipe: $S_1 = 0.9$

$\therefore$ Density of liquid: $\rho_1 = S_1 \times \rho_{water} = 0.9 \times 1000 = 900$ kg/m³

Density of mercury: $\rho_2 = 13600$ kg/m³

Difference of mercury level:

$$h_2 = 50 \text{ cm} = 0.50 \text{ m}$$

Height of liquid in left limb:

$$h_1 = 20 \text{ cm} = 0.2 \text{ m}$$

Let p_A = gauge pressure of liquid at the centre of pipe:

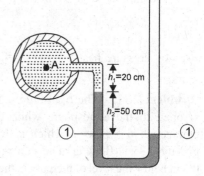

Fig. 2.24: Schematic for Problem 2.12

Gauge pressure of liquid in left limb above the datum line = gauge pressure of liquid in right limb above the datum line (1)-(1).

$$p_A + \rho_1 g h_1 + \rho_2 g h_2 = 0$$

or $p_A = - (\rho_1 g h_1 + \rho_2 g h_2)$

$$= - (900 \times 9.81 \times 0.2 + 13600 \times 9.81 \times 0.5)$$

$$= - 68473.8 \text{ N/m}^2 \text{ or Pa} = - \textbf{68.473 kPa}$$

The −ve sign indicates that the gauge pressure (*i.e.*, pressure due to liquid) is below atmospheric pressure. In other words, vacuum pressure is 68.473 kPa.

Problem 2.13: A U-tube manometer is used to measure the pressure of water in a pipe line, which is in excess of atmospheric pressure. The right limb of the manometer contains mercury and is open to atmosphere. The contact between water and mercury is in the left limb. Determine the pressure of water in the main line, if the difference in level of mercury in the limbs of U-tube is 20 cm and the free surface of mercury is in level with the centre of the pipe. If the pressure of water in pipe line is increased to 50%; calculate the new difference in the level of mercury. Sketch the arrangement in both cases.

Solution: Case-I: As shown in Fig. 2.25.

Difference of mercury level: $h = 20$ cm = 0.2 m

Density of water: $\rho_1 = 1000$ kg/m³

Density of mercury: $\rho_2 = 13600$ kg/m³

Let p_A = pressure of water in pipe line (*i.e.,* at point A)

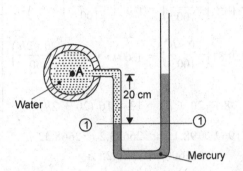

Fig. 2.25: Schematic for case - I, Problem 2.13

As we know, the gauge pressure above the datum line (1)-(1) is same in left limb and is right limb.

$$\therefore \qquad p_A + \rho_1 gh = \rho_2 gh$$

or
$$p_A = \rho_2 gh - \rho_1 gh$$
$$= gh\,(\rho_2 - \rho_1)$$
$$= 9.81 \times 0.2 \times (13600 - 1000) = \mathbf{24721.2 \ N/m^2}$$

Case-II: As shown in Fig. 2.26.

The pressure of water in pipe line is increased to 50%.

i.e.,
$$p_A = 24721.2 + 0.5 \times 24721.2 = 37081.8 \ N/m^2$$

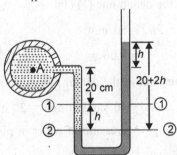

Fig. 2.26: Schematic for case - II, Problem 2.3

In this case, the mercury level in left limb will fall. The fall of mercury level in left limb will be equal to the rise of mercury level in right limb as the total volume of mercury remains same.

Let h = fall of mercury level in left limb in cm;

The rise of mercury level in right limb = x cm.

Now, the gauge pressure in left limb is equal the gauge pressure in right limb at the new datum line (2)-(2):

we get,

$$p_A + \rho_1 g \,(20 + h) = \rho_2 g \,(20 + 2h) \qquad (20 + h) \to \text{cm}$$

$$p_A + \rho_1 g \left(\frac{20+h}{100}\right) = \rho_2 g \left(\frac{20+2h}{100}\right) \qquad \left(\frac{20+h}{100}\right) \to \text{m}$$

$$37081.8 + 1000 \times 9.81 \left(\frac{20+h}{100}\right) = 13600 \times 9.81 \times \left(\frac{20+2h}{100}\right) \qquad \text{Similarly}$$

$$37081.8 + 98.1\,(20 + h) = 1334.16\,(20 + 2h) \qquad \left(\frac{20+h}{100}\right) \to \text{m}$$

$$37081.8 + 1962 + 98.1\,h = 26683.2 + 2668.32\,h$$

or $\qquad\qquad\qquad 12360.6 = 2570.22\,h$

or $\qquad\qquad\qquad\quad h = 4.809 \text{ cm}$

∴ New difference of mercury level

$$= 20 + 2h \text{ cm}$$
$$= 20 + 2 \times 4.809 = \mathbf{29.618 \text{ cm}}$$

Problem 2.14: A simple manometer having an inclined limb is used to measure intensity of pressure in a pipeline connected to the manometer as shown in Fig. 2.27. If mercury is used as the manometric liquid, determine the pressure of water in kPa.

Solution: Given data:

Height of pipe centre above the datum line (1)-(1):

$$h_1 = 60 \text{ mm}$$
$$= 0.06 \text{ m}$$

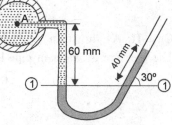

Fig. 2.27: Schematic for Problem 2.14

Vertical height of the mercury above the datum line in right limb:

$$h_2 = 40 \sin 30° \text{ mm} = 20 \text{ mm} = 0.02 \text{ m}$$

Let $\qquad\qquad\qquad p_A = $ pressure of water in pipeline

Density of water: $\qquad \rho_1 = 1000 \text{ kg/m}^3$

Density of mercury: $\qquad \rho_2 = 13600 \text{ kg/m}^3$

The gauge pressure above the datum line (1)-(1) is equal in left limb and in right limb: we get

$$p_A + \rho_1 g h_1 = \rho_2 g h_2$$
or $\qquad\qquad p_A = \rho_2 g h_2 - \rho_1 g h_1$
$$= 13600 \times 9.81 \times 0.02 - 1000 \times 9.81 \times 0.06$$
$$= 2079.72 \text{ N/m}^2 \text{ or Pa} = \mathbf{2.0792 \text{ kPa}}$$

Problem 2.15: The pressure of water flowing through a pipeline is measured by a manometer shown in Fig.2.8. The manometric liquid is mercury in all the tubes and water is enclosed between mercury columns. Find the pressure of water in the pipeline.

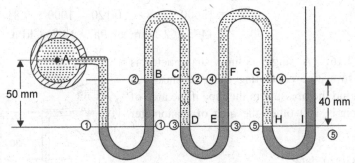

Fig. 2.28: Schematic for Problem 2.15

Solution:

As we know, the gauge pressure on the datum line (5)-(5) at points H and I is same.

$$\therefore \qquad p_H = p_I = \rho_{Hg} g h$$
$$= 13600 \times 9.81 \times 0.04 = 5336.64 \text{ N/m}^2$$

The gauge pressure below the datum line (4)-(4) at points F and G is same

$$\therefore \qquad p_F = p_G = p_H - \rho_{H_2O} g h$$

$$= 5336.64 - 1000 \times 9.81 \times 0.04 = 4944.24 \text{ N/m}^2$$

The gauge pressure above the datum line (3)-(3) at points D and E is same.

$$\therefore \qquad p_D = p_E = p_F + \rho_{Hg} g h$$
$$= 4944.24 + 13600 \times 9.81 \times 0.04 = 10280.88 \text{ N/m}^2$$

The gauge pressure below the datum line (3)-(3) at points B and C is same.

$$\therefore \qquad p_B = p_C = p_D - \rho_{H_2O} g h$$
$$= 10280.88 - 1000 \times 9.81 \times 0.04 = 9888.48 \text{ N/m}^2$$

The gauge pressure above the datum line (1)-(1) is same.

$$p_A + \rho_{H_2O} g h_1 = \rho_{Hg} g h + p_B$$

or $\qquad p_A + 1000 \times 9.81 \times 0.05 = 13600$
$\times 9.81 \times 0.04 + 9888.48$

$$p_A =$$

14734.62 N/m^2 or Pa = **14.734 kPa**

OR

Total height of mercury above datum lines (1)-(1), (3)-(3) and (5)-(5) = $3 \times 40 = 120$ mm

The equivalent simple manometer will be as shown in Fig. 2.29.

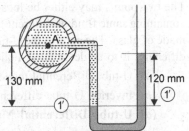

Fig. 2.29: Equivalent Simple manometer

The gauge pressure above the datum line $(1')$-$(1')$ is same:

$\therefore \qquad p_A + \rho_{H_2O}\, g h_1 = \rho_{Hg} g h_2$

$$p_A = \rho_{Hg} g h_2 - \rho_{H_2O}\, g h_1$$
$$= 13600 \times 9.81 \times 0.120 - 1000 \times 9.81 \times 0.130$$
$$= 14734.62 \text{ N/m}^2 \text{ or Pa} = \textbf{14.734 kPa}$$

Problem 2.16: A single column manometer is connected to a pipe containing water as shown in Fig. 2.30. Find the pressure in the pipe if the area of the reservoir is 100 times the area of manometer tube.

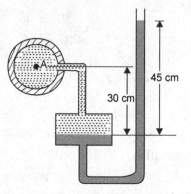

Solution: Given data:

$$\frac{\text{Area of reservoir}}{\text{Area of right limb}} = \frac{A}{a} = 100$$

Height of water: $\quad h_1 = 30 \text{ cm} = 0.30 \text{ m}$

Rise of mercury in right limb:

$$h_2 = 45 \text{ cm} = 0.45 \text{ m}$$

Fig. 2.30: Schematic for Problem 2.16

As we know, the gauge pressure at point A:

$$p_A = \rho_2 g h_2 - \rho_1 g h_1 + (\rho_2 - \rho_1) g \Delta h$$

$$= \rho_2 g h_2 - \rho_1 g h_1 + h_2 \times \frac{a}{A} (\rho_2 - \rho_1) g$$

where $\qquad \Delta h = h_2 \times \dfrac{a}{A}$

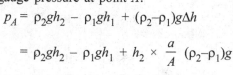

$\therefore \qquad p_A = 13600 \times 9.81 \times 0.45 - 1000 \times 9.81 \times 0.30 + 0.45 \times \dfrac{1}{100}$

$$\times (13600 - 1000) \times 9.81$$

$$p_A = 57650.42 \text{ N/m}^2 \text{ or Pa} = \textbf{57.65 kPa}$$

2.7.2 Differential Manometers

A differential manometer is used to measure the difference in pressure at two points. The two points may either be located in the same pipe or vessel, or in different pipes containing same fluid or different fluids. A differential manometer consists of a U-tube made of glass. The ends of the U-tube are connected to two gauge points whose pressure difference is to be determined. Following types of differential manometers are used:

    (*a*) U-tube differential manometer, and

    (*b*) Inverted U-tube differential manometer.

    (*a*) **U-tube Differential Manometer: (Fig. 2.31)**

Let $\quad p_A =$ pressure of the liquid in pipe A

$\quad\quad p_B =$ pressure of the liquid in pipe B

ρ_1 = density of the liquid in pipe A

ρ_2 = density of the liquid in pipe B

ρ_{Hg} = density of the manometric liquid (mercury)

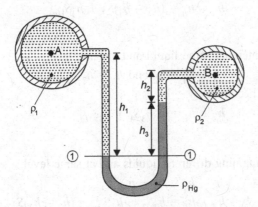

Fig. 2.31: U-tube differential manometer.

Let (1)-(1) be the datum line.

h_1 = height of the centre of the pipe A above
the datum line (1)-(1)

h_2 = height of the centre of the pipe B above the surface of

manometric liquid in the Left limb of the U-tube

h_3 = difference of mercury level in the U-tube.

Pressure in the left limb at the datum line (1)-(1) = Pressure in the right limb at the datum line (1)-(1)

$$p_A + \rho_1 g h_1 = p_B + \rho_2 g h_2 + \rho_{Hg} g h_3$$
$$p_A - p_B = \rho_2 g h_2 + \rho_{Hg} g h_3 - \rho_1 g h_1 \qquad \ldots(i)$$

Above Eq. (i), the pressure difference between pipe A and pipe B.

Let ρ_w = density of water,

Dividing Eq. (i) by $\rho_w g$.

$$\frac{p_A}{\rho_w g} - \frac{p_B}{\rho_w g} = \frac{\rho_2 g h_2}{\rho_w g} + \frac{\rho_{Hg}}{\rho_w g}.g h_3 - \frac{\rho_1 g h_1}{\rho_w g}$$

$$h_A - h_B = h_2 S_2 + h_3 S_{Hg} - h_1 S_1$$

where S_1, S_2 and S_{Hg} are specific gravity of liquid in pipe A, pipe B and manometric liquid (mercury) respectively.

If h_1, h_2 and h_3 in m, then pressure head difference between pipe A and pipe B in m of water.

i.e., $\qquad h_A - h_B = h_2 S_2 + h_3 S_{Hg} - h_1 S_1$ **m of water** $\qquad \ldots(ii)$

Some useful results for different conditions may be derived as follows:

Case-I

Two pipes containing same liquid,

i.e., $\qquad\qquad\qquad S_1 = S_2$

The above Eq. (*ii*), becomes

$$h_A - h_B = (h_2 - h_1)S_2 + h_3 S_{Hg}$$

Case-II

Two pipes containing same liquid and at same level,

i.e., $\qquad\qquad S_1 = S_2$ and $h_1 = h_2 + h_3$

The Eq. (*ii*), becomes

$$h_A - h_B = - h_3 S_1 + h_3 S_{Hg}$$

Case-III

Two pipes containing different liquids and at same level

i.e., $\qquad\qquad\qquad h_1 = h_2 + h_3$

$$\begin{aligned} h_A - h_B &= h_2 S_2 + h_3 S_{Hg} - (h_2 - h_3)S_1 \\ &= h_2 S_2 + h_3 S_{Hg} - h_2 S_1 + h_3 S_1 \\ &= (S_2 - S_1)h_2 + (S_{Hg} + S_1)h_3 \end{aligned}$$

U-tube differential manometer is used to measure the difference of pressure between two points of the same pipe: (Fig. 2.32)

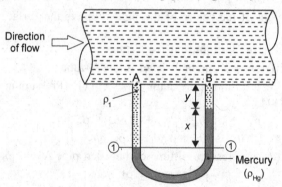

P_1

Mercury (ρ_{Hg})

Fig. 2.32: Schematic of U-tube differential manometer is used on the same pipe

Let (1)-(1) be the datum line.

Pressure in the left limb at datum line (1)-(1) = pressure in the right limb at the datum line (1)-(1).

$$P_A + (x + y)\,\rho_1 g = P_B + \rho_1 g y + \rho_{Hg} g x$$

or $\qquad\qquad P_A - P_B = \rho_1 g y + \rho_{Hg} g x - (x + y)\,\rho_2 g$

$$P_A - P_B = \rho_1 g y + \rho_{Hg} g x - \rho_1 g x - \rho_1 g y$$

$$P_A - P_B = \rho_{Hg} g x - \rho_1 g x = x[\rho_{Hg} - \rho_1]g \qquad\qquad …(iii)$$

Dividing the above Eq. (*iii*) by specific gravity ($\rho_1 g$) of liquid flowing through pipe, we get

$$\frac{p_A - p_B}{\rho_1 g} = \frac{x[\rho_{Hg} - \rho_1]g}{\rho_1 g} = x\left[\frac{\rho_{Hg}}{\rho_1} - 1\right]$$

$$h = x\left[\frac{\rho_{Hg}}{\rho_1} - 1\right]$$

where

$$h = \frac{p_A}{\rho_1 g} - \frac{p_B}{\rho_1 g}$$

= pressure head difference between points A and B in the pipes

$$h = x\left[\frac{\rho_{Hg}}{\rho_1} - 1\right] \text{ m of liquid flow through pipe.}$$

where x is manometer reading in m.

In general:

$$h = x\left[\frac{\rho_{mano}}{\rho_{pipe}} - 1\right]$$

ρ_{mano} = density of liquid used in manometer

ρ_{pipe} = density of liquid flow through pipe

(b) Inverted U-tube Differential Manometer: The type of manometer consists of an inverted U-tube, and is used to measure low pressure difference between two pipes or vessels (Fig. 2.33). The density of manometric liquid is less than that of the liquids flow through pipes.

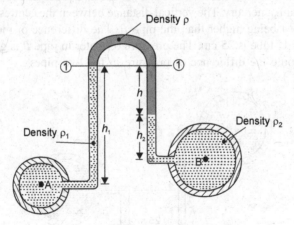

Fig. 2.33: Inverted U-tube differential manometer.

Let (1)-(1) be the datum line.

h_1 = height of liquid in left limb below the datum line (1)-(1)

h_2 = height of liquid in right limb below manometric

liquid

h = difference of manometric liquid.

ρ_1 = density of liquid at A

ρ_2 = density of liquid at B.

ρ = density of manometric liquid.

Pressure in the left limb below the datum line (1)-(1) = pressure in the right limb below the datum line (1)-(1).

$$p_A - \rho_1 g h_1 = p_B - \rho_2 g h_2 - \rho g h$$

$$p_A - p_B = \rho_1 g h_1 - \rho_2 g h_2 - \rho g h \qquad \ldots(iv)$$

Eq. (iv) is dividing by $\rho_w g$ both sides, we get   [ρ_w = Density of water]

$$\frac{p_A}{\rho_w g} - \frac{p_B}{\rho_w g} = \frac{\rho_1}{\rho_w} h_1 - \frac{\rho_2}{\rho_w} h_2 - \frac{\rho}{\rho_w} h$$

Pressure head difference between pipe A and pipe B,

$$h_A - h_B = S_1 h_1 - S_2 h_2 - Sh \quad \text{m of water}$$

where h_1, h_2 and h are in m.

S_1 = specific gravity of liquid at A

S_2 = specific gravity of liquid at B

and S = specific gravity of manometric liquid.

Problem 2.17: To measure the difference of pressure of water flowing through two pipes A and B, a U-tube is used to connect the two pipes as shown in Fig. 2.34. The U-tube contains mercury. The vertical distance between the centres of the pipeline is 2 m, the pipe A being higher than the pipe B. The difference of mercury between two limbs of the U-tube is 25 cm. The pressure of water in pipe B is greater than that in pipe A. Compute the difference of pressure in the two pipes.

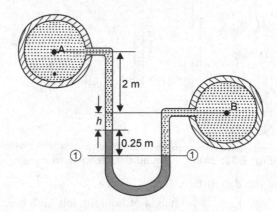

Fig. 2.34: Schematic for Problem 2.17

Solution. Given data:

The vertical distance between the centres of the pipeline:

$$h_1 = 2 \text{ m}$$

As we know, the gauge pressure in the left limb is equal to right limb above the datum line (1)-(1).

$$p_A + \rho_2 g h_1 + \rho_1 g h + \rho_2 g \times 0.25 = p_B + \rho_2 g (h + 0.25)$$
$$p_A + \rho_1 g h_1 + \rho_1 g h + \rho_2 g \times 0.25 = p_B + \rho_1 g h + \rho_1 g \times 0.25$$

or
$$p_A - p_B = \rho_1 g h_1 + \rho_2 g \times 0.25 - \rho_1 g \times 0.25$$
$$= 1000 \times 9.81 \times 2 + 13600 \times 9.81 \times 0.25$$
$$- 1000 \times 981 \times 0.25$$
$$= 50521.5 \text{ N/m}^2 \text{ or Pa} = \textbf{50.521 kPa}$$

Problem 2.18: A differential manometer is used to measure the difference of pressure of oil contained in two pipes at the same level. If the deflection of manometric liquid which is mercury, be 100 mm, determine the difference of pressure of oil in the two pipes in kPa.

Solution: Given data:

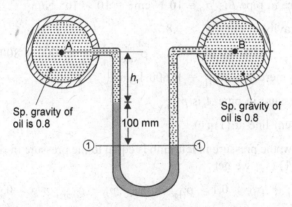

Fig. 2.35: Schematic for Problem 2.18

Specific gravity of oil: $S = 0.8$

$\therefore$ Density of oil: $\rho_1 = 0.8 \times \rho_{water} = 0.8 \times 1000 = 800 \text{ kg/m}^3$

As we know, the gauge pressure above the datum line (1)-(1) is same in left limb and in right limb.

$$p_A + \rho_1 g h + \rho_2 g \times 0.1 = p_B + \rho_1 g h_2$$

or
$$p_B - p_A = \rho_1 g h + \rho_2 g \times 0.1 - \rho_1 g h_2 \qquad \because h_2 = h + 0.1$$
$$p_B - p_A = \rho_1 g h + \rho_2 g \times 0.1 - \rho_1 g (h + 0.1)$$
$$= \rho_1 g h + \rho_2 g \times 0.1 - \rho_1 g h - \rho_1 g \times 0.1$$
$$= \rho_2 g \times 0.1 - \rho_1 g \times 0.1$$
$$= 13600 \times 9.81 \times 0.1 - 800 \times 9.81 \times 0.1$$
$$= 12556.8 \text{ N/m}^2 \text{ or Pa} = \textbf{12.556 kPa}$$

Problem 2.19: A differential mercury manometer is connected in the two pipes A and B as shown in Fig. 2.36. At pipe B, air pressure is 10 N/cm^2 (absolute), find the absolute pressure in pipe A.

Solution: Given data:

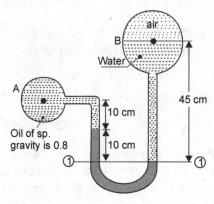

Fig. 2.36: Schematic for Problem 2.19

Air pressure in pipe B: $p_B = 10$ N/cm$^2 = 10 \times 10^4$ N/m^2

Specific gravity of oil: $S_1 = 0.8$

∴ Density: $\rho_1 = 0.8 \times \rho_{\text{water}} = 0.8 \times 1000 = 800$ kg/m^3

Density of mercury: $\rho_{\text{Hg}} = 13600$ kg/m^3

Let the pressure in pipe A is p_A

Taking datum line at (1)-(1).

As we know, the pressure in left limb is equal to the pressure in right limb above the datum line (1)-(1); we get

$$p_A + \rho_1 g \times 0.1 + \rho_{\text{Hg}} g \times 0.1 = p_B + \rho_{\text{water}} \times g \times 0.45$$

$$p_A + 800 \times 9.81 \times 0.1 + 13600 \times 9.81 \times 0.1 = 10 \times 10^4 + 1000 \times 9.81 \times 0.45$$

or

$$p_A = 90288.1 \text{ N/m}^2$$

$$= 90288.1 \times 10^{-4} \text{ N/cm}^2 = \mathbf{9.028 \text{ N/cm}^2}$$

∴ Absolute pressure in pipe A is **9.028 N/cm^2**

Problem 2.20: A differential manometer is connected at two points A and B of two pipes as shown in Fig. 2.37. The pipe A contains a liquid of specific gravity 1.6 while pipe B contains a liquid of specific gravity 0.8. The pressure at A and B are 1.2 kgf/cm^2 and 2 kgf/cm^2 respectively. Find the difference in mercury level in the differential manometer.

Solution: Given data:

Specific gravity of liquid A:

$$S_1 = 1.6$$

∴ Density: $\rho_1 = S_1 \times \rho_{water} = 1.6 \times 1000 = 1600 \text{ kg/m}^3$

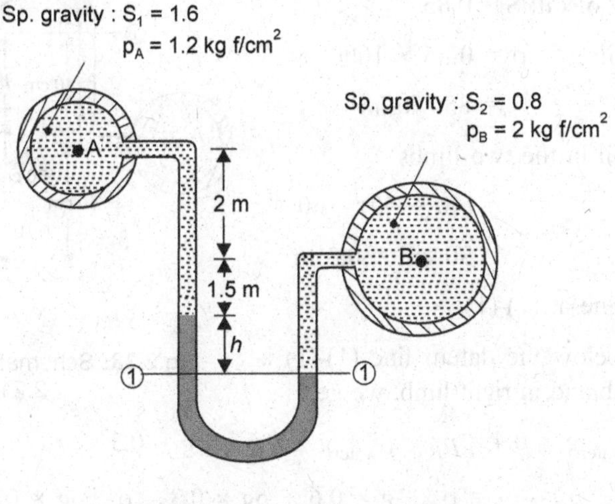

Sp. gravity : $S_1 = 1.6$
$p_A = 1.2 \text{ kg f/cm}^2$

Sp. gravity : $S_2 = 0.8$
$p_B = 2 \text{ kg f/cm}^2$

2 m

1.5 m

h

①   ①

Fig. 2.37: Schematic for Problem 2.20

Specific gravity of liquid B:

$$S_2 = 0.8$$

∴ Density: $\rho_2 = S_2 \times \rho_{water} = 0.8 \times 1000 = 800 \text{ kg/m}^3$

Pressure at point A: $p_A = 1.2 \text{ kg f/cm}^2$

$= 1.2 \times 10^4 \text{ kg f/m}^2$ ∵ 1 kg f = 9.81 N

$= 1.2 \times 10^4 \times 9.81 \text{ N/m}^2 = 11.772 \times 10^4 \text{ N/m}^2$

Pressure at point B: $p_B = 2 \text{ kg f/cm}^2$

$= 2 \times 9.81 \times 10^4 \text{ N/m}^2 = 19.62 \times 10^4 \text{ N/m}^2$

As we know, the gauge pressure in left limb is equal to the right limb above the datum line (1)-(1), we get

$p_A + \rho_1 g \times (2 + 1.5) + \rho_{Hg} gh = p_B + \rho_2 g (1.5 + h)$

$11.772 \times 10^4 + 1600 \times 9.81 \times 3.5$

$+ 13600 \times 9.81 \times h = 19.62 \times 10^4 + 800 \times 9.81 \times (1.5 + h)$

$172656 + 133416 h = 19.62 \times 10^4 + 800 \times 9.81 \times 1.5 + 800 \times 9.81 \times h$

$172652 + 133416 h = 207972 + 7848 h$

or $133416h - 7848 h = 207972 - 172652$

or $125568 h = 35320$

or $h = 0.28128 \text{ m} = \textbf{28.125 cm}$

Problem 2.21: An inverted differential manometer is connected to two pipes A and B which convey water. The fluid in manometer is oil of specific gravity 0.85. For the manometer reading shown in Fig. 2.38, find the pressure difference between A and B.

Solution: Given data:

Specific gravity of oil: $S = 0.85$

$\therefore$ Density of oil: $\rho = 0.85 \times 1000$

                                    $= 850 \text{ kg/m}^3$

Difference of oil in the two limbs

                        $= (40 + 50) - 60$

                        $= 30 \text{ cm} = 0.3 \text{ m}$

Taking datum line at (1)-(1).

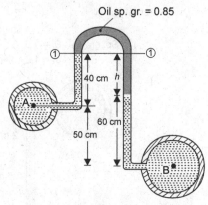

Oil sp. gr. = 0.85

Pressure below the datum line (1)-(1) is same in left limb and in right limb, we get

Fig.2.38: Schematic for Problem 2.21

$$p_A - \rho_{water} g \times 0.4 = p_B - \rho_{water} g \times 0.6 - \rho g \times 0.3$$

or $$p_B - p_A = \rho_{water} g \times 0.6 + \rho g \times 0.3 - \rho_{water} g \times 0.4$$

$$= 1000 \times 9.81 \times 0.6 + 850 \times 9.81 \times 0.3 - 1000 \times 9.81 \times 0.4$$

$$= 4463.55 \text{ N/m}^2 \text{ or Pa} = \textbf{4.463 kPa}$$

Problem 2.22: Find out the differential reading 'h' of an inverted U-tube manometer containing oil of specific gravity 0.7 as the manometric fluid when connected across pipes A and B shown in Fig. 2.39 below, conveying liquids of specific gravities 1.2 and 1.0 and immiscible with manometric fluid. Pipes A and B are located at the same level and assume the pressures at A and B to be equal.

Solution: Given data:

Specific gravity of liquid in pipe A:
                $S_1 = 1.2$

$\therefore$ Density of liquid in pipe A:

                $\rho_1 = 1.2 \times 1000$

                      $= 1200 \text{ kg/m}^3$

Specific gravity of liquid in pipe B:

                $S_2 = 1$

$\therefore$ Density of liquid in pipe B:

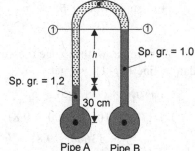

Sp. gr. = 0.7

Sp. gr. = 1.0

Sp. gr. = 1.2

Fig.2.39: Schematic for Problem 2.22

                $\rho_1 = 1 \times 1000 = 1000 \text{ kg/m}^3$

Specific gravity of oil: $S = 0.7$

$\therefore$ Density of oil: $\rho = 0.7 \times 1000 = 700 \text{ kg/m}^3$

As we know, the pressure below the datum line (1)-(1) is same in left limb and in right limb, we get

$$p_A - \rho_1 g \times 0.3 - \rho g h = p_B - \rho_2 g \, (h + 0.3)$$

But $\qquad\qquad p_A = p_B \text{ (given)}$

$\therefore \qquad\qquad - \rho_1 g \times 0.3 - \rho g h = - \rho_2 g \, (h + 0.3)$

$$-1200 \times 9.81 \times 03 \times 700 \times 9.81 \times h = - 1000 \times 9.81 \, (h + 0.3)$$

$$- 3531.6 - 6867 \, h = - 9810 \, h - 2943$$

or $\qquad\qquad 9810 \, h - 6867 \, h = 9810 \, h - 2943$

$$2943 \, h = 588.6$$

or $\qquad\qquad h = 0.2 \text{ m} = \textbf{20 cm}$

2.8 MECHANICAL GAUGES

Mechanical guages are used to measure high fluid pressure and where high precision is not required. Some of common types of mechanical gauges are given below:

    (a) Diaphragm pressure gauge

    (b) Bourdon tube pressure gauge

    (c) Dead-weight pressure gauge and

    (d) Bellows pressure gauge.

2.8.1 Diaphragm Pressure Gauge

The pressure responsive element in this gauge is an elastic sheet corrugated diaphragm. The diaphragm gets deflection being towards the low pressure side. When the fluid enters into the diaphragm, it causes its elastic deformation under pressure to be transmitted to a pointer through link and hinge joints as shown in Fig. 2.40. A pointer is moved on a graduated circular dial in pressure units. However, this pressure gauge is used to measure relatively low pressure. The Aneroid barometer operates on a similar principle.

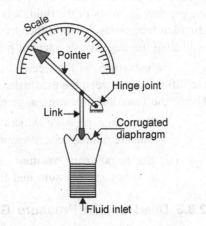

Fig. 2.40: Diaphragam pressure guage

2.8.2 Bourdon Tube Pressure Gauge

The pressure responsive element in Bourdon tube pressure gauge is a tube of steel or bronze which is elliptical cross-section and curved into a circular arc, called Bourdon tube. The outer end of the tube is closed and free to move. The other end of the tube, through which the fluid enters, is rigidly fixed to the frame as shown in Fig. 2.41.

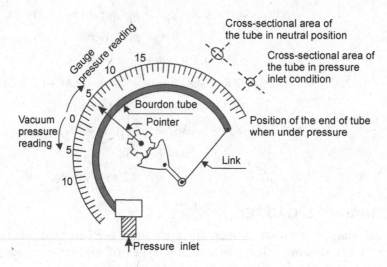

Fig. 2.41: Bourdon tube pressure gauge.

The pressure gauge is connected to the vessel containing fluid under pressure. Due to increase in internal pressure, the elliptical cross-section of the tube tends to become circular, thus causing the tube to straighten out slightly. The outward movement of the free end of the tube is transmitted, through a link, quadrant and pinion, to a pointer which moving clockwise over the graduated circular dial indicates the pressure intensity of the fluid. When a gauge is connected to a partial vacuum, the Bourdon tube tends to close, thereby moving the pointer in anti-clockwise duration, indicating the negative or vacuum pressure.

The movement of the free end of the Bourdon tube in directly proportional to the difference between the external atmospheric pressure and internal fluid pressure. Hence the Bourdon pressure gauge records—

(*i*) the gauge pressure which is the difference between fluid pressure and outside atmospheric pressure, and

(*ii*) the negative or vacuum pressure which is difference between outside atmospheric pressure and fluid pressure.

2.8.3 Dead-weight Pressure Gauge

It consists of placing a dead weight on the top of a plunger fitted in a vertical cylinder. Oil is used as the working fluid in dead weight pressure gauge. We know that intensity

of pressure for any load W on the plunger is given by $p = \dfrac{W}{A}$,

where $A = \dfrac{\pi}{4} D^2$, cross-sectional area of the plunger

                 D = diameter of the plunger.

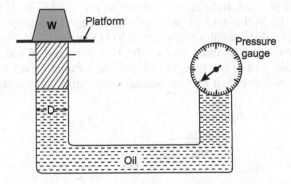

Fig. 2.42: Dead weight pressure gauge.

According to Pascal's law; the intensity of pressure (p) at any point in a fluid at rest is same in all direction. So same pressure (p) is transmitted to the pressure gauge to be calibrated and the pointer of the pressure gauge moves and takes up a steady position on the dial.

That position is marked as 'p'. In this way, by loading the plunger by loads of other magnitudes, other intensities of pressures are marked on the dial.

2.8.4 Bellows Pressure Gauge

In this pressure gauge the pressure responsive element is made of a thin metallic tube having deep circumferential corrugations. In response to the pressure changes this elastic element expands or contracts, thereby moving the pointer on a graduated circular dial as shown in Fig. 2.43.

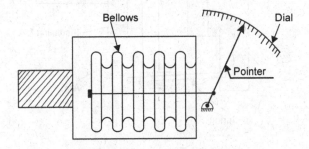

Fig. 2.43: Bellows pressure gauge.

2.9 PRESSURE TRANSDUCER

A pressure transducer is a transducer that converts the pressure into an electric signal. Pressure transducers also called pressure sensors, use various techniques to convert the pressure effect to an electrical effect such as change in voltage, capacitance, or resistance. Pressure transducers are small in size and are faster than mechanical pressure measurement devices. A wide variety of pressure transducers are available to measure gauge, absolute and differential pressure in various applications.

One typical pressure transducer is shown in Fig. 2.44. This is a bourdon tube pressure transducer. In this type of pressure transducer a bourdon tube is connected to a linear variable differential transformer (LVDT) as shown in Fig. 2.44. Core of LVDT is connected to the closed end of the Bourdon tube. When the pressure is applied at the other end of the tube, the closed end of tube moves the core through the coil and due to movement of core through coil output voltage develops.

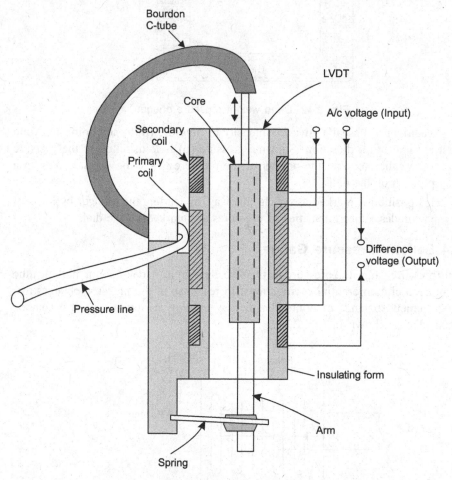

Fig. 2.44: Pressure transducer

This voltage is a linear function of applied pressure and could be recorded on an oscillograph or processing on a computer. Since bourdon tube in this pressure transducer is a elastic element, it is suitable for measuring pressure that are changing slowly. Another type of pressure transducer uses a elastic diaphragm as sensing element instead of bourdon tube, because bourdon tube pressure transducers are not well suited for measuring pressure that are changing rapidly. In pressure transducers using diaphragm as sensing element, diaphragm is in contact with the fluid. Due to change in pressure the diaphragm deforms (deflects) and this deformation is sensed and converted to electric voltage. One type of pressure transducer using diaphragm as sensing element is strain gage pressure transducer. It works by having diaphragm deflect between two chambers open to the pressure input. As the diaphragm streches in responce to a change in pressure difference across it, the strain gauge stretches and a wheatstone bridge circuit amplifies the output. One another type of pressure transducer is capacitance transducer which works in a similar manner as strain gauge pressure transducer, the difference is that in later capacitance change is measured instead of resistance change as the diaphragm deflects.

SUMMARY

1. **Pressure** is defined as normal force exerted by a fluid per unit area. It is also known as the **intensity of pressure.**

   Pressure: $p = \dfrac{\text{Force} : F}{\text{Area} : A}$

   SI units of pressure is N/m^2 or Pa.

2. Pressure head: $\quad h = \dfrac{p}{\rho g} = \dfrac{p}{w}$

   where $\qquad\qquad p$ = intensity of pressure

   $\qquad\qquad\qquad w$ = specific weight = ρg

3. **Pascal's law.** It states that the intensity of pressure at any point in a fluid at rest is same in all direction.

4. Absolute pressure (p_{abs}) = atmospheric pressure (p_{atm}) + gauge pressure (p_g)

   i.e., $\qquad\qquad\qquad p_{abs} = p_{atm} + p_g$

5. Absolute pressure (p_{abs}) = atmospheric pressure (p_{atm}) – vacuum pressure (p_{vac})

   $\qquad\qquad\qquad p_{abs} = p_{atm} - p_{vac}$

6. Measurement of pressure: The pressure of a fluid is measured by using the following two principles:

   (*a*) Principle-I $\qquad\qquad\qquad$ (*b*) Principle-II

   (*a*) **Principle-I:** According to the principle-I, the instruments which are used to measure the pressure of a fluid, are called manometers.

   Manometers are classified as:

   (*i*) Simple manometers. $\qquad\qquad$ (*ii*) Differential manometers.

   (*b*) **Principle-II:** According to the principle-II, the instruments which are used to measure the pressure of a fluid, are called mechanical gauges. Mechanical gauges are classified as:

   (*i*) Diaphragm pressure gauge. (*ii*) Bourdon tube pressure gauge.

   (*iii*) Dead-weight pressure gauge, and

   (*iv*) Bellows pressure gauge.

Contd...

7. Simple manometers are further classified as:
   (*i*) Piezometer   (*ii*) U-tube manometer, and
   (*iii*) Simple column manometer.
8. Differential manometers are also classified as:
   (*i*) U-tube differential manometer, and
   (*ii*) Inverted U-tube differential manometer.
9. Pressure Transducer. A pressure transducer is a device that converts the presure into an electric signal.

ASSIGNMENT - 1

1. What do you mean by intensity of pressure? State its dimension and units.
2. Explain pressure head.
3. What do you understand by hydrostatic law?
4. State and prove the Pascal's law.
5. Describe the various types of barometers.
6. Explain atmospheric pressure. What is the value of atmospheric pressure head in terms of mercury column and in terms of water column?
7. What is the basic principle of measurement of pressure?
8. What do you mean by gauge and absolute pressure heads?
9. What do you mean by vacuum pressure?
10. What is a manometer? How are they classified?
11. Explain the difference between simple and differential manometers.
12. What is the difference between U-tube differential manometers and inverted U-tube differential manometers? Where are they used?
13. Differentiate between a manometer and a mechanical gauge.
14. What are the different types of mechanical pressure gauges? With the help of a neat sketch explain the working of a Bourdon's gauge.
15. Write short notes on:
    (*i*) Aneroid barometer.   (*ii*) Bourdon gauge.
    (*iii*) Diaphragm gauge.

ASSIGNMENT - 2

1. Determine the intensity of pressure in water at a depth 6 m below the free surface of water.   **Ans.** 58.86 kPa
2. Convert intensity of pressure of 4 MPa into equivalent pressure head of oil of specific gravity 0.82.   **Ans.** 497.25 m
3. A hydraulic press has a ram of 500 mm diameter and a plunger of 70 mm diameter. Find the weight lifted by the hydraulic press when the force applied at the plunger is 1000 N.   **Ans.** 51.02 kN

4. An open tank contains water up to a depth of 3 m and above it an oil of specific gravity 0.9 for a depth of 1 m. Find the intensity of pressure:

   (i) at the interface of the low liquids and

   (ii) at the bottom of the tank.   **Ans.** (i) 8.829 kPa (ii) 38.259 kPa.

5. An open tank contains mercury upto a depth 0.5 m and above it water upto a depth 1 m and above water an oil of specific gravity 0.89 for a depth of 0.5 m. Find the intensity of pressure:

   (i) at the interface of the oil and water

   (ii) at the interface of the water and mercury

   (iii) at the bottom of the tank.   **Ans.** (i) 4.365 kPa (ii) 14.175 kPa (iii) 80.88 kPa

6. The diameters of the plunger and ram of a hydraulic press are respectively 60 mm and 140 mm. If a load of 6 kN is applied on the plunger to lift a load W on the ram, determine the value of W when:

   (i) the plunger and the ram are at the same level.

   (ii) the plunger is 400 mm above the ram assuming that water is used in the hydraulic press.   **Ans.** (i) 32.666 kN (ii) 32.729 kN

7. Determine the vacuum pressure in metre of water when the absolute pressure is 0.5434 bar. Assume atmospheric pressure as 10.30 metre of water.

Ans. 4.76 m of water

8. At a point 8 m below free surface of water, determine the pressure at the point in:

   (i) kPa   (ii) mm of Hg.

   If atmospheric pressure in 760 mm of Hg, determine the absolute pressure in:

   (iii) m of water   (iv) bar and   (v) mm of Hg.

   Ans. (i) 78.48 kPa, (ii) 588.23 mm of Hg (iii) 18.336 m of water

   (iv) 1.798 bar, (v) 1348.23 mm of Hg

9. A manometer is fitted to a pipe line containing water. The centre of the pipe line is at a height of 25 cm from the free surface of mercury in the right hand limb of the U-tube and the deflection of mercury is 8 cm in the right hand limb of the U-tube. Determine the pressure of water in the pipe line.

Ans. 7.436 kPa

10. The right limb of a simple U-tube manometer containing mercury is open to the atmosphere while the left limb is connected to a pipe in which a liquid of specific gravity 0.85 is flowing. The centre of the pipe is 0.2 m below the level of mercury in the right limb. Find the pressure of liquid in the pipe if the difference between the level of mercury in the two limbs is 0.4 m.

Ans. 51.698 kPa

11. A simple U-tube manometer containing mercury is connected to a pipe in which a liquid of specific gravity 0.86 and having vacuum pressure is

flowing. The other end of the manometer is open to atmosphere. If the difference of mercury level in the two limbs is 60 cm and the height of liquid in the left limb from the centre of pipe is 18 cm below, find the vacuum pressure in pipe.   **Ans.** −81.568 kPa

12. A U-tube manometer is used to measure the pressure of water in a pipe line, which is in excess of atmospheric pressure. The right contact between water and mercury is in the left limb. Determine the pressure of water in the main line, if the difference in level of mercury in the limbs of U-tube is 10 cm and the free surface of mercury is in level with the centre of the pipe. If the pressure of water in the pipe line is reduced to 9810 N/m², calculate the new difference in the level of mercury. Sketch the arrangements in both the cases.

                                                       Ans. 12.36 kPa, 8.016 cm

13. Determine the intensity of pressure of oil contained in the pipe line shown in Fig. 2.45. The relative density of oil is 0.8 and mercury is used as the gauge liquid. Determine the intensity of pressure of oil in the pipe line in: (*i*) metre of oil (*ii*) metre of water (*iii*) Pa.

    Ans. (*i*) 3.35 m of oil; (*ii*) 2.68 m of water; (*iii*) 26290.8 Pa

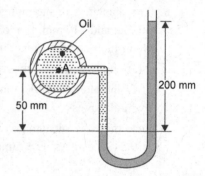

Fig. 2.45: Q.13

14. A compound manometer is used to measure the intensity of pressure of water flowing through a pipe line shown in Fig. 2.46. Determine the intensity of pressure of water in kPa. Assume that mercury is used as the gauge liquid.

                                                       Ans. 21.85668 kPa

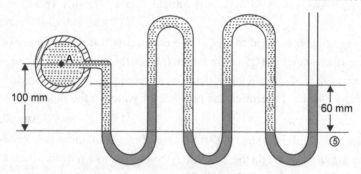

Fig. 2.46: Q.14

15. A single column manometer is connected to a pipe containing a liquid of specific gravity 0.9 as shown in Fig. 2.47. Find the pressure in the pipe if the area of the reservoir is 100 times the area of the tube for the manometer reading as shown in Fig. 2.47. The specific gravity of mercury is 13.6.

                                                       Ans. 52.098 kPa

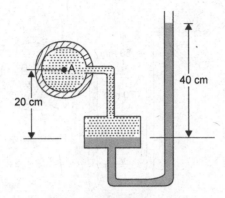

Fig. 2.47 : Q.15

16. Find the difference in pressure in the pipes A and B containing water and connected to an inverted differential manometer as shown in Fig. 2.48. The gauge liquid is oil of specific gravity 0.9. **Ans.** 9.07 kPa

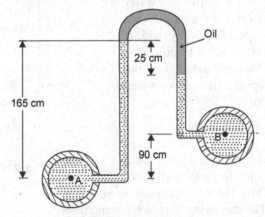

Fig. 2.48 : Q.16

17. Two points in a horizontal pipeline are connected by a U-tube differential manometer. Oil having specific gravity of 0.8 flows through the pipeline and mercury is used as the gauge liquid. If the difference of level of mercury in the two limbs of the U-tube be 3000 mm, determine the difference of pressure between the two points in:

(*i*) metre of oil (*ii*) metre of water (*iii*) Pa and (*iv*) kgf/cm².

Ans (*i*) 4.8 m of oil; (*ii*) 3.84 m of water; (*iii*) 37670.4 Pa;
(*iv*) 0.384 kgf/cm²

Hint. 1 kgf = 9.81 N

18. Determine the difference of pressure between two pipes A and B. The pipe A contains a liquid of specific gravity 1.5 and pipe B carries another liquid having specific gravity 0.9. The arrangement of pipes has been shown in Fig. 2.49. Mercury is used as the manometric liquid. **Ans.** −16.09 kPa

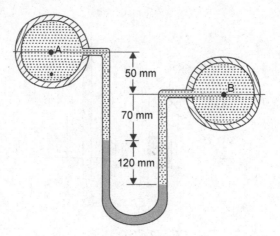

Fig. 2.49: Q.18

19. Fig. 2.50 shows an inverted differential manometer connected to pipes A and B containing water under pressure. The gauge liquid is an oil of specific gravity 0.75. If the pressure of water in pipe B is 1.2 m of water, determine the pressure in pipe A in metre of water.
    Ans. 1.17 m

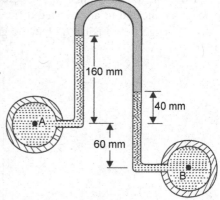

Fig. 2.50: Q-19

20. Pipes A and B containing water and a liquid of specific gravity 1.2 are connected to a U-tube manometer as shown in Fig. 2.51. The measuring fluid in the manometer is oil of specific gravity 0.80. Determine the difference of pressure between the centres of the fluids in the pipes.
    Ans. 549.36 N/m^2

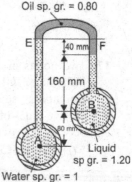

Oil sp. gr. = 0.80

Water sp. gr. = 1

Fig. 2.51: Q.20

21. An inverted U-tube manometer is connected to two horizontal pipes A and B through which water is flowing. The vertical distance between the axes of these pipes is 30 cm. When an oil of specific gravity 0.8 is used as a gauge fluid, the vertical heights of water column in the two limbs of the inverted manometer (when measured from the respective centre lines of the pipes) are found to be same and equal to 35 cm. Determine the difference of pressure between the pipes.
    Ans. 2354.4 N/m^2

□□□

Hydrostatic Forces on Surface

3.1 INTRODUCTION

The term "Hydrostatic" means the study of pressure exerted by liquid at static condition *i.e.*, at rest. The direction of pressure is always perpendicular to the surface on which it acts. The surface may be either plane or curved. The forces acting on the fluid particles will be normal to the surface due to pressure of fluid and due to gravity (or self weight of fluid particles).

3.2 TOTAL PRESSURE AND CENTRE OF PRESSURE

3.2.1 Total Pressure (F)

It is defined as the resultant force exerted by a static fluid on a surface (either plane or curved). When the fluid comes in contact with the surface, this force always acts at right angle (*i.e.*, normal) to the surface. It is denoted by F. The SI unit of total pressure (F) is Newton (*i.e.*, N).

3.2.2 Centre of Pressure (CP)

It is defined as the point of application of the total pressure (F) on the surface.

3.3 HYDROSTATIC PRESSURE

Hydrostatic pressure means force acting on a surface due to weight of a liquid. For example: (Fig. 3.1)

Hydrostatic pressure at point a : $p_a = 0$

Hydrostatic pressure at point b : $p_b = \rho g h_1$

Hydrostatic pressure at point c : $p_c = \rho g h_2$

Hydrostatic pressure at point d : $p_d = \rho g h_3$

© The Author(s) 2023
S. Kumar, *Fluid Mechanics (Vol. 1)*,
https://doi.org/10.1007/978-3-030-99762-5_3

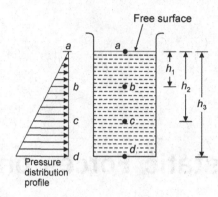

Fig. 3.1: Hydrostatic Pressure

3.3.1 Hydrostatic Law

It states that the rate of increase of pressure in vertical direction is equal to weight density (or specific weight). The rate of increase of pressure in vertical direction: $\dfrac{dp}{dh} = \rho g$

or $$\int dp = \rho g \int dh$$

$$p = \rho g h$$

where p = hydrostatic pressure above atmospheric pressure.

                   h = depth of the point from the free surface.

3.4 TOTAL PRESSURE (F) AND CENTRE OF PRESSURE (CP) FOR SUBMERGED SURFACES

The submerged surfaces may be:

   (*a*) Vertical plane surface,

   (*b*) Horizontal plane surface,

   (*c*) Inclined plane surface, and

   (*d*) Curved surface.

3.4.1 Vertical Plane Surface Submerged in Liquid

Consider a vertical plane surface of arbitrary shape immersed in a liquid as shown in Fig. 3.2.

   Let A = total area of the plane surface.

                   G = centre of gravity of plane surface.

                   $\bar{h}$ = distance of CG of the area from free surface of liquid.

                   CP = centre of pressure

                   h^* = distance of centre of pressure (CP) from free surface of liquid.

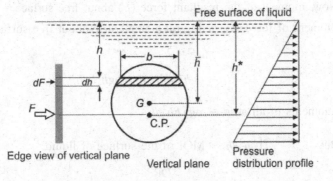

Fig. 3.2: Immersed vertical plane.

Now, we find out two parameters which are:

(a) Total pressure (F) and

(b) Position of the centre of pressure from free surface of liquid (h^*).

(a) **Total pressure (F):** Consider a small strip of thickness dh and width b at a depth of h from free surface of liquid as shown in Fig. 3.2.

Pressure intensity on the small strip: $p = \rho gh$

Area of the small strip: $\qquad dA = b \times dh$.

Pressure force on small strip: $\qquad dF = p \times dA = \rho gh \times bdh$

Total pressure on the whole surface:

$$F = \int dF = \int \rho gh \cdot bdh$$

$$F = \rho g \int bhdh$$

$$= \rho g \int h \cdot dA \qquad\qquad \because dA = bdh$$

But $\qquad\qquad \int hdA$ = Moment of surface area about free surface of liquid.

OR

First moment of area about free surface.

= area of surface × distance of CG from free surface.

$$= A\bar{h}$$

∴ Total pressure: $\qquad F = \rho g A\bar{h}$

(b) **Position of the centre of pressure from free surface of liquid (h^*):** It is calculated by using the principle of moments, which states that the moment of the resultant force about an axis is equal to the sum of moments of the component forces about the same axis.

The resultant force F is acting at CP at a distance h^* from free surface of the liquid as shown in Fig. 3.2.

Now, moment of the resultant force (F) about free surface $= F \times h^* \dots$ (i)

Moment of force dF, acting on a small strip about free surface of liquid

$$= dF \times h$$

$$= \rho g h b dh \times h = \rho g h^2 b dh$$

Sum of moments of all strip $= \rho g \int b h^2 dh$

But $\qquad \int b h^2 dh =$ MOI of the surface of liquid.

OR

Second moment of area about free surface of liquid

$$= I_0$$

$\therefore$ Sum of moment of all strip $= \rho g I_0$ $\qquad\qquad \dots$ (ii)

Equating Eqs. (i) and (ii), we get

$$F h^* = \rho g I_0$$

$$\rho g A \bar{h} h^* = \rho g I_0$$

$\therefore \qquad\qquad h^* = \dfrac{I_0}{A \bar{h}}$

$$= \dfrac{\text{2nd moment of area about free surface of liquid}}{\text{1st moment of area about free surface of liquid}} \dots \text{(iii)}$$

By parallel axis theorem: $I_0 = I_G + A \bar{h}^2$

Substituting the value of I_0 in Eq. (iii), we get

$\therefore \qquad\qquad h^* = \dfrac{I_G + A \bar{h}^2}{A \bar{h}} = \dfrac{I_G}{A \bar{h}} + \bar{h}$

$$\boldsymbol{h^* = \bar{h} + \dfrac{I_G}{A \bar{h}}} \qquad\qquad \dots \text{(iv)}$$

Hence the Eq. (iv) gives the following information:

(a) Centre pressure lies below the centre of gravity of the vertical plane surface.

(b) The distance of centre of pressure from free surface of liquid is independent of the density of the liquid.

Table 3.1: Geometric Properties of Some Plane Surfaces.

S. No.	Plane Surface	Area	C.G. from the Base	C.G. from the Top	Moment of Inertia About an Axis Passing Through C.G. and Parallel to Base Line, I_G	Moment of Inertia About Base (I_o)
1.	Rectangle	bd	$\dfrac{d}{2}$	$\dfrac{d}{2}$	$\dfrac{bd^3}{12}$	$\dfrac{bd^3}{3}$
2.	Square	b^2	$\dfrac{b}{2}$	$\dfrac{b}{2}$	$\dfrac{b^4}{12}$	$\dfrac{b^4}{3}$
3.	Triangle	$\dfrac{bh}{2}$	$\dfrac{h}{3}$	$\dfrac{2h}{3}$	$\dfrac{bh^3}{36}$	$\dfrac{bh^3}{12}$
4.	Circle	$\dfrac{\pi d^2}{4}$	$\dfrac{d}{2}$	$\dfrac{d}{2}$	$\dfrac{\pi d^4}{64}$	$I_G + A\bar{h}^2$
5.	Semi-circle	$\dfrac{\pi d^2}{8}$	$\dfrac{4r}{3\pi}$	$1.80r$ $\dfrac{}{\pi}$	$\dfrac{d^4}{144}$	$\dfrac{\pi d^4}{128}$
6.	Trapezium	$\dfrac{(a+b)h}{2}$	$\left(\dfrac{b+2a}{b+a}\right)\cdot\dfrac{h}{3}$	$\left(\dfrac{a+2b}{a+b}\right)\cdot\dfrac{h}{3}$	$\left(\dfrac{a^2+b^2+4ab}{a+b}\right)$ $\times\dfrac{h^3}{36}$	$I_G + A\bar{h}^2$

3.4.2 Horizontal Plane Surface Submerged in Liquid

Consider a horizontal plane surface immersed in a static fluid the depth $\bar{h}$ below the free surface is same at every point on the surface (Fig. 3.3). The intensity of pressure at every point on a horizontal plane surface is the same and is equal to $\rho g h$.

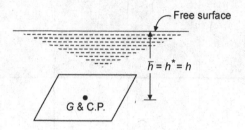

Fig. 3.3: Immersed horizontal plane.

where h is depth of horizontal plane surface

Total pressure: $F = p \times \text{area} = \rho g h A$

where A = total area of horizontal surface.

$$F = \rho g A h = \rho g A \bar{h} = \rho g A h^*$$

where $\bar{h}$ = depth of CG from free surface of liquid = h

and h^* = depth of centre of pressure (CP) from free surface

                                          of liquid = $\bar{h}$

$\therefore$ $\bar{h} = h^* = h$

3.4.3 Inclined Plane Surface Submerged in Liquid

Consider a plane surface of arbitrary shape immersed in a liquid and making an angle θ with the free surface of the liquid as shown is Fig. 3.4.

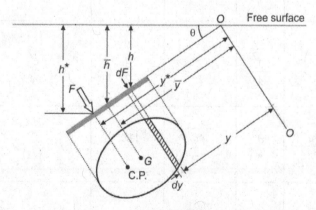

Fig. 3.4: Immersed inclined plane.

Let A = total area of inclined surface.

                                          $\bar{h}$ = depth of CG of inclined area from free surface

                                          h^* = depth of CP from free surface of liquid.

Let the plane of the surface is produced to meet the free liquid surface at point O. Then O–O is the axis perpendicular to the plane of the surface.

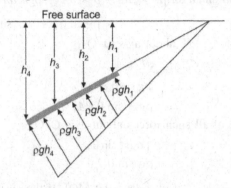

Fig. 3.5: Pressure distribution profile on an inclined plane.

Let $\bar{y}$ = distance of the CG of the inclined surface from O–O

y^* = distance of the CP of the inclined surface from O–O

Let a small strip of area dA at a depth h from free surface and at a distance y from the axis O–O.

Pressure intensity on the small strip: $p = \rho gh$

∴ Pressure force on the small strip: $dF = p \times$ Area of small strip

$= \rho gh dA$

∴ Total pressure force on the whole area: $F = \int dF = \int \rho gh \cdot dA$

From Fig. 3.6,

$$\sin \theta = \frac{h}{y} = \frac{\bar{h}}{\bar{y}} = \frac{h^*}{y^*}$$

or $\quad \sin \theta = \dfrac{h}{y}$

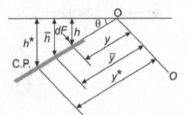

Fig. 3.6: Total Pressure

or $\quad h = y \sin \theta$

∴ $\quad F = \rho g \int y \sin \theta \, dA$

$= \rho g \sin \theta \int y \, dA$

$= \rho g \sin \theta \, A\bar{y} \qquad \because \int y \, dA = A\bar{y}$

$= \rho g \bar{h} A \qquad\qquad \because \bar{h} = \bar{y} \sin \theta$

Position of the Centre of Pressure from Free Surface of Liquid (h^*)

It is calculated by using the principle of moments, which states that the moment of the resultant force about an axis is equal to the sum of moments of the components about the same axis.

The resulting force F is acting at CP at a distance y^* from axis O–O.

∴ Moment of the resulting force (F) about axis O–O

$= F \times y^* \qquad\qquad\qquad ... (i)$

Pressure intensity on a small strip: $p = \rho g h$

Pressure force on small strip: $dF = p \times dA = \rho g h \cdot dA$

$$= \rho g \sin \theta \, dA \quad \because \sin \theta = \frac{h}{y} \text{ or } h = y \sin \theta$$

Moment of the force, dF about axis O–O

$$= dF \times y$$

$$= \rho g \sin \theta \, dA \, y$$

$$= \rho g y^2 \sin \theta \, dA$$

Sum of moment of all such forces about axis O–O

$$= \int \rho g y^2 \sin \theta \, dA$$

$$= \rho g \sin \theta \, I_0 \qquad\qquad\qquad \dots (i)$$

$$\because \quad \int y^2 dA = I_0, \text{ MOI of the surface about axis O–O}$$

Equating Eqs. (i) and (ii), we get

$$F \times y^* = \rho g \sin \theta \, I_0$$

$$y^* = \frac{\rho g \sin \theta \, I_0}{F} \qquad\qquad\qquad \dots (iii)$$

Substituting $F = \rho g \bar{h} A$ and $y^* = \dfrac{h^*}{\sin \theta}$ in Eq. (iii), we get

$$\frac{h^*}{\sin \theta} = \frac{\rho g \sin \theta \, I_0}{\rho g \bar{h} A}$$

$$h^* = \frac{I_0 \sin^2 \theta}{\bar{h} A}$$

By parallel axis theorem,

$$I_0 = I_G + A\bar{y}^2 \qquad\qquad\qquad \Bigg| \qquad \because h^* = \frac{I_0 \sin^2 \theta}{\bar{h} A}$$

$$\frac{h^* \bar{h} A}{\sin^2 \theta} = I_G + \frac{A \times \bar{h}^2}{\sin^2 \theta} \qquad \Bigg| \qquad \text{or } I_0 = \frac{h^* \bar{h} A}{\sin^2 \theta}$$

or $$h^* = \frac{I_G \sin^2 \theta}{\bar{h} A} + \bar{h} \qquad\qquad \Bigg| \qquad \text{and } \sin \theta = \frac{\bar{h}}{y}$$

or $$h^* = \bar{h} + \frac{I_G}{\bar{h} A} \sin^2 \theta \qquad\qquad \Bigg| \qquad \text{or } \bar{y} = \frac{\bar{h}}{\sin \theta}$$

where I_G = MOI of the inclined plane surface about an axis passing through G and parallel to axis O–O.

If $\theta = 90°$, $\sin^2 90° = 1$

Then $$h^* = \bar{h} + \frac{I_G}{\bar{h} A}$$

which is same as vertical plane surface submerged in liquid.

If $\theta = 0°$, $\sin^2 0° = 1$

Then $$h^* = \bar{h}$$

which is same as horizontal plane surface submerged in liquid.

Problem 3.1: A vertical rectangular plane surface of width 2 m and depth 3.4 m is placed in water so that its upper edge is at a depth 1 m below the free water surface. Determine the total pressure and position of the centre of pressure on the plane surface.

Solution: Given data:

Width of rectangular plane surface: $b = 2$ m

Depth of rectangular plane surface: $d = 3.4$ m

Now, area of rectangular plane surface,
$$A = b \times d = 2 \times 3.4$$
$$= 6.8 \text{ m}^2$$

Centre of gravity from free surface of the liquid:

$$\bar{h} = 1 + \frac{d}{2}$$

$$= 1 + \frac{3.4}{2} = 2.7 \text{ m}$$

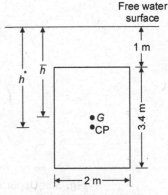

Fig. 3.7: Schematic for Problem 3.1

Moment of inertia about an axis passing through the centroid:

$$I_G = \frac{bd^3}{12} = \frac{2 \times (3.4)^3}{12} = 6.55 \text{ m}^4$$

(a) Total pressure (F) on the plane surface:

$$F = \rho g \bar{h} A \qquad \because \rho = 1000 \text{ kg/m}^3 \text{ for water.}$$
$$= 1000 \times 9.81 \times 2.7 \times 6.8 = \textbf{180111.60 N}$$

(b) Position of the centre of pressure from free surface of liquid: h^*

$$h^* = \bar{h} + \frac{I_G}{A\bar{h}}$$

[This is formula used when vertical plane surface submerged in liquid]

$$= 2.7 + \frac{6.55}{6.8 \times 2.7} = \textbf{3.056 m}$$

Problem 3.2: A rectangular plane surface is 1.5 m wide and 4 m deep. It lies vertically in water. Determine the total pressure and position of the centre of pressure on the plane surface when upper edge is horizontal and (a) coincides with free water surface (b) 3 m below the free water surface.

Solution: Given data:

Width of rectangular plane surface: $b = 1.5$ m

Depth of rectangular plane surface: $d = 4$ m

Now,

Area of rectangular plane surface: $A = b \times d = 1.5 \times 4 = 6$ m^2

Moment of inertia about an axis passing through the centroid:

$$I_G = \frac{bd^3}{12} = \frac{1.5 \times (4)^3}{12} = 8 \text{ m}^4$$

(a) Upper edge coincides with free water surface: As shown in Fig. 3.8

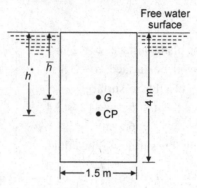

Fig. 3.8: Upper edge coincides with free water surface.

Total pressure: $F = \rho g A \bar{h}$

where: $\rho = 1000 \text{ kg/m}^3$

              $g = 9.81 \text{ m/s}^2$

              $A = 6 \text{ m}^2$

and $\bar{h} = \frac{d}{2} = \frac{4}{2} = 2 \text{ m}$

∴ Total pressure: $F = 1000 \times 9.81 \times 6 \times 2 = \textbf{117720 N}$

Position of the centre of pressure from free surface of water: h^*

∴ $h^* = \bar{h} + \frac{I_G}{A\bar{h}} = 2 + \frac{8}{6 \times 2} = \textbf{2.666 m}$

(b) Upper edge is 3 m below the free water surface: As shown in Fig. 3.9

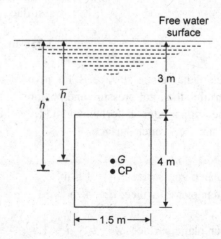

Fig. 3.9: Upper edge is 3 m below the free water surface.

Total pressure: $\qquad F = \rho g \overline{h} A$

where $\qquad \rho = 1000$ kg/m^3

$\qquad g = 9.81$ m/s^2

$\qquad A = 6$ m^2

and $\qquad \overline{h} = 3 + \dfrac{d}{2} = 3 + \dfrac{4}{2} = 5$ m

$\therefore$ Total pressure: $\qquad F = 1000 \times 9.81 \times 5 \times 6 =$ **294300 N**

Position of the centre of pressure from free surface of water: h^*

$\therefore \qquad\qquad\qquad h^* = \overline{h} + \dfrac{I_G}{A\overline{h}} = 5 + \dfrac{8}{6 \times 5} =$ **5.266 m**

Problem 3.3: A circular plate of diameter 1.2 m is placed vertically in water in such a way that the centre of the plate is 2.5 m below the free surface of water. Determine the total pressure and the position of the centre of pressure.

Solution: Given data:

Diameter of plate: $d = 1.2$ m

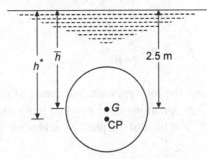

Fig. 3.10: Schematic for Problem 3.3

$\therefore$ Area: $\qquad A = \dfrac{\pi}{4}d^2 = \dfrac{3.14}{4} \times (1.2)^2 = 1.130$ m^2

MOI about centroid: $\quad I_G = \dfrac{\pi}{64}d^4 = \dfrac{3.14}{64}(1.2)^4 = 0.1017$ m^4

$\qquad\qquad\qquad\qquad \overline{h} = 2.5$ m

For water, Density: $\qquad \rho = 1000$ kg/m^3

and $\qquad\qquad\qquad g = 9.81$ m/s^2

$\therefore$ Total pressure: $\qquad F = \rho g \overline{h} A = 1000 \times 9.81 \times 2.5 \times 1.130 =$ **27713.25 N**

And position of the centre of pressure:

$$h^* = \overline{h} + \dfrac{I_G}{A\overline{h}} = 2.5 + \dfrac{0.1017}{1.13 \times 2.5} = 2.5 + 0.036 = \textbf{2.536 m}$$

Problem 3.4: A circular plate of diameter d is submerged in a liquid and the depth of centroid of the plate is 'p' m below the free surface of liquid. Prove that the depth of the centre of pressure is equal to $\left(p + \dfrac{d^2}{16p} \right)$ metre.

Solution: Given data:

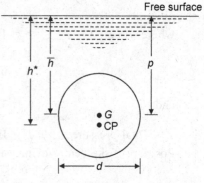

Diameter of plate   = d m

Depth of CG from free surface:

$$\bar{h} = p \text{ m}$$

Now, area of circular plate: $A = \dfrac{\pi}{4}d^2$

MOI about centroid: $I_G = \dfrac{\pi}{64}d^4 \text{ m}^4$

Depth of the centre of pressure from free surface of liquid: h^*

Fig.3.11: Schematic for Problem 3.4

$\therefore$

$$h^* = \bar{h} + \frac{I_G}{A\bar{h}}$$

$$h^* = p + \frac{\pi \cdot d^4}{64 \times \frac{\pi}{4}d^2 \times p}$$

$$h^* = p + \frac{d^2}{16\,p} \text{ m}$$

Problem 3.5: Determine the total pressure and centre of pressure on an isosceles triangular plate of base 3.5 m and attitude 3.5 m when it is submerged vertically in an oil of specific gravity 0.85. The base of the plate coincides with the free surface of oil.

Solution: Given data:

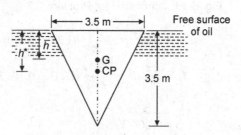

Fig. 3.12: Schematic for Problem 3.5

Base of plate: $b = 3.5$ m

Height of plate: $h = 3.5$ m

$\therefore$ Area of plate: $A = \dfrac{bh}{2} = \dfrac{3.5 \times 3.5}{2} = 6.125 \text{ m}^2$

Specific gravity of oil: $S = 0.85$

$\therefore$ Density of oil: $\rho = \rho_{\text{water}} \times S = 1000 \times 0.85 = 850 \text{ kg/m}^3$

Depth of CG from free surface of oil:

$$\bar{h} = \frac{h}{3} = \frac{3.5}{3} = 1.166 \text{ m}$$

MOI of triangular plane about its CG:

$$I_G = \frac{bh^3}{36} = \frac{3.5 \times (3.5)^3}{36} = 4.168 \text{ m}^4$$

Total pressure: $\quad\quad F = \rho g \bar{h} A = 850 \times 9.81 \times 1.166 \times 6.125 = \textbf{59551.48 N}$

and depth of CP from free surface of oil: h^*

$$h^* = \bar{h} + \frac{I_G}{A\bar{h}} = 1.166 + \frac{4.168}{6.125 \times 1.166} = \textbf{1.749 m}$$

Problem 3.6: A triangular plane having 3 m width and 3 m height is submerged vertically in an oil of specific gravity 0.8. The vertex of the triangular surface is above its base and just touches the free surface of oil. Find:

(*i*) the force acting on the surface due to oil pressure.

(*ii*) the depth of the centre of pressure from free surface.

Solution: Given data:

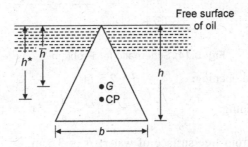

Fig. 3.13: Schematic for Problem 3.6

Width of the plane: $\quad\quad b = 3$ m

Height of the plane: $\quad\quad h = 3$ m

∴ Area of plane: $\quad\quad A = \dfrac{bh}{2} = \dfrac{3 \times 3}{2} = 4.5 \text{ m}^2$

Depth of CG from free surface of oil:

$$\bar{h} = \frac{2h}{3} = 2 \times \frac{3}{3} = 2 \text{ m}$$

MOI about the centroid: $\quad\quad I_G = \dfrac{bh^3}{36} = \dfrac{3 \times 3^3}{36} = 2.25 \text{ m}^4$

Specific gravity of oil: $\quad\quad S = 0.8$

∴ Density of oil: $\quad\quad \rho = \rho_{\text{water}} \times 0.8 = 1000 \times 0.8 = 800 \text{ kg/m}^3$

(*i*) The force acting on the surface due to oil pressure:

$$F = \rho g \bar{h} A = 800 \times 9.81 \times 2 \times 4.5 = \textbf{70632 N}$$

(*ii*) The depth of the centre of pressure from free surface:

$$h^* = \bar{h} + \frac{I_G}{A\bar{h}} = 2 + \frac{2.25}{4.5 \times 2} = \mathbf{2.25 \ m}$$

Problem 3.7: A circular opening, 2.5 m diameter, in a vertical side of a tank is closed by a disc of 2.5 m diameter which can rotate about horizontal diameter. Find

(*i*) the force on the disc, and

(*ii*) the torque required to maintain the disc in equilibrium in vertical position when the head of water above the horizontal diameter is 3.5 m.

Solution: Given data:

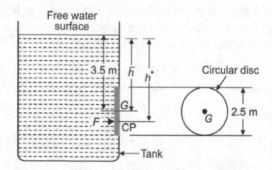

Fig. 3.14: Schematic for Problem 3.7

Diameter of the opening: $\qquad d = 2.5$ m

$\therefore$ Area of opening: $\qquad A = \frac{\pi}{4}d^2 = \frac{\pi}{4}(2.5)^2 = 4.906 \ m^2$

Depth of CG from free surface of water: $\bar{h} = 3.5$ m

MOI of circular disc about its CG:

$$I_G = \frac{\pi}{64}d^4 = \frac{3.14}{64} \times (2.5)^4 = 1.916 \ m^4$$

(*i*) Force on the disc: $\qquad F = \rho g A\bar{h} = 1000 \times 9.81 \times 4.906 \times 3.5$

$$= \mathbf{168447.51 \ N}$$

(*ii*) Torque required to maintain the disc in equilibrium: T

The depth of the centre of pressure from free surface:

$$h^* = \bar{h} + \frac{I_G}{A\bar{h}} = 3.5 + \frac{1.916}{4.906 \times 3.5} = \mathbf{3.611 \ m}$$

Taking moment of the force (F) about CG

$$= F(h^* - \bar{h}) = 168447.51 \times (3.611 - 3.5)$$
$$= \mathbf{18697.67 \ Nm} \qquad \text{(anti clockwise)}$$

Hence a torque (T) of 18697.67 Nm must be applied on the disc in the clockwise direction to maintain the disc in equilibrium position.

Problem 3.8: A vertical sluice gate is used to cover an opening in a dam. The opening is 3 m wide and 3.5 m high. On the upstream of the gate, the liquid of specific gravity 1.5, lies upto a height of 2 m above the top of the gate, whereas on the downstream side the water is available upto a height touching the top of the gate. Find:

(*i*) Resultant force acting on the gate.

(*ii*) Position of the centre of pressure from free surface.

(*iii*) Position of the centre of pressure of resultant force.

(*iv*) The force acting horizontally at the top of the gate which is capable of opening it.

Assume that the gate is hinged at the bottom.

Solution: Given data:

Width of gate:	$b = 3$ m
Depth of gate:	$d = 3.5$ m
∴ Area:	$A = b \times d = 3 \times 3.5 = 10.5$ m^2
Specific gravity of liquid:	$S_1 = 1.5$ (on the upstream of the gate)
∴ Density of liquid:	$\rho_1 = \rho_{water} \times S_1 = 1000 \times 1.5 = 1500$ kg/m^3

Let F_1 = force exerted by the fluid of specific gravity 1.5 on gate.

F_2 = force exerted by water on the gate.

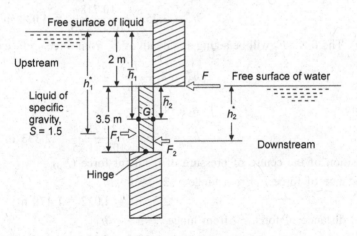

Fig. 3.15: Schematic for Problem 3.8

The force F_1 is given by: $F_1 = \rho_1 g A \overline{h_1}$

where

$\rho_1 = 1500$ kg/m^3

$g = 9.81$ m/s^2

$A = 10.5$ m^2

$$\overline{h_1} = 2 + \frac{d}{2} = 2 + \frac{3.5}{2} = 3.75 \text{ m}$$

∴ $F_1 = 1500 \times 9.81 \times 10.5 \times 3.75 = 579403.125$ N

Similarly $\qquad\qquad F_2 = \rho g A \bar{h}_2$

where $\qquad\qquad\qquad \rho = 1000 \text{ kg/m}^3$

$\qquad\qquad\qquad\qquad g = 9.81 \text{ m/s}^2$

$\qquad\qquad\qquad\qquad A = 10.5 \text{ m}^2$

and $\qquad\qquad\qquad \bar{h}_2 = \dfrac{d}{2} = \dfrac{3.5}{2} = 1.75 \text{ m}$

$\therefore \qquad\qquad F_2 = 1000 \times 9.81 \times 10.5 \times 1.75 = 180258.75 \text{ N}$

(*i*) Resultant force acting on the gate:

$$F_R = F_1 - F_2$$
$$= 579403.125 - 180258.75 = \mathbf{399144.375\,N}$$

(*ii*) Position of the centre of pressure from free surface h_1^* and h_2^*

(*a*) The force F_1 will be acting at a depth of h_1^* from free surface of liquid,

$$h_1^* = \bar{h}_1 + \frac{I_G}{A\bar{h}_1}$$

where $\qquad\qquad\qquad \bar{h}_1 = 3.75 \text{ m}$

$\qquad\qquad\qquad\qquad A = 10.5 \text{ m}^2$

and $\qquad\qquad\qquad I_G = \dfrac{bd^3}{12} = \dfrac{3 \times (3.5)^3}{12} = 10.718 \text{ m}^4$

$\therefore \qquad\qquad h_1^* = 3.75 + \dfrac{10.718}{10.5 \times 3.75} = \mathbf{4.022\ m}$

(*b*) The force F_2 will be acting at a depth of h_2^* from force surface of liquid:

$$h_2^* = \bar{h}_2 + \frac{I_G}{A\bar{h}_2}$$

where $\qquad\qquad\qquad \bar{h}_2 = 1.75 \text{ m}$

$$h_2^* = 1.75 + \frac{10.718}{10.5 \times 1.75} = \mathbf{2.333\ m}$$

(*iii*) Position of the centre of pressure of resultant force (F_R)

Distance of force F_1 from hinge $= 2 + 3.5 - h_1^*$

$\qquad\qquad\qquad\qquad\qquad = 2 + 3.5 - 4.022 = 1.478 \text{ m}$

and distance of force F_2 from hinge $= 3.5 - h_2^*$

$\qquad\qquad\qquad\qquad\qquad = 3.5 - 2.333 = 1.167 \text{ m}$

Let h is distance of resultant force from hinge; taking the moments of resultant force (F_R), F_1 and F_2 about the hinge, we get

$$F_R \times h = 579403.125 \times 1.478 - 180258.75 \times 1.167$$

or $\qquad\qquad h = \dfrac{579403.125 \times 1.478 - 180258.75 \times 1.167}{F_R}$

$$= \frac{645995.8575}{399144.375} = \mathbf{1.6184\ m\ above\ the\ hinge}$$

(*iv*) The force acting horizontally at the top of the gate, which is capable of opening it. Let F = force required on the top of the gate to open it as shown in Fig. 3.15.

Taking the moments of forces F, F_1 and F_2 about the hinge, we get

$$F \times 3.5 + F_2 \times 1.167 = F_1 \times 1.478$$
$$F \times 3.5 + 180258.75 \times 1.167 = 579403.125 \times 1.478$$

or $F = \textbf{184570.245 N}$

Problem 3.9:   A square orifice in the vertical side of a tank has one diagonal vertical and is completely covered by a plane hinged along one of the upper sides of the orifice. The diagonals of the orifice are 1.8 m long and the tank contains a liquid of specific gravity 0.9. The centre of orifice is 2 m below the free surface. Find the resultant force exerted on the plate by the liquid and position of the centre of pressure.

Solution: Given data:

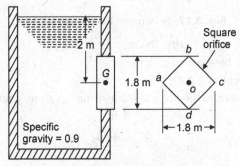

Fig. 3.16: Schematic for Problem 3.9

Diagonals of square orifice: $ac = bd = 1.8$ m.

∴ Area of square orifice: A = area of Δacb + area of Δacd

$$= \frac{ac \times ob}{2} + \frac{ac \times od}{2} = \frac{1.8 \times 0.9}{2} + \frac{1.8 \times 0.9}{2} = 1.62 \text{ m}^2.$$

Specific gravity of liquid: $S = 0.9$

∴ Density of liquid: $\rho = S \times$ density of water = $0.9 \times 1000 = 900$ kg/m^3

Depth of the centre of orifice from free surface: $\bar{h} = 2$ m

∴ The resultant force exerted on the plate by the liquid:

$$F = \rho g \bar{h} A$$
$$= 900 \times 9.81 \times 2 \times 1.62 = \textbf{28605.96 N}$$

Position of the centre of pressure from free surface:

$$h^* = \bar{h} + \frac{I_G}{A\bar{h}}$$

where I_G = MOI of $abcd$ about diagonal ac

= MOI of Δabc about ac + MOI of Δacd about ac

$$= \frac{ac \times ob^3}{12} + \frac{ac \times ad^3}{12} = \frac{1.8 \times 0.9^3}{12} + \frac{1.8 \times 0.9^3}{12}$$

$$= 0.2187 \ m^4$$

$$\therefore \qquad h^* = 2 + \frac{0.2187}{1.62 \times 2} = 2.0675 \ m$$

Problem 3.10: A sluice gate is placed across a trapezoidal channel 10 m wide at the top and 4 m wide at a depth of 3 m. Determine the total pressure on the gate and position of centre of pressure when the depth of water acting on the gate is 2 m.

Solution: Given data as shown in Fig. 3.17.

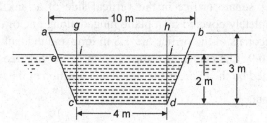

Fig. 3.17: Schematic for Problem 3.10

Width of gate at top: $ab = 10$ m
Width of gate at bottom: $cd = 4$ m
            Depth of gate = 3 m.
Now cg and dh are perpendiculars drawn on ab from points c and d.
$$\therefore \qquad\qquad gh = cd = 4 \ m$$

and $\qquad\qquad ag = bh = \dfrac{10-4}{2} = 3\,m$

Now the water is filled upto 2 m height, Therefore, water level ef is drawn which cuts the perpendiculars cg and dh at points i and j respectively. Determine width ef, from similar triangles $\triangle agc$ and $\triangle eic$,

$$\frac{ag}{ei} = \frac{gc}{ic}$$

or $\qquad\qquad ei = ag \times \dfrac{ic}{gc} = \dfrac{3 \times 2}{3} = 2 \ m$

Similarly from triangles $\triangle hbd$ and $\triangle jfd$,

$$\frac{hb}{jf} = \frac{hd}{jd}$$

or $\qquad\qquad jf = hb \times \dfrac{jd}{hd}$

$$= \frac{3 \times 2}{3} = 2 \ m$$

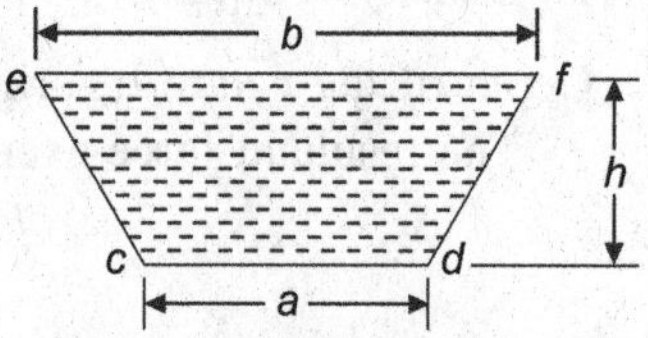

Fig. 3.18: Trapezoidal cross-section of water

$$\therefore \qquad ef = ei + ij + jf =$$
$2 + 4 + 2 = 8$ m

As we know that the total pressure on gate due to depth of the water 2 m: $F = \rho g \overline{h} A$
where $\rho = 1000 \ kg/m^3$ for water

$$g = 9.81 \text{ m/s}^2$$

$\bar{h}$ = depth of centroid from free water surface

$$= \left(\frac{b+2a}{a+b}\right)\cdot\frac{h}{3} = \left(\frac{8+2\times4}{4+8}\right)\times\frac{2}{3} = 0.8888 \text{ m}$$

and A = area of gate up to depth of water 2 m

$$= (a+b)\frac{h}{2} = (4+8)\frac{2}{2} = 12 \text{ m}^2.$$

$\therefore$ Total pressure: F $= 1000 \times 9.81 \times 0.8888 \times 12$

$$= \mathbf{104629.536 \ N}$$

Depth of the centre of pressure from free surface of water:

$$h^* = \bar{h} + \frac{I_G}{A\bar{h}}$$

where I_G = MOI of trapezoidal efcd upto depth of water 2 m about its centroid.

$$= \left(\frac{a^2+b^2+4ab}{a+b}\right)\times\frac{h^3}{36}$$

$$I_G = \frac{(4^2+8^2+4\times4\times8)}{(4+8)}\times\frac{2^3}{36} = 3.8518 \text{ m}^4$$

$\therefore$ $$h^* = 0.8888 + \frac{3.8518}{12\times0.8888} = \mathbf{1.2499 \ m}$$

Problem 3.11:   The dimensions of a cubical tank are 2 m × 2 m × 2 m. It contains water in the lower 0.8 m depth. The upper remaining part is filled with oil of specific gravity 0.85. Find (i) the total pressure on one side on the tank, and (ii) the position of the centre of pressure.

Solution: Given data:

The dimensions of a cubical tank are 2 m × 2 m × 2 m.

Depth of water in lower part of the tank: $h_2 = 0.8$ m

Depth of oil in upper remaining part: $h_1 = 2 - 0.8 = 1.2$ m

Specific gravity of the oil: $S = 0.85$

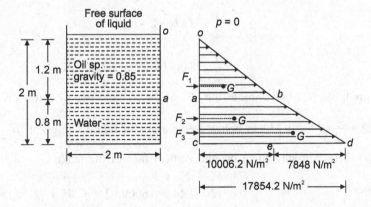

Fig. 3.19: Schematic for Problem 3.11

∴ Density of oil: $\qquad \rho_1 = S \times$ density of water

$$= 0.85 \times 1000 = 850 \text{ kg/m}^3$$

Density of water: $\qquad \rho_2 = 1000 \text{ kg/m}^3$

Now let us draw the pressure diagram for the vertical face of the tank

Intensity of pressure at o: $\qquad p_o = 0$

(at free surface of liquid)

Intensity of pressure at a : $\qquad p_a = \rho_1 gh_1 = 850 \times 9.81 \times 1.2 = 10006.2 \text{ N/m}^2$

Intensity of pressure at c : $\qquad p_c = \rho_1 gh_1 + \rho_2 gh_2$

$$= p_a + \rho_2 gh_2$$
$$= 10006.2 + 1000 \times 9.81 \times 0.8$$
$$= 10006.2 + 7848 = \textbf{17854.2 N/m}^2$$

The pressure diagram can be split into the components oab, $abec$ and bde. The total pressure on one vertical face consists of the following components:

(a) Force: $\qquad F_1 = $ area of triangle $oab \times$ width of the tank

$$= \frac{1}{2} \times ab \times oa \times 2 = \frac{1}{2} \times 10006.2 \times 1.2 \times 2 = 12007.44 \text{ N}$$

This force will be acting at the CG of the triangle oab i.e., at a distance of

$\frac{2}{3} \times 1.2 = 0.8$ m below free surface of liquid i.e., point 'o'

(b) Force: $\qquad F_2 = $ area of rectangle $abec \times$ width of the tank

$$= ab \times ac \times 2$$
$$= 10006.2 \times 0.8 \times 2 = 16009.92 \text{ N}$$

This force will be acting at the CG of the rectangle $obec$ i.e., at a distance of

$$1.2 + \frac{ab}{2} = 1.2 + \frac{0.8}{2}$$

$= 1.6$ m below free surface of liquid i.e., point 'o'

(c) Force: $\qquad F_3 = $ area of triangle $dbe \times$ width of the tank

$$= \frac{1}{2} \times ed \times be \times 2$$

$$= \frac{1}{2} \times 7848 \times 0.8 \times 2 = 6278.4 \text{ N}$$

This force will be acting at the CG of the triangle dbe i.e., at a distance of

$1.2 + \frac{2}{3} \times 0.8 = 1.733$ m below free surface of liquid i.e., point 'o'

(i) Total pressure force on one vertical face of the tank:

$$F = F_1 + F_2 + F_3$$
$$= 12007.44 + 16009.92 + 6278.4$$
$$= \textbf{34295.76 N}$$

(ii) The position of the centre of pressure: h^*

Let the total pressure force F is acting at a depth of h^* from the free surface of liquid i.e., from point 'o'.

Taking the moments of all forces about point 'o' on the free surface of liquid, we get

$$F \times h^* = F_1 \times 0.8 + F_2 \times 1.6 + F_3 \times 1.733$$

or
$$h^* = \frac{0.8F_1 + 1.6F_2 + 1.733F_3}{F}$$

$$= \frac{0.8 \times 12007.44 + 1.6 \times 16009.92 + 1.733 \times 6278.4}{34295.76} = 1.3442 \text{ m}$$

Problem 3.12: A circular plate of 4.0 m diameter is immersed in water in such a way that its greatest and least depth below the free surface are 6 m and 3 m respectively. Find total pressure on the face of the plate and position of the centre of pressure.

Solution: Given data:

Diameter of plate: $d = 4$ m

∴ Area of plate: $A = \dfrac{\pi}{4}d^2 = \dfrac{\pi}{4} \times (4)^2 = 12.56$ m^2

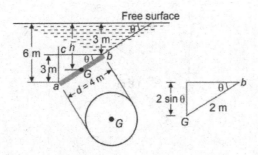

Fig. 3.20: Schematic for Problem 3.12

In $\triangle abc$,

$$\sin \theta = \frac{ac}{ab} = \frac{3}{4}$$

$$\sin \theta = 0.75$$

$$\theta = \sin^{-1}(0.75) = 48.59°$$

Depth of the CG of the plate from the free surface of water:

$$\bar{h} = 3 + 2 \times \sin 48.59° = 4.5 \text{ m}$$

MOI of the circular plate about its centroid and parallel to free surface of water:

$$I_G = \frac{\pi}{64}d^4 = \frac{3.14}{64} \times (4)^4 = 12.56 \text{ m}^4$$

Total pressure: $F = \rho g \bar{h} A = 1000 \times 9.81 \times 4.5 \times 12.56 = \mathbf{554461.2\ N}$

Position of the CP from free surface of water:

$$h^* = \bar{h} + \frac{I_G}{A\bar{h}} \sin^2\theta = 4.5 + \frac{12.56 \times \sin^2 48.59°}{12.56 \times 4.5}$$

$$= \mathbf{4.625\ m\ below\ the\ free\ surface}$$

Problem 3.13:   A rectangular plane 1.2 m by 1.8 m is submerged in water and makes an angle of 30° with the horizontal, the 1.2 m side being horizontal. Calculate the magnitude of the net force on one face and the position of centre of pressure when the top edge of the plane is (*i*) at the water surface and (*ii*) 30 m below the water surface.

Solution: Given data:

Width of the plane : $b = 1.2$ m

Depth of the plane : $d = 1.8$ m

Inclination of plane : $\theta = 30°$

Area of the plane : $A = b \times d = 1.2 \times 1.8 = 2.16$ m^2

Moment of inertia: $I_G = \dfrac{bd^3}{12} = \dfrac{1.2 \times (1.8)^3}{12} = 0.5832$ m^4

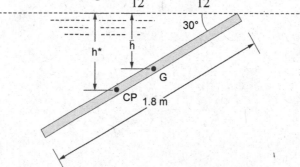

Fig. 3.21: Schematic of Case-I for Problem 3.13

Case-I: Top edge of the plane is at the water surface.

$$\bar{h} = \frac{1.8}{2} \times \sin 30° = 0.45\ m$$

$$F = \rho g \bar{h} A$$

$$= 1000 \times 9.81 \times 0.45 \times 2.16 = 9535.32\ N$$

$$= \mathbf{9.535\ kN}$$

Position of centre of pressure: $h^* = \bar{h} + \dfrac{I_G}{A\bar{h}} \sin^2\theta$

$$= 0.45 + \frac{0.5832}{2.16 \times 0.45} \times \sin^2 30°$$

$$= 0.60 \text{ m}$$

Case-II: Top edge of the plane is 30 m below the water surface.

$$\bar{h} = 30 + \frac{1.8}{2} \times \sin 30° = 30.45 \text{ m}$$

$$F = \rho g \bar{h} A$$

$$= 1000 \times 9.81 \times 30.45 \times 2.16 = 645223.32 \text{ N}$$

$$= 645.223 \text{ kN}$$

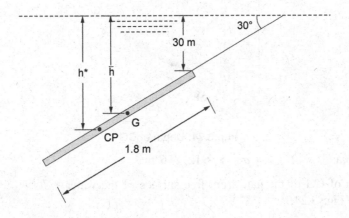

Fig. 3.22: Schematic of Case-II for Problem 3.13

Position of centre of pressure,

$$h^* = \bar{h} + \frac{I_G \sin^2 \theta}{A \bar{h}}$$

$$= 30.45 + \frac{0.5832 \times \sin^2 30°}{2.16 \times 30.45} = 30.45 \text{ m}$$

Problem 3.14: An inclined rectangular sluice gate AB, 1.2 m by 5 m size as shown Fig. 3.23 is installed to control the discharge of water, the end A is hinged. Determine the force normal to the gate applied at B to open it.

Solution: Given data:

    Width of the gate: $b = 5$ m

    Depth of the gate: $d = 1.2$ m

    Inclination of gate: $\theta = 45°$

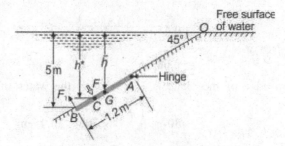

Fig. 3.23: Schematic for Problem 3.14

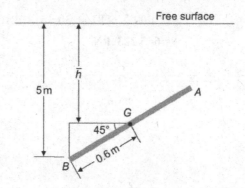

Fig. 3.24: Depth of CG

∴ Area: $A = b \times d = 5 \times 1.2 = 6 \ \text{m}^2$

Depth of CG of the gate from free surface of the water = $\overline{h}$
From Fig. 3.24,

$$\overline{h} = 5 - 0.6 \sin 45° = 4.5757 \ \text{m}$$

The total pressure force (F) acting on the gate:

$$F = \rho g A \overline{h} = 1000 \times 9.81 \times 6 \times 4.5757 = 269325.702 \ \text{N}$$

The position of the centre of pressure from free surface of liquid: h^* (Fig. 3.25)

$$h^* = \overline{h} + \frac{I_G \sin^2 \theta}{A\overline{h}}$$

where I_G = MOI of rectangular gate
about centroid

$$= \frac{bd^3}{12} = \frac{5 \times 1.2^3}{12} = 0.72 \ \text{m}$$

∴ $h^* = 4.5757 + \dfrac{0.72 \times \sin^2 45°}{6 \times 4.5757}$

$$= 4.5888 \ \text{m}$$

From $\triangle COC'$; we get

$$\sin 45° = \frac{CC'}{CO}$$

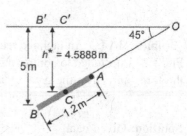

Fig. 3.25: Position of the
centre of pressure

or
$$CO = \frac{CC'}{\sin 45°} = \frac{4.5888}{0.7071}$$
$$CO = 6.4896 \text{ m}$$

and from $\triangle BOB'$, we get
$$\sin 45° = \frac{BB'}{BO}$$

or
$$BO = \frac{BB'}{\sin 45°} = \frac{5}{0.7071}$$
$$BO = 7.0711 \text{ m}$$

∴
$$BC = BO - CO \text{ from Fig. 3.25}$$
$$= 7.0711 - 6.4896 = 0.5815 \text{ m}$$

and
$$CA = BA - BC$$
$$= 1.2 - 0.5815 = 0.6185 \text{ m}$$

Now taking the moments about the hinge A, we get
$$F_1 \times BA = F \times CA$$
$$F_1 = \frac{F \times CA}{BA} = \frac{269325.702 \times 0.6185}{1.2} = \mathbf{138814.95 \text{ N}}$$

Problem 3.15: A rectangular gate 4 × 2 m is hinged at its base and inclined at 62° to the horizontal as shown in Fig. 3.24. To keep the gate in a stable position, a counter weight of 4500 kgf is attached at the upper end of the gate as shown in Fig. 3.26. Find the depth of water at which the gate begins to fall. Neglect the weight of the gate and friction at the hinge and pulley.

Solution: Given data:

Length of gate: $AB = 4$ m
Width of gate $= 2$ m
Inclination of gate: $\theta = 62°$
Counter weight: $W = 4500$ kgf $= 4500 \times 9.81$ N $= 44145$ N

As the pulley is frictionless, the force acting at point $B = 44145$ N. First find the total force F acting on the gate AB for a given depth (h) of water.

From $\triangle A'DA$, we get

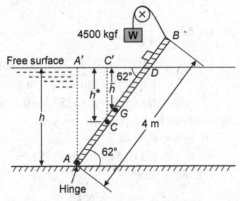

Fig. 3.26: Schematic for Problem 3.15

$$\sin 62° = \frac{AA'}{AD} = \frac{h}{AD}$$

or $$AD = \frac{h}{\sin 62°} = 1.13257\,h$$

∴ Area of gate immersed in water:

$$A = AD \times \text{width of gate} = 1.13257\,h \times 2 = 2.26514\,h\ \text{m}^2$$

Position of the CG from free surface of water: $\bar{h}$

$$\bar{h} = \frac{h}{2} = 0.5\,h\ \text{m}.$$

∴ Total pressure force: $F = \rho g \bar{h} A$

$$= 1000 \times 9.81 \times 0.5\,h \times 2.26514\,h$$
$$= 11110.51\,h^2\ \text{N}$$

And the position of the centre of pressure free surface of water:

$$h^* = \bar{h} + \frac{I_G \sin^2 \theta}{A\bar{h}}$$

where I_G = MOI of gate immersed in water about centroid.

∴ $$I_G = \frac{b \times (AD)^3}{12}$$

$$= \frac{2 \times (1.13257)^3}{12} = 0.24212\,h^3\ \text{m}^4$$

∴ $$h^* = 0.5\,h + \frac{0.24212\,h^3 \times \sin^2 62°}{2.26514\,h \times 0.5\,h}$$

$$= 0.5\,h + 0.16666\,h\ \text{m}$$
$$= 0.66666\,h\ \text{m}$$

From $\Delta C'DC$, we get $\sin 62° = \dfrac{C'C}{CD}$

or $$CD = \frac{C'C}{\sin 62°} = \frac{h^*}{\sin 62°} = \frac{0.66666\,h}{0.88294} = 0.75504\,h\ \text{m}$$

and $$AC = AD - CD = 1.13257\,h - 0.75504\,h = 0.37753\,h\ \text{m}.$$

Now, taking moments about hinge point A; we get,

$$W \times 4 = F \times AC$$
$$44145 \times 4 = 11110.51\,h^2 \times 0.37753\,h$$
$$176580 = 4194.55\,h^3$$

or $$h^3 = \frac{176580}{4194.55}$$

$$h^3 = 42.097$$

or $\qquad\qquad\qquad\qquad h = (42.097)^{1/3} = \textbf{3.478 m}$

Problem 3.16: A circular plate, 4 m diameter has circular hole of 1 m diameter, its centre 1 m above the centre of the plate as shown in Fig. 3.27. The plate is immersed in water at an angle of 30° to the horizontal and with its top edge 2 m below the free water surface. Find

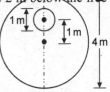

    (*i*) the total pressure, and

    (*ii*) the depth of the centre of pressure.

    Solution: Given data:

Diameter of circular plate: $D = 4$ m

Fig. 3.27: Circular Plate

∴ Area of the whole plate: $A = \dfrac{\pi}{4}D^2$

$$= \dfrac{3.14}{4} \times (4)^2 = 12.56 \text{ m}^2$$

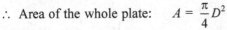

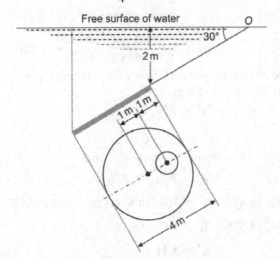

Fig. 3.28: Schematic for Problem 3.16

Diameter of the hole: $d = 1$ m

∴ Area of the hole: $a = \dfrac{\pi}{4}d^2 = \dfrac{3.14}{4} \times 1^2 = 0.785 \text{ m}^2$

$$\bar{h}_1 = 2 + 2\sin 30° = 3 \text{ m}$$

and $\qquad\qquad \bar{h}_2 = 2 + 1\sin 30° = 2.5$ m

Total pressure on whole plate:

$$F_1^* = \rho g \bar{h}_1 A = 1000 \times 9.81 \times 3 \times 12.56 = 369640.8 \text{ N}$$

Total pressure on the small circular plate equal to the hole:

$$F_2^* = \rho g \bar{h}_2 a = 1000 \times 9.81 \times 2.5 \times 0.785 = 19252.125 \text{ N}$$

Total pressure on the plate with hole:

$$F^* = F_1^* - F_2^* = 369640.8 - 19252.125 = \textbf{350388.675 N}$$

Now to determine the centre of pressure of the whole plate without hole:

$$h_1^* = \bar{h}_1 + \frac{I_{G1} \sin^2 \theta}{A\bar{h}_1}$$

$$= 3 + \frac{\dfrac{\pi}{64} D^4 \times \sin^2 30°}{12.56 \times 3}$$

$$= 3 + \frac{3.14 \times 4^4 \times \sin^2 30°}{64 \times 12.56 \times 3} = 3 + 0.083 = 3.083 \text{ m}$$

Centre of pressure of small circular plate equal to the hole:

$$h_2^* = \bar{h}_2 + \frac{I_G \sin^2 \theta}{a\bar{h}_2}$$

$$= 2.5 + \frac{\dfrac{\pi}{64} d^4 \times \sin^2 30°}{0.785 \times 2.5}$$

$$= 2.5 + \frac{3.14 \times 1^4 \times \sin^2 30°}{64 \times 0.785 \times 2.5} = 2.5 + 0.006 = 2.506 \text{ m}$$

To determine centre of pressure (h^*) of the circular plate with hole, taking moments about the axis O–O, we get

$$F^* y^* = F_1^* y_1^* - F_2^* y_2^*$$

$$F^* \frac{h^*}{\sin \theta} = F_1^* \frac{h_1^*}{\sin \theta} - F_2^* \frac{h_2^*}{\sin \theta}$$

$$F^* h^* = F_1^* h_1^* - F_2^* h_2^*$$

$$350388.675 \times h^* = 369640.8 \times 3.083 - 19252.125 \times 2.506$$

$$350388.675 \times h^* = 1091356.761$$

or $h^* = 3.11 \text{ m}$

3.4.4 Curved Surface Submerged in Liquid

Case-I: Liquid in the concave curved surface.

Consider a curved surface AB, submerged in a static fluid as shown in Fig. 3.29.

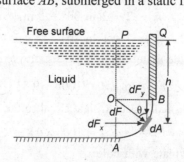

Fig. 3.29: Submerged concave curved surface.

Let dA = area of small strip at depth of h from free surface

The pressure intensity on the area dA: $p = \rho gh$

and pressure force: $dF = p \cdot dA$

$$= \rho ghdA$$

This force dF acts normal to the surface.

Hence total pressure force on the curved surface:

$$F = \int dF = \int \rho ghdA$$

$$= \int \rho ghdA \qquad\qquad ...\ (i)$$

But the direction of force on the surface changes from point to point. Hence integration of Eq. (i) for curved surface is impossible. The problem can be solved by resolving the force dF in two components dF_x and dF_y in x-direction and y-direction respectively.

Resolving the force dF,

$\qquad\qquad dF_x = dF\sin\theta \qquad\qquad \because\ dF = \rho ghdA$

$\therefore \qquad\qquad dF_x = \rho ghdA \sin\theta.$

$\qquad\qquad dF_y = dF \cdot \cos\theta$

$\qquad\qquad\quad\ = \rho ghdA \cos\theta$

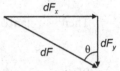

Fig.3.30: Components of dF

Total force in x-direction and y-direction

$$F_x = \int \rho ghdA \sin\theta = \rho g \int hdA \sin\theta \quad ...\ (ii)$$

and $\qquad F_y = \int \rho ghdA \cos\theta = \rho g \int hdA \cos\theta \quad ...\ (iii)$

From area diagram,

Let $\qquad (dA)_V = dA \sin\theta$ = vertical projection of area dA

and $\qquad (dA)_H = dA \cos\theta$

$\qquad\qquad\qquad = $ horizontal projection of area dA

From Eq. (ii),

$\qquad dA \sin\theta = (dA)_V = $ vertical projection area of the area dA and

$\therefore \qquad\qquad F_x = \rho g \int h \cdot (dA)_V$

$\qquad\qquad\ \boldsymbol{F_x = \rho g A_V \bar{h}} \qquad\qquad \because \int h(dA)_H = A_V \bar{h}$

where $\qquad A_V = \int (dA)_V = $ vertical projection of curved surface

$\qquad\qquad\qquad = $ area of OA.

$\qquad\qquad \bar{h} = $ depth of centroid of vertically projected area (A_V) from free surface of the liquid.

From Eq. (iii)

$\qquad dA \cos\theta = (dA)_H = $ horizontal projection of the area dA and

$\therefore \qquad\qquad F_y = \rho g \int h \cdot (dA)_H$

$$F_y = \rho g V \qquad\qquad \because \int h \cdot (dA)_H = V$$

where V = volume of the liquid $PQBA$.

$$F_x = \rho g A_V \overline{h}$$

   = total pressure force on the projected area (A_V) of the curved
     surface on vertical plane OA

and $F_y = \rho g V$

   = weight of liquid supported by the curved surface upto free
     surface of liquid

   = weight of liquid in volume $PQBA$.

Resultant force: $F = \sqrt{F_x^2 + F_y^2}$

and let the resultant force F makes angle ϕ with the horizontal,

Then $\tan \phi = \dfrac{F_y}{F_x}$

or $\phi = \tan^{-1}\left(\dfrac{F_y}{F_x}\right)$

Fig. 3.32:
Components of
resultant force
F

Case-II: Liquid in the convex side of the curved surface.

If the liquid is below the curved surface AB as shown in Fig.
3.31, the same expression are obtained and vertical components F_x and F_y as in the
previous case

$$F_x = \rho g A_V \overline{h} \, .$$

where A_v is the vertical projection of the curved AB i.e., area of OA

   $\overline{h}$ = depth of centroid of vertically projected area (A_V) from free
        surface of the liquid

and $F_y = \rho g V$

where V = the volume of the imaginary liquid $PQBA$.

The results are same in both the cases. But the directions of F_x and F_y are
different. F_y acts downwards in case-I and it acts upwards in case-II. F_x acts toward
right in case-I and it acts toward left in case-II.

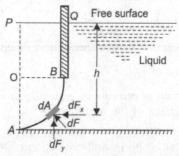

Fig. 3.33: Submerged convex side of the curved surface.

Problem 3.17: The gate *OA* shown in figure below is hinged at *O* and is in the form of a quadrant of a circle of radius 1 m. If supports water on one side. If the width of the gate is 3 m.

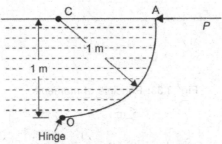

Fig. 3.34: Schematic for Problem 3.17

Calculate the force required to hold the gate in position.

Solution: Given data:

$$\text{Width of gate} = 3 \text{ m}$$
$$\text{Radius of gate} = 1 \text{ m}$$

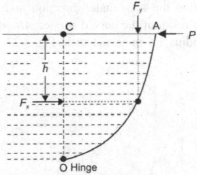

Fig. 3.35: Horizontal and vertical forces act on the gate

Horizontal force F_x exerted by water on gate is

$$F_x = \rho g A_V \bar{h}$$

where A_V = vertical projection area of curve surface OA.

= area of OC = OC × width of gate = 1 × 3 = 3 m²

$\bar{h}$ = depth of *CG* of *OC* from free surface of water

$$= \frac{1}{2} = 0.5 \text{ m}$$

$\therefore$
$$F_x = 1000 \times 9.81 \times 3 \times 0.5 = 14715 \text{ N}$$

and vertical force F_y exerted by water is

$$F_y = \rho g V$$

where V = volume of water COA

$$= \frac{\pi}{4} \times (CA)^2 \times 3 = \frac{3.14}{4} \times 1 \times 3 = 2.355 \text{ m}^3$$

$\therefore$
$$F_y = 1000 \times 9.81 \times 2.355 = 23102.55 \text{ N}$$

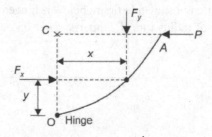

Fig. 3.36: Principle of moments

The distance $\qquad\qquad x = 0.5$ m

and $\qquad\qquad\qquad y = \dfrac{1}{3}(CO) = \dfrac{1}{3} = 0.333$ m

Taking moments about the hinge O, we get

$$P \times 1 = F_y \times x + F_x \times y$$
$$P = 14715 \times 0.5 + 23102.15 \times 0.333 = \mathbf{15065.15 \ N}$$

Problem 3.18: Determine the magnitude, direction and point of action of the hydrostatic force per unit width of the curved surface AB which is in the form of a quadrant of a circle of radius 1.5 m as shown in Fig. 3.37.

Solution: Given data:

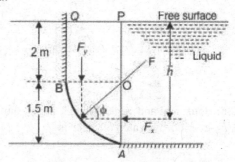

Fig. 3.37: Schematic for Problem 3.18

Width of gate = 1 m

Radius of gate = 1.5 m

$\therefore \ BO = OA = 1.5$ m

Horizontal force F_x exerted by water on gate is

$$F_x = \rho g A_V \overline{h}$$

where $\qquad\qquad A_V$ = vertical projection area of curve surface AB

$\qquad\qquad\qquad$ = area of OA

$\qquad\qquad\qquad$ = $OA \times$ width of the gate = $1.5 \times 1 = 1.5$ m^2

$\qquad\qquad \overline{h}$ = depth of CG of OA from free surface of water

$\qquad\qquad\qquad = 2 + \dfrac{1.5}{2} = 2.75$ m

$\therefore \qquad\qquad F_x = 1000 \times 9.81 \times 1.5 \times 2.75$

$$= 40466.25 \text{ N}$$

and vertical force F_y exerted by water is

$$F_y = \rho g V$$

where V = volume of liquid $PQBA$

$$F_y = \rho g V$$

$$= \rho g \text{ [volume of } PQBO + \text{volume of } OBA]$$

$$= 1000 \times 9.81 \left[2 \times 1.5 \times 1 + \frac{\pi}{4} \times (OB)^2 \times 1 \right]$$

$$= 9810 \left[3.0 + \frac{3.14}{4} \times (1.5)^2 \times 1 \right] = \mathbf{46756.91 \text{ N}}$$

Resultant hydrostatic force: $F = \sqrt{F_x^2 + F_y^2} = \sqrt{(40466.25)^2 + (46756.91)^2}$

$$= \mathbf{61836.28 \text{ N}}$$

Its direction with horizontal: $\phi = \tan^{-1}\left(\dfrac{F_y}{F_x}\right)$

$$= \tan^{-1}\left(\frac{46756.91}{40466.25}\right) = \tan^{-1}(1.15545) = \mathbf{49.12°}$$

The line of action of the resultant hydrostatic force (F) is such that it passes through the centre of curvature O.

Problem 3.19: Determine the magnitude, direction and location of total pressure exerted by water on the curved surface AB, which is the quadrant of a circle (Fig. 3.38). Given that the radius of the surface is 2.5 m and width perpendicular to the paper is 3 m.   (*GGSIP University Delhi, Dec. 2008*)

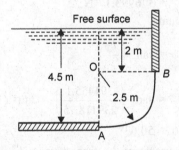

Fig. 3.38: Schematic for Problem 3.19

Solution: Given data:

Radius of surface   = 2.5 m

∴ $OA = OB = 2.5$ m

Width of curve $AB = 3$m

Horizontal force F_x exerted by water on curved surface AB:

$$F_x = \rho g A_v \overline{h}$$

where A_V = vertical projections are of curved surface AB

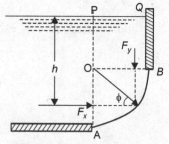

$$= \text{area of } OA$$

$$= OA \times \text{width of curve } AB$$

$$= 2.5 \times 3 = 7.5 \text{ m}^2$$

and $\bar{h}$ = depth of CG of OA from free surface of water.

Fig. 3.39: Horizontal and vertical forces act on curved surface

$$= 2 + \frac{2.5}{2} = 2 + 1.25$$

$$= 3.25 \text{ m}$$

$\therefore$ $F_x = 1000 \times 9.81 \times 7.5 \times 3.25 \text{ N} = 239118.75 \text{ N}$

and vertical force F_y exerted by water is

$$F_y = \rho g V$$

where V = volume of water $PQBA$

$$= \text{volume of } PQBO + \text{volume of } OBA$$

$$= 2 \times 2.5 \times 3 + \frac{\pi}{4} \times (OB)^2 \times 3$$

$$= 15 + \frac{3.14}{4} \times (2.5)^2 \times 3 = 15 + 14.71 = 29.71 \text{ m}^3$$

$\therefore$ $F_y = 1000 \times 9.81 \times 29.71 \text{ N} = 291455.1 \text{ N}$

Total pressure: $F = \sqrt{F_x^2 + F_y^2} = \sqrt{(239118.75)^2 + (291455.1)^2}$

$$= \mathbf{376993.17 \text{ N}}$$

Direction with horizontal:

$$\phi = \tan^{-1}\left(\frac{F_y}{F_x}\right)$$

$$= \tan^{-1}\left(\frac{291455.1}{239118.75}\right) = \tan^{-1}(1.2188)$$

$$\phi = 50.63°$$

Position of total pressure from the free surface: h^*

$$h^* = \bar{h} + \frac{I_G}{A_V \bar{h}}$$

where I_G = MOI of OA about its CG

$$= \frac{bd^3}{12} = \frac{3 \times (2.5)^3}{12} = 3.906 \text{ m}^4$$

$\therefore$ $h^* = 3.25 + \frac{3.906}{7.5 \times 3.25} = 3.25 + 0.16 = \mathbf{3.41 \text{ m}}$

Problem 3.20: Determine the horizontal and vertical component of water pressure acting on the segmental gate of radius 2 m as shown in Fig. 3.40. Take width of the gate 3 m.

Solution: Given data:

Radius of the gate: $r = 2$ m

Width of the gate $= 3$ m (not shown in Fig. 3.40)

Horizontal force on the gate:

$$F_x = \rho g A_V \bar{h}$$

where A_V = vertical projected area of gate ACB

= area of plane ADB

= $2AD$ × width of the gate

= $2 × 2 \sin 45° × 3$

= 8.485 m²

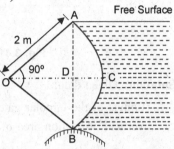

Fig. 3.40: Schematic for Problem 3.20

and $\bar{h}$ = depth of CG of the plane ADB from free surface of water.

$$= \frac{AB}{2}$$

$$= AD$$

$$= 2 \sin 45° = 1.4142 \text{ m}$$

$\therefore$ $F_x = 1000 × 9.81 × 8.485 × 1.4142$

= **117714.96 N**

$AOB = 2AD$
$\because AD = DB$

and

Vertical force on the gate:

$$F_y = \rho g V$$

where V = volume of water under the $ACBDA$

= [area of sector $ACBOA$ – area of $\triangle ABO$] × width of the gate

$$= \left[\frac{\pi r^2}{4} - \frac{OB × OA}{2} \right] × 3$$

$$= \left[\frac{3.14 × 2^2}{4} - \frac{2 × 2}{2} \right] × 3 = 3.42 \text{ m}^3$$

$\therefore$ $F_y = 1000 × 9.81 × 3.42 = $ **33550.2 N**

Problem 3.21: Find the resultant force due to water per metre length, acting on the circular gate of radius 3 m as shown in Fig. 3.41. Find also the angle at which the resultant force will act.

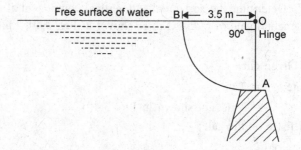

Fig. 3.41: Schematic for Problem 3.21

Solution: Given data:

Radius of the circular gate: r = 3.5 m

Vertical projection area of curved surface AB:

$$A_V = \text{area of } OA = OA \times \text{width of the gate}$$
$$= 3.5 \times 1 = 3.5 \text{ m}^2$$

Depth of CG of OA from free surface of water:

$$\overline{h} = \frac{3.5}{2} = 1.75 \text{ m}$$

∴ Horizontal force exerted by water on the circular gate AB:

$$F_x = \rho g A \overline{h} = 1000 \times 9.81 \times 3.5 \times 1.75$$
$$= 60086.25 \text{ N} = \mathbf{60.08 \text{ kN}}$$

and vertical force F_y, exerted by water is

$$F_y = \rho g V = \rho g \text{ [volume of } OBA]$$

$$= 1000 \times 9.81 \times \frac{\pi r^2}{4} \times 1 = 9810 \times \frac{3.14}{4} \times (3.5)^2$$

$$= 94335.41 \text{ N} = \mathbf{94.33 \text{ kN}}$$

∴ Resultant force due to water:

$$F = \sqrt{F_x^2 + F_y^2} = \sqrt{(60.08)^2 + (94.33)^2} = \mathbf{111.83 \text{ kN}}$$

The angle at which the resultant force will act with horizontal: ϕ

$$\tan \phi = \frac{F_y}{F_x} = \frac{94.33}{60.03}$$

$$\tan \phi = 1.57$$

or $\phi = \tan^{-1}(1.57) = \mathbf{57.50°}$

Problem 3.22: A roller gate of cylindrical form 3 m in diameter has a span of 10 m. Find the magnitude and direction of resultant force acting on the gate, when it is placed on the dam and the water level is such that it is going to spill.

Solution: Given data:

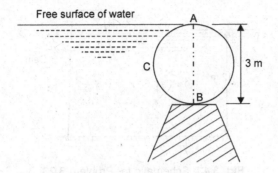

Fig. 3.42: Schematic for Problem 3.22

Diameter of gate: $d = 3$ m, Span: $l = 10$ m and Depth of water $= 3$ m.
Vertical projection area of gate: $A_V = 3 \times 10 = 30$ m^2
and depth of CG of the vertical projection area, AB:

$$\bar{h} = \frac{3}{2} = 1.5 \text{ m}$$

Horizontal force on the roller gate: $F_x = \rho g \bar{h} A_V = 1000 \times 9.81 \times 1.5 \times 30$
$$= 441450 \text{ N} = 441.45 \text{ kN}$$

Vertical force acting on the roller gate: F_y
$$F_y = \rho g V$$
$$= \rho g \times \text{ volume of imaginary water lying in the curved portion } ACB$$
$$= 9.81 \times 1000 \times \frac{1}{2} \times \frac{\pi}{4} \times d^2 \times l = 9810 \times \frac{1}{2} \times \frac{3.14}{4} \times 3^2 \times 10$$
$$= 346538.25 \text{ N} = 346.538 \text{ kN}$$

∴ Resultant force acting on the gate: F

$$F = \sqrt{F_x^2 + F_y^2} = \sqrt{(441.45)^2 + (346.538)^2} = \mathbf{561.22 \text{ kN}}$$

Direction of resultant force: ϕ

Let ϕ = angle which the resultant force makes with the horizontal.

$$\tan \phi = \frac{F_y}{F_x} = \frac{346.538}{441.45}$$

$$\tan \phi = 0.7849$$

or $\phi = \tan^{-1}(0.7849) = \mathbf{38.12°}$

Problem 3.23: The curved of a dam, retaining water, is shaped according to the relationship $y = \dfrac{x^2}{4}$ as shown in Fig. 3.43.

The height of water retained by the dam is 12 m. Find the magnitude and direction of the resultant force on the dam. Consider width of the dam as unity.

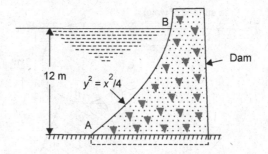

Fig. 3.43: Schematic for Problem 3.23

Solution: Given data:

Relation of water face curve: $y = \dfrac{x^2}{4}$

or $\qquad\qquad x^2 = 4y$

or $\qquad\qquad x = \sqrt{4y} = 2\sqrt{y}$

Height of water: $h = 12$ m.

Width: $\qquad b = 1$ m

The horizontal force exerted by water on

curve AB: $\qquad F_x = \rho g \bar{h} A_V$

where $\qquad \bar{h} = \dfrac{h}{2} = \dfrac{12}{2} = 6$ m

$\qquad\qquad A_V = 12 \times 1 = 12 \text{ m}^2$

$\therefore \qquad\qquad F_x = 1000 \times 9.81 \times 6 \times 12$

$\qquad\qquad\quad = 706320\,\text{N} = 706.32\,\text{kN}$

Vertical component, F_y is given by

$\qquad F_y$ = weight of water
supported by the
curve AB

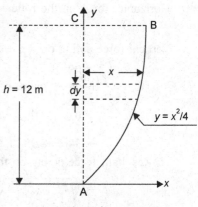

Fig. 3.44: Curved Surface

$\qquad$ = weight of water in the portion ABC

$\qquad$ = $Mg = \rho V g$

$\qquad$ = $\rho g \times$ volume of liquid under curve ABC

$\qquad$ = $\rho g \times$ (area of ABC) $\times$ width of dam

$\qquad$ = $\rho g \left[\displaystyle\int_0^{12} 2\sqrt{y}\,dy \right] \times 1$ $\qquad$ Area of strip = xdy

$\qquad\qquad\qquad\qquad\qquad\qquad\qquad\qquad$ Area of $ABC = \displaystyle\int_0^{12} xdy$

$\qquad$ = $1000 \times 9.81 \displaystyle\int_0^{12} 2\sqrt{y}\,dy = 9810 \times 2 \left[\dfrac{y^{3/2}}{3/2} \right]_0^{12}$

$$= 9810 \times \frac{4}{3} \left[(12)^{3/2} - 0 \right] = 543725.38 \text{ N} = 543.72 \text{ kN}$$

∴ Resultant force on dam: F

$$F = \sqrt{F_x^2 + F_y^2} = \sqrt{(706.32)^2 + (543.72)^2} = \sqrt{794519.38}$$

$$= \textbf{891.35 kN}$$

Direction of the resultant force: ϕ

Let ϕ = angle made by resultant force with horizontal direction

$$\tan \phi = \frac{F_y}{F_x} = \frac{543.72}{706.32} = 0.7697$$

$$\phi = \tan^{-1}(0.7697) = \textbf{37.58°}$$

3.5 TOTAL PRESSURE AND CENTRE OF PRESSURE ON LOCK GATES

Lock gates are the devices used for changing the water level in a river or a canal for navigation or boating. It is constructed between two different levels of water. The upstream means higher water level and the downstream means lower water level.

In order to transfer a boat from upstream to downstream, the upstream, gates are opened while the downstream gates are closed and water level in the chamber rises up to the upstream water level. The boat is then admitted in the chamber. Then the upstream gates are closed and downstream gates are opened and the water level in the chamber is lowered to the downstream water level. Now the boat can enter in downstream. If the boat is to be transferred from downstream to upstream side, the above explained procedure is reversed.

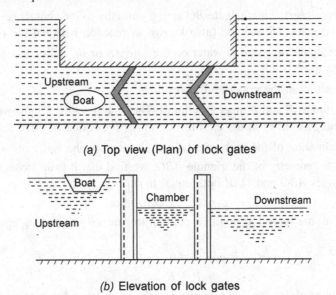

(a) Top view (Plan) of lock gates

(b) Elevation of lock gates

Fig. 3.45: Transfer of boat at upstream to downstream.

Now consider a set of lock gates AB and BC hinged at the top and bottom at A and C respectively as shown in Fig. 3.46. In the closed position, the gates meet at point B.

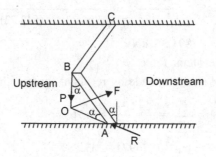

(a) Top view (plan) of lock gate

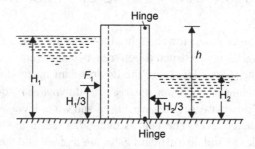

(b) Elevation of lock Gate

Fig. 3.46: Lock gate

Let P = force exerted by the gate BC acting normally to the contact surface of the two gates AB and BC (also known as reaction between the two gates)

     F = resultant force due to water on the gate AB or BC acting at right angle to the gate.

     R = reaction at the upper and lower hinge.

Let the forces P and F meet at point O. Then the reaction R must pass through point O as the gate AB is in the equilibrium under the action of three forces.

Let α = Inclination of the lock gate with the normal to the walls of the lock.

From the geometry of the triangle ABO, we find that it is an isosceles triangle having its angles ABO and OAB both equal to α.

    i.e., $\angle ABO = \angle OAB = \alpha$

Resolving the forces, at point O, parallel to gate AB and putting equal to zero, we get

$$R \cos \alpha - P \cos \alpha = 0$$

    or $P = R$

and now resolving the forces normal to the gate AB and putting equal to zero, we get

$$R \sin \alpha + P \sin \alpha - F = 0$$

or
$$F = R \sin \alpha + P \sin \alpha$$
$$F = P \sin \alpha + P \sin \alpha \qquad \because R = P$$
$$F = 2P \sin \alpha$$

$\because$
$$P = \frac{F}{2 \sin \alpha} \qquad \qquad \text{... (i)}$$

or
$$R = \frac{F}{2 \sin \alpha}$$

(a) Find P or R.

From equation (i), P can be calculated if F and α are known. The value of α is calculated from the angle between the lock gates. The angle between the two lock gates is equal to $(180 - 2\alpha)$. Hence α can be calculated. The value of F is calculated as:

Let H_1 = height of water on the upstream side

H_2 = height of water on the downstream side

F_1 = total pressure of the water on the gate on upstream side

F_2 = total pressure of the water on the gate on downstream side

l = width of gate.

We know,

$$F_1 = \rho g A_1 \overline{h_1}$$

where A_1 = wetted area of each gate on the upstream side

$$= H_1 l$$

and $$\overline{h_1} = \frac{H_1}{2}$$

$\therefore$
$$F_1 = \rho g \times H_1 l \times \frac{H_1}{2} = \frac{\rho g l H_1^2}{2}$$

Similarly, $$F_2 = \rho g A_2 \overline{h_2}$$

where A_2 = wetted area of each gate on the downstream side

$$= H_2 l$$

and $$\overline{h_2} = H_2/2$$

$\therefore$
$$F_2 = \rho g \times H_2 l \times H_2/2 = \rho g l \frac{H_2^2}{2}$$

Since the direction of F_1 and F_2 are in the opposite direction, therefore the resultant force: F

$$F = F_1 - F_2$$

$$F = \rho g l \frac{H_1^2}{2} - \rho g l \frac{H_2^2}{2}$$

Substituting the value of α and F in Eq. (i), the value of P or R can be calculated.

(b) Find reaction at the top and bottom hinges:

Let R_T = reaction of the top hinge, and

R_B = reaction of the bottom hinge.

Since the total reaction (R) will be shared by the two hinges (R_T and R_B),

$\therefore$ $R = R_T + R_B$

We know that the total pressure F_1 will act through its centre of pressure, which is at a height of $H_1/3$ from the bottom of the gate. Similarly, the total pressure F_2 will also act through its centre of pressure, which is at a height of $H_2/3$ from the bottom of the gate.

The resultant of F_1 and F_2 acts normal to the gate. Half of the value of F will be resisted by the hinges of one lock gate and other half will be resisted by the hinges of other lock gate.

Taking moment, about the lower hinge,

$$R_T \sin \alpha \times h = \frac{F_1}{2} \times \frac{H_1}{3} - \frac{F_2}{2} \times \frac{H_2}{3}$$

where h is the distance between the two

$$\boldsymbol{R_T \, h \, \sin \, \alpha = \frac{F_1 H_1}{6} - \frac{F_2 H_2}{6}} \qquad \qquad \dots (ii)$$

Also resolving the forces horizontally,

$$\boldsymbol{R_T \sin \, \alpha + R_B \sin \, \alpha = F_1 - F_2} \qquad \qquad \dots (iii)$$

From Eqs. (ii) and (iii), we can calculate R_T and R_B

Problem 3.24: Two lock gates are provided in a canal of 16 m width meeting at an angle 120°. Find the force acting on each gate, when the depth of water on upstream side is 5 m.

Solution: Given data:

Width of lock gates = 16 m

Inclination of gates: $(180° - 2\alpha) = 120°$

or $2\alpha = 180° - 120° = 60°$

$\alpha = 30°$

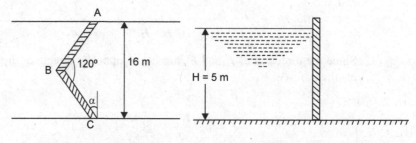

(a) Top view of lock gates   (b) Elevation of lock gates

Fig. 3.47: Schematic for Problem 3.24

Width of each gate: $l = \dfrac{16/2}{\cos\alpha} = \dfrac{8}{\cos 30°} = \dfrac{8}{0.866} = 9.237$ m

∴ Wetted area of each gate:

$$A = 5 \times 9.237 = 46.185 \text{ m}^2$$

Now force acting on each gate:

$$F = \rho g A \bar{h} = \rho g A \dfrac{H}{2}$$

$$= 1000 \times 9.81 \times 46.185 \times \dfrac{5}{2}$$

$$= 1132687.125 \text{ N} = \textbf{1132.68 kN}$$

Problem 3.25: The lock gates 12 m wide make an angle of 130° when closed. The water levels on the upstream and downstream sides are 7 m and 5 m respectively as shown in Fig. 3.48. Each gate is carried by two hinges placed at the top and bottom of the gate. Find:

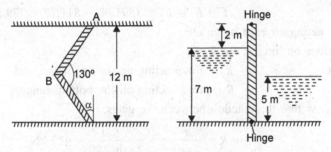

Fig 3.48: Schematic for Problem 3.25

(*i*) Resultant force on each gate, and

(*ii*) Force acting on each hinge.

Solution: Given data:

Width of the lock gates = 12 m

Angle at which the gates make when closed = 130°

∴ $180° - 2\alpha = 130°$

or $2\alpha = 180° - 130° = 50°$

$\alpha = 25°$

$H_1 = 7$ m

$H_2 = 5$ m

Distance between hinges: $h = 9$ m

Width of each gate $= \dfrac{12/2}{\cos\alpha} = \dfrac{6}{\cos 25°} = 6.62$ m

∴ Wetted area of each gate on the upstream side: A_1

$$A_1 = H_1 \times 6.62 = 7 \times 6.60 = 46.34 \text{ m}^2$$

and wetted area of each gate on the downstream side: A_2

$$A_2 = H_2 \times 6.62 = 5 \times 6.62 = 33.1 \text{ m}^2$$

We know that force acting on each gate on the upstream side: F_1

$$F_1 = \rho g \bar{h}_1 A_1 = 1000 \times 9.81 \times \frac{H_1}{2} \times 46.34 \; \because \; \bar{h}_1 = \frac{H_1}{2}$$

$$= 9810 \times \frac{7}{2} \times 46.34 = 1591083.9 \text{ N} = 1591.08 \text{ kN}$$

and force acting on each gate on the downstream side: F_2

$$F_2 = \rho g \bar{h}_2 A_2 = 1000 \times 9.81 \times \frac{H_2}{2} \times 33.1 \; \because \; \bar{h}_2 = \frac{H_2}{2}$$

$$= 9810 \times \frac{5}{2} \times 33.1 = 811777.5 \text{ N} = 811.77 \text{ kN}$$

(*i*) Resultant force on each hinge: F

∴ Resultant force on each gate:

$$F = F_1 - F_2 = 1591.08 - 811.77 = \mathbf{779.31 \text{ kN}}$$

(*ii*) Force acting on each hinge OR

Reactions on hinges: R_T, R_B

where $\qquad\qquad R_T$ = force acting on the top hinge, and

$\qquad\qquad\qquad R_B$ = force acting on the bottom hinge.

We know that the reaction between the gates,

$$R = \frac{F}{2 \sin \alpha} = \frac{779.31}{2 \times \sin 25°} = 922 \text{ kN}$$

We also know from the equation of forces acting on the hinges that;

$$R_T h \sin\alpha = \frac{F_1 H_1}{6} - \frac{F_2 H_2}{6}$$

$$R_T \times 9 \times \sin 25° = \frac{1591.08 \times 7}{6} - \frac{811.77 \times 5}{6}$$

$$3.80 \; R_T = 1856.26 - 676.47$$

or $\qquad\qquad\qquad R_T = \mathbf{310.47 \text{ kN}}$

and $\qquad\qquad\qquad R = R_B + R_T$

$$922 = R_B + 310.47$$

or $\qquad\qquad\qquad \mathbf{R_B = 611.53 \text{ kN}}$

3.6 PRESSURE DISTRIBUTION IN A LIQUID MASS SUBJECTED TO UNIFORM ACCELERATION

When a liquid is contained in a tank and there is no relative motion between the liquid particles or between the liquid particles and the boundary of the tank, the liquid is said to be in state of absolute rest or absolute equilibrium. Such a liquid obeys the hydrostatic laws. On the other hand when a tank is made to move with a constant acceleration. At the beginning, the fluid particles will move relative to each other and to the boundaries of the tank. However, after some time, there will not be any relative motion between the liquid particles and the boundaries of the tank. The liquid will come to rest in this new position relative to the tank. The entire liquid mas moves as a single unit. Liquid in such a motion is said to be in relative equilibrium. Such liquid also obeys the hydrostatic laws. Because of no relative motion between the liquid particles, hence the shear stresses and shear forces between liquid particles will be zero. Further, the liquid pressure acts normal to the surface in contact with it. Under these conditions, the following are the important cases:

(*i*) Liquid mass in a continer subjected to constant acceleration in the horizontal direction.

(*ii*) Liquid mass in a container subjected to uniform acceleration in the vertical direction.

3.6.1 Liquid Mass in a Container Subjected to Contant Accteration in the Horizontal Direction

Consider a tank filled with liquid is being accelerated horizontally from left to right with uniform acceleration a as shown in Fig. 3.49 (b). Thus the free surface of liquid due to constant horizontal acceleration will become a downward sloping inclined plane with the liquid rising at the back side and the liquid falling at the front side.

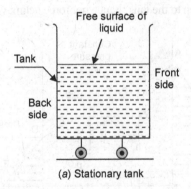

(a) Stationary tank

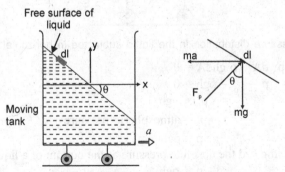

(b) Tank moving with constant acceleration in horizontal direction

Fig. 3.49: Liquid mass in a container

The equation for the free surface of liquid can be derived by considering the equilibrium of a fluid element dl lying on the free surface of liquid. The forces acting on the fluid element dl are:

(a) Pressure force F_p extered by the fluid on the element dl. This force is normal to the free surface.

(b) Weight of the fluid element: mg acting vertically downward

(c) Inertia force: ma acting horizontal and apposite direction of the moving tank.

Resolving the forces horizontally,

We get

$$F_p \sin \theta = ma \qquad (i)$$

and resolving the forces vertically, we get

$$F_p \cos \theta = mg \qquad (ii)$$

Dividing Eq. (i) by Eq. (ii), we get

$$\tan \theta = \frac{a}{g} \qquad (iii)$$

The term a/g is constant at all points on the free surface of liquid, hence $\tan \theta$ is constant and consequently the free surface is a straight plance inclined down at angle θ along the direction of acceleration.

Considering the equilibrium of a liquid element at depth h from the free surface. There is no vertical acceleration given to the tank, hence net force acting vertically should be zero.

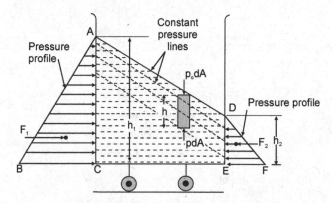

Fig. 3.50: Pressure distribution in the liquid subjected in horizontal acceleration

$\therefore \quad p \, dA - p_0 \, dA - \rho \, ghdA = 0$

or $\qquad\qquad p - p_0 - \rho gh = 0$

or $\qquad\qquad p - p_0 = \rho gh$

where $\qquad\qquad p_0 = $ atmospheric pressure

or $\qquad\qquad p = \rho gh + p_0 \qquad (iv)$

Eq. (iv) is indicated the absolute pressure at the bottom of a liquid element.

$$p = \rho gh \qquad (v)$$

or Pressure head: $\qquad \dfrac{p}{\rho g} = h$

Equation (v) is indicated the gauge pressure at the bottom of a liquid element. It is clear that pressure head at any point in a liquid subjected to a constant horizontal acceleration is equal to the height of the liquid column above that point. Therefore the pressure distribution in a liquid subjected to a constant acceleration is same as hydrostatic pressure distribution. The lines of constant pressure are parallel to the free surface of liquid. The constant pressure lines and the variation in liquid pressure on back side and front side of the tank as shown in Fig. 5.50.

Let h_1 = depth of liquid at the back side of the tank.

            h_2 = depth of liquid at the front side of the tank.

F_1 = total pressure force exerted by liquid on the back side of the tank.

        = area of $\triangle$ ABC × width

$$= \frac{1}{2} \times BC \times AC \times b$$

$$= \frac{1}{2} \rho g h_1 \times h_1 \times b$$

$$= \frac{\rho g b h_1^2}{2} = \rho g A_1 \bar{h_1}$$

where b = width of tank perpendicular to the plane of the paper, not shown in figure.

$$A_1 = h_1 b, \quad \bar{h_1} = \frac{h_1}{2}$$

F_2 = total pressure force exerted by liquid on the front side of the tank.

        = area of $\triangle$ DFE × width

$$= \frac{1}{2} \times EF \times DE \times b$$

$$= \frac{1}{2} \times \rho g h_2 \times b$$

$$= \frac{\rho g b h_2^2}{2} = \rho g \bar{h_2} A_2 \quad \text{where} \quad A = h_2 b \quad \text{and} \quad \bar{h_2} = \frac{h_2}{h}$$

Thereforce net pressure force,

$$F = F_1 - F_2$$

According to Newton's second law of motion, the net force is equal to the product of the mass of liquid and acceleration in the same direction.

Mathematically,

$$F = ma$$

or $F_1 - F_2 = ma$

where m = mass of the liquid contained in the tank.

                a = *horizontal acceleration of the tank.*

3.6.2 Liquid Mass in a Container Subjected to Uniform Accteration in the Vertical Direction

Consider a tank open at the top, containing a liquid and moving vertically upward with a uniform acceleration. The tank is subjected to an acceleration in vertical direction only, therefore the liquid will remain horizontal.

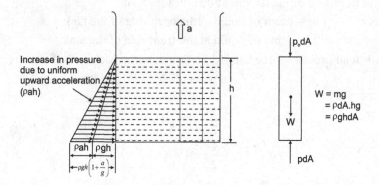

Fig. 3.51: Liquid subjected to constant vertical upward acceleration

Consider a small column of the liquid of height h and cross-sectional area dA in the tank as shown in Fig. 3.51.

Let dA = cross-sectional area of a small column of the liquid.

$\quad$ h = height of a small column of the liquid

$\quad$ P_0 = atmaspheric pressure acting on the top of column.

$\quad$ p = pressure exerted by liquid at the bottom of column.

According to Newton is second low of molion, the net force acting on a small column of the liquid is equal to the product of the mass of liquid and acceleration in the same direction.

Mathematically,

$$\Sigma F_{upward} = ma$$

$$pdA - p_0dA - \rho ghdA = \rho\, dA\, ha \qquad\qquad \therefore\ m = \rho\, dA\, h$$

or $\qquad\qquad\qquad p = p_0 - \rho gh = \rho gh$

or $\qquad\qquad\qquad p - p_0 = \rho gh + \rho ha$

or $\qquad\qquad\qquad p_g = \rho gh + \rho ha$

$$p_g = \rho gh \left[1 + \frac{a}{g}\right] \qquad\qquad (i)$$

where $p_g = p - p_0$, gauge pressure.

Similarly, the liquid mass is subjected to uniform acceleration in the vertical downward direction, acceleration a shall be -ve and then Eq. (i) becomes to

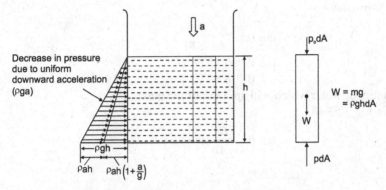

Fig. 3.52: Liquid subjectd to contant vertical downward acceleration

$$p_g = \rho g h \left[1 - \frac{a}{g} \right] \qquad (ii)$$

If the tank is moved vertical downward at the gravitational acceleration, then $a = g$ and Eq. (*ii*) becomes $p_g = 0$. Hence, the pressure in uniform and equivalent to surrounding atmospheric pressure, and there is no force on the base and walls of the tank.

Problem 3.26:   An open rectangular tank 3 m long, 2.5 m wide and 1.25 m deep is completely filled with water. If the tank is moved with an acceleration of 1.5 m/s², find the slop of the free surface of water and the quantity of water which will spill out of the tank.

Solution: Given data

Length of the tank: $l = 3$ m
Width of the tank : $b = 2.5$ m
Depth of water in tank : $h = 1.25$ m
Acceleration of the tank : $a = 1.5$ m/s²

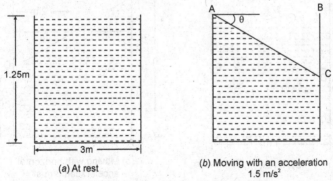

(a) At rest

(b) Moving with an acceleration
1.5 m/s²

Fig. 3.53: Schematic for Problem 3.26

Slope of the free surface of water:
Let θ = angle which the free surface of water makes with the horizontal.

We know that

$$\tan \theta = \frac{a}{g} = \frac{1.5}{9.81}$$

$$\tan \theta = 0.1529$$

or $\qquad\qquad\qquad \theta = 8.69°$

Quantity of water which will spill out of the tank:

From Fig 3.53 (b),

$$\tan \theta = \frac{BC}{AB}$$

$$\tan 8.69° = \frac{BC}{3}$$

or $\qquad\qquad\qquad BC = 3 \times \tan 8.69° = 3 \times 0.1529 = 0.4587$

∴ Quantity of water which will spill out of the tank

$$= \text{are of } \Delta ABC \times \text{width of tank}$$

$$= \frac{1}{2} \times AB \times BC \times 2.5 \;\; = \frac{1}{2} \times 3 \times 0.4587 \times 2.5 \; = \mathbf{1.72 \ m^3}$$

Problem 3.27: A rectangular tank 4 m long, 2 m wide and 3 m deep contains water to a depth of 2 m. If the tank is accelerated horizontally at 3 m/s², find the total pressure at the back and front sides of the tank.

Solution : Given data

Length of the tank: $\qquad\qquad l = 4$ m

Width of the tank : $\qquad\qquad b = 2$ m

Depth of water in tank : $\qquad\quad H = 3$ m

Depth of water in tank : $\qquad\quad h = 2$ m

Horizontial acceleration of the tank : $a = 3$ m/s²

Let θ = angle which the free surface of water makes with the horizontal.

We know that

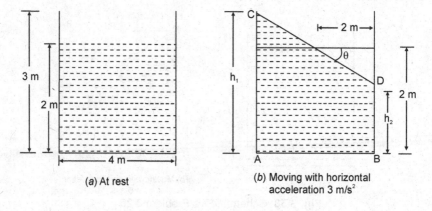

(a) At rest   (b) Moving with horizontal acceleration 3 m/s²

Fig. 3.54: Schematic for Problem 3.27

$$\tan \theta = \frac{a}{g}$$

$$\tan \theta = \frac{3}{9.81}$$

$$\tan \theta = 0.3058$$

or $\qquad\qquad \theta = 17°$

From Fig. 3.54 (*b*),

Depth of water at the front side,

$$h_2 = 2\text{--}2 \tan \theta = 2\text{--}2 \times \tan 17° = 2\text{--}2 \times 0.3058$$
$$= 1.388 \text{ m}$$

Depth of water at the back side,

$$h_1 = 2+2 \tan \theta = 2+2 \tan 17° = 2+2 \times 0.3058 = 2.61 \text{ m}$$

We know that the total pressure at the back side of the bank,

$$F_1 = \rho g A_1 h_1$$

where $\qquad A_1 = AC \times$ with of tank

$$= h_1 \times b$$
$$= 2.61 \times 2 = 5.22 \text{ m}^2$$

and $\qquad \bar{h_1} = \frac{h_1}{2} = \frac{2}{2} = 1\, m$

∴ $\qquad\qquad F_1 = 1000 \times 9.81 \times 5.22 \times 1 = 5120\ 8.2 \text{ N} = \textbf{51.208 kN}$

and total pressure at the front side of the tank,

$$F_2 = \rho g A_2 \bar{h_2}$$
$$= 5$$

where $\qquad A_2 = BD \times$ width of tank

$$= h_2 \times b = 1.388 \times 2 = 2.77 \text{ m}$$

and $\qquad \bar{h_2} = \frac{h_2}{2} = \frac{1.388}{2} = 0.694 \text{ m}$

∴ $\qquad\qquad F_2 = 1000 \times 9.81 \times 2.77 \times 0.694 = 18858.5 \text{ N}$
$$= \textbf{18.858 kN}$$

Problem 3.28: An open tank 10 m long, 4 m wide and 2 m deep is filled with water upto a depth of 1.5 m. The tank is uniformly accelerated from rest to 12.5 m/s. Find the shortest time in which the tank may be accelerated without the water spilling over the edge.

Solution: Given data

Length of the tank: $\qquad l = 10$ m

Width of the tank : $\qquad b = 4$ m

Depth of water in tank : $\quad H = 2$ m

Depth of water in tank : $\quad h = 1.5$ m

Initial velocity of the tank: $u = 0$

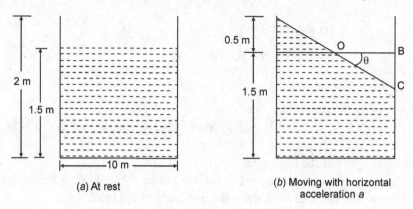

(a) At rest

(b) Moving with horizontal acceleration a

Fig. 3.55: Schematic for Problem 3.28

Final velocity of the tank : v = 12.5 m/s

Let t = time required for

θ = angle which the free surface of water makes with the horizontal.

a = uniform acceleration of the tank.

From Fig. 3.55 (b),

$$\tan \theta = \frac{BC}{OB}$$

$$\tan \theta = \frac{0.5}{5} = \frac{1}{10}$$

also

$$\tan \theta = \frac{a}{g}$$

∴

$$\frac{1}{10} = \frac{a}{9.81}$$

or

$$a = \frac{9.81}{10} = 0.981 \text{ m/s}^2$$

We know that final velocity of the tank,

$$v = u + at$$

$$12.5 = 0 + 0.981 \times t$$

or

$$t = \frac{12.5}{0.981} = 12.74 \text{ s}$$

Problem 3.29: A closed oil tanker 2 m deep, 1.8 m wide and 3.5 m long has been filled with an oil of specific gravity 0.85 upto a depth of 1.5 m. Find the acceleration which may be imparted to the tank in the direction of its length so that bottom front side of the tank is just exposed. Also find the net horizontal force acting on the tanker sides and show that this equals the forces necessary to accelerate the liquid mass in the tanker.

Solution: Given data:

Depth of the tank: $H = 2$ m
Width of the tank : $b = 1.8$ m
Length of the tank : $l = 3.5$ m
Specific gravity of oil: $S = 0.85$
∴ Density of oil : $\rho = S \times 1000$ kg/m$^3 = 0.85 \times 1000$ kg/m$^3 = 850$ kg/m^3
Depth of oil in tank : $h = 1.5$ m

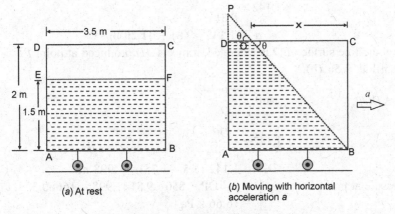

(a) At rest

(b) Moving with horizontal acceleration a

Fig. 3.56: Schematic for Problem 3.29

Since the top of the tank is closed, and the liquid cannot spill from it under any acceleration imparted to it. Hence, the quantity of oil inside the tanker remains the same. The oil surface which was initally horizontal EF as shown in Fig 3.56 (a) and inclined OB when the front bottom end B is just exposed.

Now equating volumes of oil before and after the motion.

Volume of rectangle ABFE = volume of trapezium ABOD

lbh = volume of rectange ABCD - volume of $\triangle$ OCB

$$lbh = lbh - \frac{xH}{2}b$$

$$lbh = bh\left(l - \frac{x}{2}\right)$$

$$lh = H\left(l - \frac{x}{2}\right)$$

or
$$3.5 \times 1.5 = 2\left(3.5 - \frac{x}{2}\right)$$

$$5.25 = 7 - x$$

or $x = 7 - 5.25 = 1.75$ m

From Fig. 3.56 (b)

$$\tan\theta = \frac{BC}{OC} = \frac{H}{x}$$

$$\tan\theta = \frac{2}{1.75} = 1.142$$

also $$\tan\theta = \frac{a}{g}$$

$\therefore$ $$1.142 = \frac{a}{9.81}$$

or a = 1.142× 9.81 = **11.20 m/s²**

Now the free surface BO is extended, it meets AD produced at point P.
From Fig. 3.56 (b),

In $\triangle DOP$, $$\tan\theta = \frac{DP}{DO}$$

$$1.142 = \frac{DP}{3.5 - x}$$

or DP = 1.142 (3.5 – 1.75) = 1.998 m

Pressure at point D : p_D = ρg×DP = 850× 9.81×1.998 = 16660.32 N/m²
                                        = 16.66 k Pa

Pressure at point A: p_A = ρg× AP = 850×9.81×(2+1.998) = 33337.32 N/m²
                                        = 33.33 kPa

$\therefore$ Force on the back side AD,

$$F_{AD} = \text{average pressure} \times \text{area} = \left(\frac{p_D + p_A}{2}\right) \times Hb$$

$$= \left(\frac{16.66 + 33.33}{2}\right) \times 2 \times 1.8 = 89.98 \text{ kN}$$

and force on the front side BC,

                     F_{BC} = 0 because no oil pressure exerted on front side BC.

$\therefore$ Net pressure force: F = F_{AD} – F_{BC}
                     = 89.99 – 0 = **89.99 kN**

The force needed to accelerate the liquid mass in the tank is

          F = mass of oil × uniform linear acceleration
             = ρ × volume of liquid × a
             = 850 × (3.5×1.8×1.5) × 11.20
             = 89964N = 89.96 kN ≈ **89.99 kN**

Hence, the difference between the forces on the two sides of the tank is equal to
the force necessary to accelerate the liquid mass in the tank.

Problem 3.30: A vessel containing 1.5 m height of water is moved vertically
downward with a constant acceleration a. If water starts boiling at absolute pressure

of 0.07 bar at the ambient temperature, find the value of acceleration a such that water just begins to boil.

Solution: Given data

Height of water in vessel: $h = 1.5$ m

Absolute pressure : $p_{abs} = 0.07$ bar $= 0.07 \times 10^5$ N/m^2

also $p_{abs} = p_g + p_{atm}$

or Gauge pressure : $p_g = p_{abs} - p_{atm} = 0.07 \times 10^5 - 1.01325 \times 10^5$

$= -0.94325 \times 10^5$ N/m$^2 = -94325$ N/m^2

We know that the vessel is moved vertically downward, the gauge pressure is

$$p_g = \rho g h \left(1 - \frac{a}{g}\right)$$

$$-94325 = 1000 \times 9.81 \times 1.5 \left(1 - \frac{a}{9.81}\right)$$

$$-94325 = 14715 \left(1 - \frac{a}{9.81}\right)$$

or $$-6.41 = 1 - \frac{a}{9.81}$$

or $$\frac{a}{9.81} = 1 + 6.41 = 7.41$$

or $$a = 7.41 \times 9.81 = \mathbf{72.69 \ m/s^2}$$

Problem 3.31: An open rectangular tank 4 m long and 2.5 m wide contains an oil of specific gravity 0.85 upto a depth of 1.5 m. Determine the total pressure on the bottom of the tank, when the tank is moving with an acceleration of 0.5g m/s^2 (*i*) vertically upwards, and (*ii*) vertically downward.

Solution : Given data

Length of the tank: $l = 4$ m

Width of the tank: $b = 2.5$ m

Specific gravity of the oil:S $= 0.85$

∴ Density of the oil: $\rho = S \times 1000$ kg/m$^3 = 0.85 \times 1000$ kg/m$^3 = 850$ kg/m^3

Depth of oil in tank : $h = 1.5$ m

Acceleration : $a = 0.5g$ m^2/s

(*i*) Total pressure on the bottom of the tank when it is moving vertically upwards: F$_1$

Intensity of pressure on the bottom of the tank :

$$p_1 = \rho g h \left(1 + \frac{a}{g}\right) = 850 \times 9.81 \times 1.5 \left(1 + \frac{0.5g}{g}\right)$$

$$= 12507.75 (1 + 0.5) = 12507.75 \times 1.5$$

$$= 18761.62 \ N/m^2 = 18.761 \ kN/m^2$$

$\therefore$ Total pressure on the bottom of the tank:
$$F_1 = p_1 A = 18.761 \times 4 \times 2.5$$
$$= \textbf{187.61 kN}$$

(*ii*) Total pressure on the bottom of the tank when it is moving vertically downwards: F_2

Intensity of pressure on the bottom of the tank:

$$p_2 = \rho g h\left(1 - \frac{a}{g}\right)$$

$$p_2 = \rho g h\left(1 - \frac{a}{g}\right) = 850 \times 9.81 \times 1.5\left(1 - \frac{0.5g}{g}\right)$$

$$= 12507.75(1 - 0.5) = 12507.75 \times 0.5$$

$$= 6253.87 \text{ N/m}^2 = 6.253 \text{ kN/m}^2$$

$\therefore$ Total pressure on the bottom of the tank:
$$F_2 = p_2 A = 6.253 \times 4 \times 2.5 = \textbf{62.53 kN}$$

SUMMARY

1. **Total pressure (F):** It is defined as the resultant force exerted by a static fluid on a surface (Surface is either plane or curved).
2. **Centre of pressure (CP):** It is defined as the point of application of the total pressure (F) on the surface.
3. **Hydrostatic pressure:** It is the force acting on a surface due to weight of a liquid.
4. **Hydrostatic law:** It states that the rate of increase of pressure on vertical direction is equal to weight density or specific weight.

   Mathematically,
   $$p = \rho g h$$
   where $p =$ hydrostatic pressure above atmospheric pressure.
   $$h = \text{depth of the point from the free surface.}$$
5. **Total pressure (F) and centre of pressure (CP) for submerged surface:**

   (*i*) Vertical plane surface submerged in liquid:

   Total pressure: $F = \rho g A \bar{h}$

   Position of the centre of pressure from free surface of liquid:

   $$h^* = \bar{h} + \frac{I_G}{A\bar{h}}$$

   (*ii*) Horizontal plane surface submerged in liquid.

Contd...

Total pressure: $F = \rho g A \bar{h}$

Position of the centre of pressure from free surface of liquid: $h^* = \bar{h}$ *i.e.*, centre of gravity and centre of pressure coincides.

(*iii*) **Inclined plane surface submerged in liquid:**

Total pressure: $F = \rho g A \bar{h}$

Position of the centre of pressure from free surface of liquid:

$$h^* = \bar{h} + \frac{I_G \sin^2 \theta}{A \bar{h}}$$

where $\bar{h}$ = position of CG from the free surface of liquid

I_G = MOI of the plane about centroid.

θ = angle of inclination of the phase with free surface of liquid.

(*iv*) **For curved surface:**

Force exerted on the curved surface along x-direction: $F_x = \rho g A_V \bar{h}$

= Total pressure force on the vertical projected area (A_V) of the curved surface

and Force exerted on the curved surface along of direction: $F_y = \rho g V$

= weight of liquid supported by the curved surface upto free surface of liquid.

Resultant force: $F = \sqrt{F_x^2 + F_y^2}$

And the resultant force (F) makes angle ϕ with the horizontal:

$$\phi = \tan^{-1}\left(\frac{F_y}{F_x}\right)$$

6. **The resultant force on a sluice gate:** F

$$F = F_1 - F_2$$

where F_1 = pressure force on the sluice gate due to upstream liquid.

F_2 = pressure force on the sluice gate due to downstream liquid.

7. **Lock gates:**

Lock gates are the devices used for changing the water level in a river or a canal for navigation or boating.

The reaction between the two gates: P

$$P = \frac{F}{2\sin\alpha}$$

Also reaction between the two gates is equal to reaction at the hinge *i.e.*,

$$P = R$$

Contd...

where F = resultant force on the gate
                               = $F_1 - F_2$
and θ = inclination of the gate with the normal to the
                               side of the lock.

ASSIGNMENT - 1

1. Define the following terms:
   (*i*) Total pressure and (*ii*) Centre of pressure.
2. Derive an expression for the force exerted by static fluid on a submerged vertical plane surface and the distance of centre of pressure from the free surface of liquid.   (GGSIP University Delhi, Dec. 2005)
3. Derive an expression for the force exerted on a submerged inclined plane surface by the static fluid and prove that the centre of pressure is always below the centre of gravity of the submerged inclined surface.
                                          (GGSIP University Delhi, Dec. 2002)
4. How would you determine the horizontal and vertical components of the resultant pressure on a sub-merged curved surfaces?
5. Explain how you would find the resultant pressure on a curved surface immersed in a liquid?
6. Find an expression for the force exerted and centre of pressure for completely sub-merged inclined plane surface. Can the same method be applied for finding the resultant force on a curved surface immersed in the liquid? If not, why?
7. For the vertical or inclined plane submerged in liquid, centre of pressure always lies below the centre of gravity, why?
8. Prove that the vertical component of the resultant force on a submerged curved surface is equal to the weight of the liquid supported by the curved surface.
9. Prove that the reaction between the gates of a lock is equal to the reaction at the hinge.
10. Derive an expression for reaction between the gates as $P = \dfrac{F}{2\sin\alpha}$

    where F = resultant force due to water on the lock gate,
            α = inclination of the gate with normal to the side of the lock.

ASSIGNMENT - 2

1. A vertical rectangular plane surface of width 2 m and depth 4 m is placed in water so that its upper edge is at a depth 1.5 m total pressure and position of the centre of pressure on the plane surface.   **Ans.** 274680 N, 3.88 m
2. A rectangular plane surface is 2 m wide and 3.5 m deep. It lies in vertical plane in water. Determine the total pressure and position of the centre of pressure

on the plane surface when its upper edge is horizontal and (*a*) Coincides with free surface. (*b*) 2 m below the free surface.

Take density of water is 999 kg/m^3.

Ans. (*a*) 120052.32 N, 2.33 m (*b*) 257254.98 N, 4.02 m

3. A circular plate of diameter 1.6 m which is placed in such a way that the centre of the plate is 3.2 m below the free surface of water. Determine the total pressure and position of the centre of pressure.

Ans. 63066.52 N, 3.25 m below the free surface

4. A rectangular plate is '*b*' m wide and '*d*' m deep. The depth of centroid of the plate is '*p*' m below the water surface. Prove that the depth of the centre

of pressure is equal to $\left(p + \dfrac{d^2}{12\,p} \right)$ metre from free surface of water.

5. Determine the total pressure and centre of pressure on an isosceles triangle plate of base 4 m and altitude 4 m when it is submerged vertically in an oil of specific gravity 0.9. The base of the plate coincides with the free surface of oil.   **Ans.** 94152.45 N, 1.998 m

[Note: $\bar{h} = \dfrac{h}{3} = \dfrac{4}{3} = 1.333$ m.

Take the value of *h* up to three decimal points for more accurate results]

6. An isosceles triangle of base 3 m and altitude 6 m, is immersed vertically in water with its axis of symmetry horizontal as shown in Fig. 3.57. If the head of water on it is 9 m, find

 (*i*) total pressure on the plate, and

 (*ii*) the position of the centre of pressure.   **Ans.** 794610 N, 9.041 m

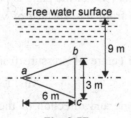

Fig. 3.57

7. A semi-circular plate 1.25 m in diameter in placed vertically in a liquid of specific gravity 1.5 with its diameter on the free liquid surface. Find the total pressure on the plate and the depth of centre of pressure.

Ans. 2395.019 N, 0.3695 m

8. A tank with vertical sides is 1.5 m × 1.5 m × 1.5 m deep. It contains water in the lower 0.6 m depth. The upper remaining part is filled with oil of specific gravity 0.9. Calculate (*i*) The total pressure on one side on the tank, and (*ii*)

The position of the centre of pressure from free surface of water.

Ans. (*i*) 15163.80 N (*ii*) 1.0052 m from free surface of water

9. A circular plate of 3 m diameter is submerged in water in such a way that its greatest and least depths below the free surface are 4 m and 3 m respectively. Find total pressure on the face of the plate and position of the centre of pressure.   **Ans.** 242569.84 N, 3.5177 m from free surface water

10. A rectangular plate 2 m wide and 4 m deep is immersed in water in such a way that its plane makes an angle of 25° with the water surface as shown in Fig. 3.58. Determine

(*i*) the total pressure on one side of the plate and

(*ii*) the position of the centre of pressure from free surface of water.

    Ans. (*i*) 207579.60 N (*ii*) 2.735 m below the free surface of water.

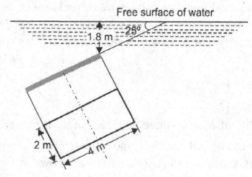

Fig. 3.58: Q-10

11. A triangular plate of base 2.5 m and 4 m which is immersed in such a way that the plan (top view) of the plate makes an angle of 30° with the free surface of the water. The base of the plate is parallel to free surface of water and at a depth of 3 m from free surface of water. Determine

(*i*) the total pressure

(*ii*) the position of the centre of pressure from free surface of water.

    Ans. (*i*) 179846.73 N (*ii*) 3.7272 m

12. Determine the magnitude and direction of the resultant force due to water acting on a roller gate of cylindrical form of 3 m diameter, when the gate is placed on the dam in such way that water is just going to spill. Take the width of the gate as 5 m.   **Ans.** 188243.11 N, 66.99

13. Determine the horizontal and vertical components of the water force exerted on the segmental gate of radius 8 m as shown in Fig. 3.59. Take width of the gate as unity.   **Ans.** F_x = 78480 N, F_y = 28352.86 N

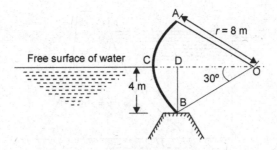

Fig. 3.59: Q-13

[Note: Take $\pi = 3.14$ and other values upto four decimal places.]

14. A tainter gate 3 m long is mounted on a spillway as shown in Fig. 3.60. Find the total pressure on the gate and angle at which it acts.

Ans. 184.85 kN, 39.3° with horizontal.

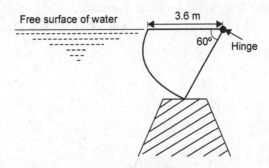

Fig. 3.60: Q-14

15. As shown in Fig. 3.61, the cross-section of a dam has a parabolic shape. Find the force exerted by the water per metre length of the dam and its inclination with the vertical. **Ans.** 8272.53 kN, 71.6°

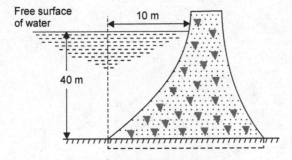

Fig. 3.61: Q-15

16. Find the magnitude and direction of the resultant water pressure force acting on a curved face of a dam which is shaped according to Fig. 3.62 The height of the water retained by the dam is 10 m. Consider the width of the dam as unity. **Ans.** 790.907 kN, 51.67°

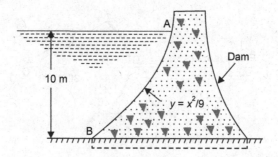

10 m

A

Dam

$y = x^2/9$

B

Fig. 3.62: Q-16

17. Each gate of a lock is 6 m high and is supported by two hinges placed on the top and bottom of the gate. When the gates are closed, they make an angle of 120°. The width of lock is 5 m. If the water levels are 4 m and 2 m on the upstream and downstream sides respectively, find the magnitude of the forces on the hinges due to water pressure.

Ans. R_T = 43.898 kN, R_B = 126.03 kN

❑❑❑

Buoyancy and Floatation

4.1 INTRODUCTION

If a body is placed over a liquid either the body sinks in the liquid or floats on the liquid, depending on following two forces:

(i) Gravitational force [acts downwards due to weight of the body].

(ii) Upward force exerted by the fluid on the body.

These two forces must be collinear, equal and opposite for equilibrium to be maintained. If upward force is equal to the weight of the body, the body will float on the liquid. If upward force is less than the weight of body, the body will sink in the liquid.

Following three cases are possible depending on the ratio of the weight (W) of the body and the buoyant force F_B.

1. If $\dfrac{W}{F_B} > 1$ $i.e.$, $W > F_B$; the body tends to move downwards and eventually sinks.

2. If $\dfrac{W}{F_B} = 1$ $i.e.$, $W = F_B$; the body floats and is only partially submerged.

3. If $\dfrac{W}{F_B} < 1$ $i.e.$, $W < F_B$; the body is lifted upward and rises to the surface.

4.2 BUOYANCY OR BUOYANT FORCE

If a body is fully or partly submerged in a fluid, the vertical upward force acting on the body is equal to the weight of the fluid displaced by the body (Fig. 4.1). The upward vertical force (F_B) is known as **force of buoyant or buoyancy or buoyant**

© The Author(s) 2023
S. Kumar, *Fluid Mechanics (Vol. 1)*,
https://doi.org/10.1007/978-3-030-99762-5_4

force.

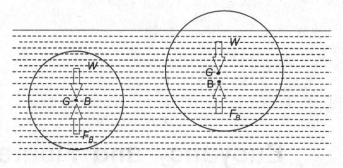

(a) Fully submerged body   (b) Partly submerged body.

Fig. 4.1: Bouyancy of fully and partly submerged bodies.

Buoyant force: F_B = weight as liquid displaced by the body

= mg

where m = mass of liquid displaced by the body

= $\rho\forall$

$\therefore$ $F_B = \rho g \forall$

where $\forall$ = volume of liquid displaced by the body (Fig. 4.2).

ρ = density of the liquid

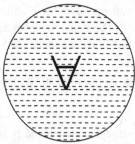
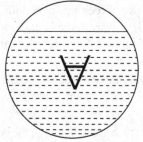

(a) Fully submerged body, volume of   (b) Partly submerged bodies,
liquid ($\forall$) is equals to the volume   volume of liquid ($\forall$) is equal to
of the whole body.   the volume of the body
                                                     submerged in liquid.

Fig. 4.2: Volume of liquid displaced by the body ($\forall$).

For the equilibrium of the body; we have

Buoyant force = weight as the body

F_B = Mg

where M = mass of the body

= $\rho_b V$

$F_B = \rho_b g V$

where ρ_b = density of the body

$$V = \text{volume of the body}$$

Buoyant force: $F_B = \text{weight of liquid displaced by the body}$

$$= \text{weight as the body}$$

$$F_B = \rho g \forall = \rho_b\, gV$$

where $\rho = \text{density of the liquid}$

$\rho_b = \text{density of body}$

$\forall = \text{volume of liquid displaced by the body}$

$V = \text{volume of the body}$

4.3 CENTRE OF BUOYANCY

The point at which the force of buoyancy (or buoyant force) acts is known as **centre of buoyancy.** It is denoted by letter B. It is always the centre of gravity of the volume of fluid displaced by the body.

4.4 PRINCIPLE OF FLOATATION (ARCHIMEDES' PRINCIPLE)

The principle of floatation states, if a body is fully or partly submerged in a fluid, the vertical upward force acts, on the body is equal to the weight of the fluid displaced by the body. The upward vertical force (F_B) is known as **buoyant force** or **buoyancy.** This principal is also known as **Archimedes' principle.**

Problem 4.1:   Find the volume of the water displaced and position of centre of buoyancy for a wooden block of width 3 m and depth 2 m, when it floats horizontally in water. The density of wooden block is 700 kg/m³ and its length 7 m.

Solution: Given data:

$$\text{Width} = 3 \text{ m}$$
$$\text{Depth} = 2 \text{ m}$$
$$\text{Length} = 7 \text{ m}$$

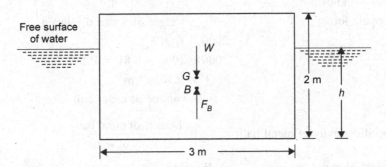

Fig. 4.3: Schematic for Problem 4.1

Volume of the block: $V = \text{width} \times \text{depth} \times \text{length}$

$$= 3 \times 2 \times 7 = 42 \text{ m}^3$$

Density of wood: $\rho = 700 \text{ kg/m}^3$

Weight of block: $W = (\rho g V)_{\text{block}} = 700 \times 9.81 \times 42 = 2.88414 \text{ N}$

For equilibrium condition:

The weight of water displaced = weight of wooden block

$$(\rho g \forall)_{water} = W$$

$$\rho_{water} \times 9.81 \times \forall = 288414$$

or Volume of water displaced: $\forall = \dfrac{288414}{1000 \times 9.81} = \textbf{29.4 m}^3$

We know, volume of water displaced: $\forall = 3 \times h \times 7$

where h = depth of wooden block submerged in water

$$2.94 = 3 \times h \times 7$$

$$h = 1.4 \text{ m}$$

$\therefore$ Centre of buoyancy: $B = \dfrac{h}{2} = \dfrac{1.4}{2} = \textbf{0.7 m from free surface or base.}$

Problem 4.2: A metal ball weighs 8000 N in air and 6000 N in water. What is its volume and specific gravity?

Solution: Given data:

Weight of metal ball in air   = 8000 N

Weight of metal ball in water   = 6000 N

We know,

Buoyant force: F_B = weight of water displaced

= weight loss of metal ball in water

= weight of metal ball in air – weight of metal ball in water

= 8000 – 6000 = 2000 N

Also we know,

Buoyant force: F_B = weight of water displaced

$$2000 = (\rho g V)_{water}$$

$$2000 = 1000 \times 9.81 \times V$$

or $V = \textbf{0.20387 m}^3$

= volume of metal ball

Specific gravity of metal ball: $S_m = \dfrac{\text{Density of metal ball}}{\text{Density of water}}$

Weight of metal in air: $W = Mg$

$$W = \rho_m V g$$

or $\rho_m = \dfrac{W}{Vg} = \dfrac{8000}{0.20387 \times 9.81}$

Density of metal ball: $\rho_m = 4000.07 \ kg/m^3$

$\therefore$ Specific gravity of metal ball:

$$S_m = \frac{\rho_m}{\rho_{water}} = \frac{4000.07}{1000} = 4$$

Problem 4.3: Find the density of metallic body which floats at the interface of mercury of specific gravity 13.6 and water such that 40% of its volume is submerged in mercury and 60% in water as shown in Fig. 4.4.

(GGSIP University; Delhi, Dec. 2001)

Solution:

Let V is the volume of a metallic body in m^3

The volume of metallic body submerged in mercury:

$$V_1 = 40\% \text{ of the volume of body} = 0.4 \ V$$

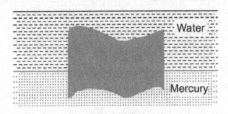

Fig. 4.4: Schematic for Problem 4.3

And the volume of metallic body submerged in water:

$$V_2 = 60\% \text{ of the volume of body}$$
$$= 0.6 \ V$$

Specific gravity of mercury: $S_1 = 13.6$

$\therefore$ Density of mercury: $\rho_1 = 13.6 \times$ density of water

$$= 13.6 \times 1000 = 13600 \ kg/m^3$$

Now,

Net buoyant force exerted on the body = buoyant force exerted by mercury + buoyant force exerted by water

$$F_B = F_{B1} + F_{B2}$$
$$= \rho_1 g V_1 + \rho_2 g V_2$$
$$= 13600 \times 9.81 \times 0.4V + 1000 \times 9.81 \times 0.6 \ V$$
$$= \textbf{59252.54 } V \ \textbf{N}$$

For equilibrium condition:

Net buoyant force exerted on the body = weight of the body

$$F_B = Mg$$
$$F_B = \rho V g$$

$\therefore$ $5925.4 \ V = \rho \times V \times 9.81$

or $\rho = \textbf{6040 } kg/m^3$

Density of the metallic body is **6040 kg/m³**

Problem 4.4: Find the density of a metallic body which floats at the interfaces of mercury of specific gravity 13.6, oil of specific gravity 0.85 and water such that 30% of its volume is submerged in mercury 60% in water and 10% in oil as shown in Fig 4.5.

Fig. 4.5: Schematic for Problem 4.4

Solution: Let V is the volume of a metallic body in m^3. The volume of metallic body submerged in mercury:

$$V_{Hg} = 30\% \text{ of the volume of body}$$

$$= 0.3 \ V$$

Volume of metallic body submerged in water:

$$V_{water} = 60\% \text{ of the volume of body}$$

$$= 0.6 \ V$$

and the volume of metallic submerged in oil:

$$V_{oil} = 10\% \text{ of the volume of body}$$

$$= 0.1 \ V$$

Specific gravity of mercury:

$$S_{Hg} = 13.6$$

$\therefore$ Density of mercury: $\rho_{Hg} = S_{Hg} \times$ density of water

$$= 13.6 \times 1000 = 13600 \text{ kg/m}^3$$

Specific gravity of oil: $S_{oil} = 0.85$

$\therefore$ Density of oil: $\rho_{oil} = 0.85 \times 1000 = 850 \text{ kg/m}^3$

Now,

Net buoyant force exerted on the body = buoyant force exerted mercury + buoyant force exerted by water + buoyant force exerted by oil

$\therefore$ $\quad\quad F_B = (F_B)_{Hg} + (F_B)_{water} + (F_B)_{oil}$

$\quad\quad\quad\quad = \rho_{Hg}\, g\, V_{Hg} + \rho_{water}\, g\, V_{water} + \rho_{oil}\, g\, V_{oil}$

$\quad\quad\quad\quad = 13600 \times 9.81 \times 0.3\, V + 1000 \times 9.81 \times 0.6 V + 850 \times 9.81 \times 0.1 V$

$\quad\quad\quad\quad = 40024.8\ V + 5886\ V + 833.85\ V$

$\quad\quad\quad\quad = 46744.65\ V$

For equilibrium condition:

Net buoyant force exerted on the body = weight of the body

$\quad\quad\quad F_B = Mg$ $\quad\quad\quad\quad\quad\quad\quad\quad\quad\quad\quad\quad \because M = \rho_m V$

$46744.65\ V = \rho_m Vg$

or $\quad\quad \rho_m = \dfrac{46744.65}{g} = \dfrac{46744.65}{9.81} = \mathbf{4765\ kg/m^3}$

Density of the metallic body is **4765 kg/m³**

Problem 4.5: A body weights 5 N in water and 6.5 N in oil of specific gravity 0.80. Find the volume and weight of the body. Find also the density and specific gravity of the material of the body.

Solution: Given data:

Weight of a body in water $\quad = 5$ N

Weight of a body in oil $\quad\quad = 6.5$ N

Specific gravity of oil: $S_{oil} = 0.8$

$\therefore$ Density of oil: $\quad\quad \rho_{oil} = 0.8 \times 1000 = 800$ kg/m³

We know,

Loss of weight of a body in liquid = weight of the liquid displaced

For water,

$\quad\quad\quad\quad W - 5 = (\rho g V)_{water}$

$\quad\quad\quad\quad W - 5 = \rho_{water} g V$

where $\quad\quad\quad\quad W =$ weight of the body

$\quad\quad\quad\quad\quad\quad V =$ volume of water displaced by body or volume of the body, because body submerged in water.

$\quad\quad\quad\quad W - 5 = 1000 \times 9.81 V$

$\quad\quad\quad\quad W - 5 = 9810\ V$ $\quad\quad\quad\quad\quad\quad\quad\quad\quad\quad ...(1)$

For oil, $\quad\quad\quad W - 6.5 = (\rho g V)_{oil}$

$\quad\quad\quad\quad W - 6.5 = 800 \times 9.81 \times V$

$\quad\quad\quad\quad W - 6.5 = 7848\ V$ $\quad\quad\quad\quad\quad\quad\quad\quad\quad ...(2)$

Eq. (1) – Eq. (2), we get

$\quad\quad\quad\quad 1.5 = 1962\ V$

or $\quad\quad\quad\quad V = \dfrac{1.5}{1962}$ m³

$\quad\quad\quad\quad V = \mathbf{7.645 \times 10^{-4}\ m^3}$

Substituting the value of V in Eq. (1), we get

$$W - 5 = 9810 \times 7.645 \times 10^{-4}$$

$$W - 5 = 7.49$$

or $\qquad W = 7.49 + 5 = \mathbf{12.49 \ N}$

also $\qquad W = Mg$

$\therefore \qquad 12.49 = M \times 9.81$

or Mass of the body: $M = 1.27$ kg

Density of the body: $\qquad \rho = \dfrac{\text{Mass:M}}{\text{Volume:V}} = \dfrac{1.27}{7.645 \times 10^{-4}} = \mathbf{1661.21 \ kg/m^3}$

Specific gravity of the body:

$$S = \dfrac{\text{Density of the body}:\rho}{\text{Density of the water}:\rho_{\text{water}}} = \dfrac{1661.21}{1000} = \mathbf{1.66}$$

Problem 4.6: A float of "the ball cock" type is required to close an opening of a supply pipe feeding. The valve should close down when 60% of volume of the spherical float is immersed in water. The valve has a diameter of 15 mm and the fulcrum of the operating lever is to be 150 mm from the valve and 0.5 m from the centre of float. Determine the minimum diameter of the float if it is required to close the valve against a pressure of 147 kN/m² gauge. Neglect the weight of the lever and the float.

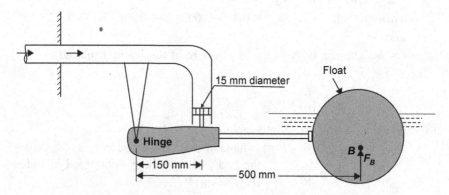

Fig. 4.6: Schematic for Problem 4.6

Solution: Given data:

60% of volume of the spherical float is immersed in water

Diameter of valve: $\qquad d = 15$ mm $= 0.015$ m

$\therefore$ Area of the valve: $\qquad a = \dfrac{\pi}{4} d^2 = \dfrac{3.14}{4} \times (0.015)^2 = 1.766 \times 10^{-4} \ m^2$

Pressure required to closed the valve:

$$p = 147 \text{ kN/m}^2 = 147 \times 10^3 \text{ N/m}^2$$

$\therefore$ Force exerted by fluid on the valve when closed:

$$F = p \times a = 147 \times 10^3 \times 1.766 \times 10^{-4} \text{ N} = 25.96 \text{ N}$$

Let $\qquad\qquad\qquad\qquad V$ = volume of float

Volume of float immersed in water:

$$\forall = 60\% \text{ of volume of float} = 0.6 \, V$$

$\therefore$ Buoyant force: $\qquad F_B$ = weight of liquid displaced by the body

$$= \rho g \forall = 1000 \times 9.8 \times 0.6 \, V = 5886 \, V \text{ N}.$$

Now taking moment about hinge point is zero

$$F_b \times 0.5 - F \times 0.150 = 0$$

$$5886 \times V \times 0.5 - 25.96 \times 0.156 = 0$$

$$5886 \times V \times 0.5 = 4.04976$$

$$V = 1.37606 \times 10^{-3} \text{ m}^3$$

Volume of float: $\qquad\qquad V = \dfrac{\pi}{6} D^3$

$\therefore \qquad\qquad 1.37606 \times 10^{-3} = \dfrac{3.14}{6} \times D^3$

$$D^3 = 2.6294 \times 10^{-3} \text{ m}^3$$

or Diameter of the float: $\quad D = 0.13802 \text{ m} = \textbf{138.02 mm}$

4.5 METACENTRE AND METACENTRIC HEIGHT

Metacentre is the point about which a body starts oscillating when the body floating in a liquid is tilted by a small angle. It is denoted by letter M. The Metacentre may also be defined as the point at which the line of action of the force of buoyancy (F_B) will meet the normal axis of the body, when a small angular displacement is given to the body.

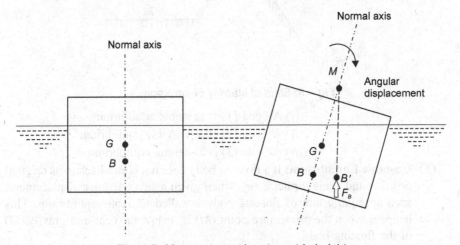

Fig. 4.7: Metacentre and metacentric height

Metacentric height: The distance between the metacentre (M) of a floating body and the centre of gravity (G) as the body is called metacentric height. It is denoted by GM (Fig. 4.7).

As a matter of fact, the metacentric height of a floating body is a direct measure of its stability or in other words, more the metacentric height of a floating body; more it will be stable. In the modern design trend, the metacentric height of a boat or ship is accurately calculated to check its stability. Some normal values of metacentric height for different types of ships are given below:

Merchant ships   : 0.3 m to 1 m

Sailing ships   : 0.45 m to 1.5 m

Battle ships   : 1 m to 2 m

River crafts   : 2 m to 3.5 m

4.6 EQUILIBRIUM OF FLOATING BODIES

A floating body has three states of equilibrium like any solid body:

(a) Stable Equilibrium

(b) Unstable Equilibrium

(c) Neutral Equilibrium

(a) **Stable Equilibrium:** When a floating body is given small angular displacement (*i.e.,* titled slightly), by some external force, and then it returns back to its original position due to internal forces [*i.e.,* weight of a body (W) and buoyant force (F_B)] such an equilibrium of floating body is called stable equilibrium. This happens when the metacentre point (M) is above the centre of gravity (G) of the floating body.

As in the case of solid body (1) is at stable equilibrium as shown in Fig. 4.8.

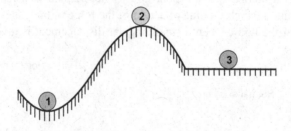

Fig 4.8: State of stability of solid body.

(i) A body (1) is at stable equilibrium.

(ii) A body (2) is at unstable equilibrium.

(iii) A body (3) is at neutral equilibrium

(b) **Unstable Equilibrium:** If a floating body does not return back to its original position and heels further away, when given a small angular displacement, such an equilibrium of floating body is called unstable equilibrium. This happens when the metacentre point (M) is below the centre of gravity (G) of the floating body.

As in case of solid body (2) is at unstable equilibrium as shown in Fig. 4.8.

(c) **Neutral Equilibrium:** If a floating body occupies a new position and remains at rest in new position, when given a small angular displacement, such an equilibrium of floating body is called neutral equilibrium. This happens when the metacentre point (M) coincides with the centre of gravity (G) of the floating body.

As in the case of solid body (3) is at neutral equilibrium as shown in Fig. 4.8.

4.7 EQUILIBRIUM OF SUBMERGED BODY

The positions of centre of gravity (G) and centre of buoyancy in case of a completely submerged body are fixed. Consider a balloon, which is completely submerged in air. Let the lower portion of the balloon contains heavier material, so that its centre of gravity (G) is lower than its centre of buoyancy (B) as shown is Fig. 4.9.(a) and at the top portion of the balloon very heavier material, so that its centre of gravity (G) is higher than its centre of buoyancy (B) as shown in Fig. 4.9 (b). As a floating body, the submerged body can have also three possible condition of equilibrium.

(a) Stable Equilibrium   (b) Unstable Equilibrium

(c) Neutral Equilibrium

(a) **Stable Equilibrium:** A body is said to be in stable equilibrium if it regains its original position when subjected to a small angular displacement. This happens when the centre of buoyancy (B) is above the centre of the gravity (G) of submerged body and $W = F_B$. If balloon is given angular displacement in clockwise direction then W and F_B restoring couple acting in the anti-clockwise direction and bring the balloon in the original position. As shown in Fig. 4.9 (a)

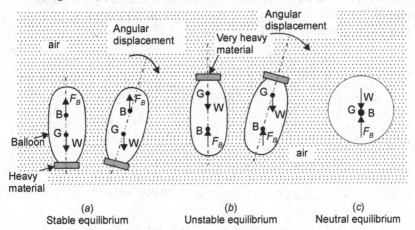

(a)   (b)   (c)

Stable equilibrium   Unstable equilibrium   Neutral equilibrium

Fig.4.9: Stabilities of submerged bodies.

(b) **Unstable Equilibrium:** A body is said to be unstable if it does not regain its original position when subjected to a small angular displacement i.e., it does not come back to its original position and heels further. This happens when the centre of buoyancy (B) is below the centre of gravity (G) of submerged body and $W = F_B$.

If balloon given angular displacement in clockwise direction, couple formed by W & FB is also clockwise. Thus the body does not return to its original position and hence the body is in unstable equilibrium as shown in Fig. 4.9. (b).

(c) **Neutral equilibrium:** A body is said to be in a state of neutral equilibrium, when subjected to a small angular displacement, neither it tends to regain its original position, nor tends to heel further but instead it keeps on its new displaced position. If $W = F_B$ and centre of buoyancy (B) and centre of gravity (G) are at the same point in neutral equilibrium as shown in Fig. 4.9 (c).

4.8 DETERMINATION OF METACENTRIC HEIGHT

The metacentric height may be determined by the following two methods:

(a) Analytical method, and   (b) Experimental method

4.8.1 Analytical Method of Determination of Metacentric Height (GM)

As shown in Fig. 4.10 (a), the position of a floating body in equilibrium. The location of centre of gravity and centre of buoyancy in this position is at G and B. When a floating body gives a small angular displacement in clockwise direction as shown in Fig. 4.10 (b), the portion EOE' as the body goes above the free surface and portion FOF', get submerged, because of this, the centre of buoyancy shifts form B to B'. The volume and weight of the two triangular prisms are same [i.e. triangular prisms EOE' and FOF']. These triangular prisms represent a gain in buoyant force on the right side and a corresponding loss of buoyant force on the left side. The gain is represented by an equal and opposite force dF_B acting vertically downward though the centroid (g') of EOE'. The couple due to buoyant force dF_B tends to rotate the ship in the counter clockwise direction. The moment caused by the displacement of the centre of buoyancy form B to B' is also in the counter clockwise direction. Thus these two couples must be equal.

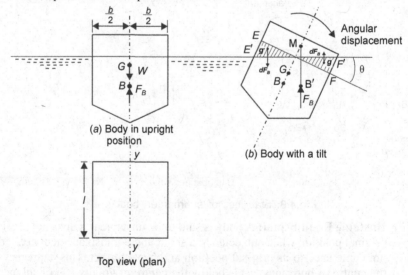

(a) Body in upright position

(b) Body with a tilt

Top view (plan)

Fig. 4.10: Metacentric height of a floating body.

Let b = width of the floating body

                        l = length of the floating body

                        θ = angle of heel or tilt

Buoyant force: = $\rho g \forall$

where ρ = density of liquid

                        $\forall$ = volume of the body submerged in liquid or volume of liquid displaced by the body.

Weight of triangular prism = $\rho g \times$ volume of the triangular prism

$$= \rho g \frac{lb^2\theta}{8}$$

Moment due to moving the triangular prism of water from g' to g = moment due to moving the buoyant force from B to B'

$$\rho g \frac{lb^2\theta}{8} \times gg' = \rho g \forall \; BB'$$

$$\frac{lb^2\theta}{8} \times gg' = \forall \times BM \sin\theta$$

$$\therefore \quad \frac{lb^2\theta}{8} \; \frac{2}{3} \; b = \forall \; BM \; \theta$$

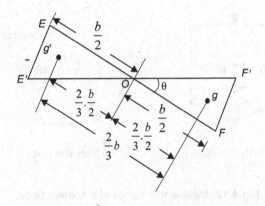

Fig. 4.11: Triangular prism of water

$$\left(\because gg' = \frac{2}{3}b \; \& \; \sin\theta \approx \theta \right)$$

$$\frac{lb^3\theta}{12} = \forall \; BM \; \theta$$

$$\frac{lb^3}{12} = \forall \; BM$$

$$I = \forall \; BM$$

Cross-sectional of triangular prism, OFF'

$$= \frac{1}{2}.\frac{b}{2}.\frac{b}{2}\tan\theta = \frac{b^2}{8}\theta \quad \because \theta \text{ is very angle}$$

$$\therefore \tan\theta \approx \theta$$

Volume of triangular prism = $\frac{b^2\theta}{8}l$

$$= \frac{lb^2\theta}{8}$$

or
$$BM = \frac{I}{\forall}$$

where
$$I = \frac{lb^3}{12}$$

= MOI of the horizontal section of the body at liquid level about its longitudinal axis (*i.e.*, y-axis).

$$BM = \frac{\text{MOI of the plan}}{\text{Volume of water displaced by the body}}$$

By definition of metacentric height (*GM*): It is distance between the centre of gravity (*G*) and metacentric point (*M*).

∴ $GM = BM - BG$ when *B* is above *G*.

and $GM = BM + BG$ when *B* is below *G*.

In general, $\boldsymbol{GM = BM \pm BG}$

4.8.2 Experimental Method of Determination of Metacentric Height (*GM*)

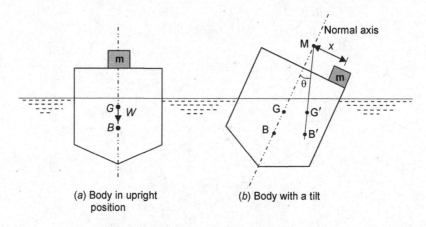

(a) Body in upright position

(b) Body with a tilt

Fig 4.12: Metacentric height of a floating body.

The metacentric height of a floating body can be determined, provided we know the centric of gravity of the floating body.

Let *m* is a known mass placed over the centre of the vessel as shown in Fig. 4.12. (*a*) and body in upright position.

Let   *W* = weight of body including weight *w* (= mg).

G = centre of gravity of the body

B = centre of buoyancy of the body.

The mass *m* is moved across the vessel toward right through a distance *x* as shown in Fig. 4.12 (*b*). The vessel will be tilted. The angle of heel (θ) is measured by means of a plumbline and a protractor attached on the vessel. The centre of gravity of the body will shift from *G* to *G'* as the mass *m* has been moved toward the right.

Also the centre of buoyancy will shift from B to B' as the body has tilted. Under equilibrium, the moment caused by the movement of the mass m though a distance x must be equal to the moment caused by the shift of the centre of gravity from G to G'. Thus the moment due to movement of mass (m) through a distance x
= the moment due to shift of centre of gravity from G to G'.

$$mg\,x = W \times GG'$$
$$mg\,x = W \times GM \sin\theta \qquad \text{∵ angle of heel } (\theta) \text{ is small}$$
$$mg\,x = W \times GN \times \theta \qquad \therefore \sin\theta \approx \theta$$

or
$$GM = \frac{mg\,x}{W\theta}$$

Problem 4.7: A solid cylinder of diameter 4 m has a height of 3 metre. Find the metacentric height of the cylinder when it is floating in water with its axis vertical. The specific gravity of the cylinder is 0.6. State whether the equilibrium is stable or unstable. *(GGSIP University, Delhi. Dec. 2004)*

Solution: Given data:

Diameter of cylinder: $d = 4$ m
Height of cylinder: $H = 3$ m
Specific gravity of the cylinder = 0.6

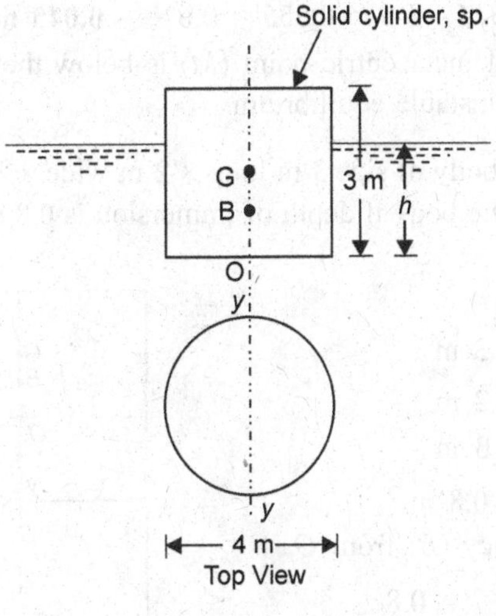

Fig. 4.13: Schematic for Problem 4.7

$\therefore$ Depth of cylinder in water: h = specific gravity × height of cylinder
$$= 0.6 \times H$$
$$= 0.6 \times 3 = 1.8 \text{ m}$$

Distance of centre of buoyancy (B) form O:

$$OB = \frac{h}{2} = \frac{1.8}{2} = 0.9 \text{ m}$$

Distance of centre of gravity (G) from O:

$$OG = \frac{H}{2} = \frac{3}{2} = 1.5 \text{ m}$$

$\therefore$
$$BG = OG - OB$$
$$= 1.5 - 0.9 = 0.6 \text{ m}$$

also
$$BM = \frac{I}{\forall}$$

where
I = MOI of the top view of the body about y–y

$$= \frac{\pi}{64} d^4 = \frac{3.14}{64} \times (4)^4 = 12.56 \text{ m}^4$$

and
$\forall$ = volume of cylinder submerged in water
= cross-sectional area × depth of cylinder in water

$$\forall = \frac{\pi}{4} d^2 \times h = \frac{3.14}{4} \times (4)^2 \times 1.8 = 22.608 \text{ m}^3$$

$\therefore$
$$BM = \frac{12.56}{22.608} = 0.555 \text{ m}$$

We know that the metacentric height:
$$GM = BM - BG = 0.555 - 0.6 = -\textbf{0.045 m}$$

The $-ve$ sign shows that metacentric point (M) is below the centre of gravity (G). Thus, the cylinder is in unstable equilibrium.

Problem 4.8:   A uniform body of size 3 m long × 2 m wide × 1 m deep floats in water. What is the weight of the body if depth of immersion is 0.8 m? Determine the meta-centric height also.

Solution: Given data:

Length of body : $l = 3$ m

Width of body : $b = 2$ m

Depth of body : $H = 3$ m

Depth of immersion : $h = 0.8$ m

Distance of centre of buoyancy (B) from O,

$$OB = \frac{h}{2} = \frac{0.8}{2}$$
$$= 0.4 \text{ m}$$

Distance of centre of gravity (G) from O,

$$OG = \frac{H}{2} = \frac{1}{2} = 0.5 \text{ m}$$

$\therefore$
$$BG = OG - OB$$
$$= 0.5 - 0.4 = 0.1 \text{ m}$$

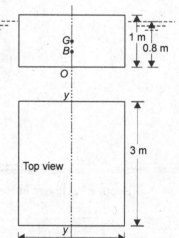

Fig. 4.14: Schematic for Problem 4.8

Let $\forall$ = volume of body submerged in water

= cross-section area × depth of body m water

$= bl \times h = blh = 2 \times 3 \times 0.8 = 4.8 \ m^3$

For equilibrium condition:

Weight of body = weight of liquid displaced by the body

$$W_{Body} = (\rho \ \forall \ g)_{water} = 1000 \times 4.8 \times 9.81 \ N = 47088 \ N$$

$$BM = \frac{I}{\forall}$$

where I is MOI of the top view of the body about $y - y$:

$$I = \frac{lb^3}{12} = \frac{3 \times (2)^3}{12} = 2 \ m^4$$

∴ $$BM = \frac{I}{\forall} = \frac{2}{4.8} = 0.416 \ m$$

We know that the metacentric height:

$$GM = BM - BG = 0.416 - 0.1 = 0.316 \ m$$

The +ve value of GM shows that metacentric point (M) is above the centre of gravity (G). Thus, the body is in stable equilibrium.

Problem 4.9: A cylindrical concrete mould 500 mm in diameter and 500 mm in height is attached to a wooden cylindrical pole of 500 mm diameter so that it floats vertically in water as a marker buoy. Determine the maximum and minimum length of the pole, such that it floats vertically above the surface of the water. Take specific gravity of concrete and wood as 3 and 0.5 respectively.

Solution: Let L = length of the wooden pole, m

Z = length of the wooden pole above the free surface of water, m

Diameters of cylindrical concrete wooden pole are same and equal to 500 mm

i.e., $d = 500 \ mm = 0.5 \ m$

Length of concrete: $\ell = 500 \ mm = 0.5 \ m$

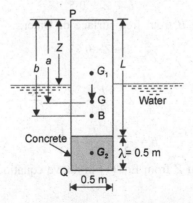

Fig. 4.15: Schematic for Problem 4.9

Specific gravity of concrete: $S_c = 3$

∴ Density: $\rho_c = 3 \times 1000 \text{ kg/m}^3 = 3000 \text{ kg/m}^3$

Specific gravity of wood:

$$S_w = 0.5$$

∴ Density: $\rho_w = 0.5 \times 1000 \text{ kg/m}^3 = 500 \text{ kg/m}^3$

For floating condition;

Bouyant force: $F_B = W$, weight of the body

$$\rho \forall g = W_{\text{wood}} + W_{\text{concretre}}$$

$$\rho \forall g = \rho_w V_w g + \rho_c V_c$$

$$\rho \forall = \rho_w V_w + \rho_c V_c$$

$$1000 \times \frac{\pi}{4} d^2 \times (L - Z + \ell) = 500 \times \frac{\pi}{4} d^2 \times L + 3000 \frac{\pi}{4} d^2 + \ell$$

$$1000 \times (L - Z + \ell) = 500 L + 3000 \ell$$

$$1000 (L - Z + 0.5) = 500 L + 3000 \times 0.5$$

$$L - Z + 0.5 = \frac{L}{2} + 3 \times 0.5$$

$$L - Z + 0.5 = \frac{L}{2} + 1.5$$

or

$$L - \frac{L}{2} + 0.5 - 1.5 = Z$$

or

$$Z = \frac{L}{2} - 1 \qquad (1)$$

As a marker buoy: $Z > 0$

∴ Minimum length: $L = 2 \ m$

Length of the body submerged in water $= (L - Z) + \ell$

$$= L - Z + 0.5$$

Centre of buoyancy B from free surface of water:

$$b - Z = \frac{L - Z + 0.5}{2}$$

or

$$b = Z + \frac{L}{2} - \frac{Z}{2} + \frac{0.5}{2}$$

$$b = \frac{Z}{2} + \frac{L}{2} + \frac{1}{4}$$

Substituting value of Z from Eq. (1) in above equation, we get

$$b = \frac{1}{2}\left[\frac{L}{2}-1\right]+\frac{L}{2}+\frac{1}{4}$$

$$b = \frac{L}{4}-\frac{1}{2}+\frac{L}{2}+\frac{1}{4}$$

$$b = \frac{3L}{4}-\frac{1}{4}$$

$$b = \frac{(3L-1)}{4}$$

The distance 'a' is found by taking moments of weights with respect to the end P of the pole:

$$Wa = W_{wood} \times \frac{L}{2} + W_{concrete} \times \left(L+\frac{\ell}{2}\right)$$

$$(W_{wood} + W_{concrete})a = W_{wood}\frac{L}{2} + W_{concrete}\left(L+\frac{0.5}{2}\right)$$

$$(\rho_w V_w g + \rho_c V_c g)a = \rho_w V_w g\frac{L}{2} + \rho_c V_c g\,(L+0.25)$$

$$(\rho_w V_w g + \rho_c V_c)a = \rho_w V_w \frac{L}{2} + \rho_c V_c\,(L+0.25)$$

$$\left(500\times\frac{\pi}{4}d^2 L+3000\times\frac{\pi}{4}d^2\ell\right)a = 500\times\frac{\pi}{4}d^2.L.\frac{L}{2}+3000\times\frac{\pi}{4}d^2\ell(L+0.25)$$

$$(500\,L+3000\times 0.5\,)\,a = 500\frac{L^2}{2}+3000\times 0.5\,(L+0.25)$$

$$(500\,L+1500)\,a = 500\frac{L^2}{2}+1500\,(L+0.25)$$

$$(L+3)\,a = \frac{L^2}{2}+3\,(L+0.25)$$

$$(L+3)\,a = \frac{L^2}{2}+3L+0.75$$

$$(L+3)\,a = \frac{L^2+6L+1.5}{2}$$

or

$$a = \frac{L^2+6L+1.5}{2L+6}$$

∴

$$GB = b - a$$

$$= \frac{3L-1}{4} - \frac{(L^2+6L+1.5)}{2L+6}$$

$$GB = \frac{(3L-1)(2L+6)-4(L^2+6L+1.5)}{4(2L+6)}$$

$$= \frac{6L^2 + 18L - 2L - 6 - 4L^2 - 24L - 6}{4(2L+6)}$$

$$= \frac{2L^2 - 8L - 12}{4(2L+6)}$$

$$= \frac{2(L^2 - 4L - 6)}{4(2L+6)}$$

$$GB = \frac{L^2 - 4L - 6}{4L+12}$$

We know that the metacentric height:

$$GM = \frac{I}{\forall} - GB$$

where $I = \frac{\pi}{64} d^4$, MOI about centroid of the top view

$\forall = \frac{\pi}{4} d^2 (L + \ell - Z)$, volume of the body submerged in water

$\forall = \frac{\pi}{4} d^2 (L + 0.5 - \frac{L}{2} + 1)$

$\forall = \frac{\pi}{4} d^2 (\frac{L}{2} + 1.5)$

$$GM = \frac{\frac{\pi}{64} d^4}{\frac{\pi}{4} d^2 (\frac{L}{2} + 1.5)} - \frac{(L^2 - 4L - 6)}{4L+12}$$

$$= \frac{d^2}{16(\frac{L}{2} + 1.5)} - \frac{(L^2 - 4L - 6)}{4L+12}$$

$$= \frac{(0.5)^2}{2(4L+12)} - \frac{(L^2 - 4L - 6)}{4L+12}$$

$$= \frac{1}{(4L+12)} \left[\frac{0.25}{2} - L^2 + 4L + 6 \right]$$

$$= \frac{1}{4L+12} [0.125 - L^2 + 4L + 6]$$

$$GM = \frac{-L^2 + 4L + 6.125}{4L+12}$$

For stable condition: $GM \geq 0$

$- L^2 + 4L + 6.125 \geq 0$

$$L = \frac{-4 \pm \sqrt{(4^2) - 4 \times (-1) \times 6.125}}{2 \times (-1)}$$

$$= \frac{-4 \pm \sqrt{16 + 24.5}}{-2}$$

$$= \frac{-4 \pm 6.364}{-2} = \frac{4 \pm 6.364}{2}$$

$$L = \frac{4 + 6.364}{2} = \textbf{5.182 m}$$

$L = \textbf{5.182 m}$ maximum height

Problem 4.10: An iceberg floats in an ocean so that one-eight of its volume is above the surface. Find the specific gravity of the iceberg with respect to ocean water.

Solution:

Let V = volume of the iceberg

Volume of the iceberg above the ocean water:

$$V_1 = \frac{1}{8} V$$

and volume of the iceberg displaced by the ocean water (or volume of iceberg immersed in water): $\forall$

$$\forall = V - V_1$$

$$= V - \frac{V}{8} = \frac{7}{8} V$$

Let ρ_{ice} = density of iceberg

$\rho_{o.w.}$ = density of ocean water

We know,

Weight of iceberg = weight of ocean water displaced by iceberg

$$Mg = (\rho\, g \forall)_{o.w.} \qquad | \text{ o.w.} = \text{ocean water}$$

$$\rho_{ice} V g = \rho_{o.w.}\, g \times \forall$$

or

$$\frac{\rho_{ice}}{\rho_{o.w.}} = \frac{\forall}{V} = \frac{7}{8} \frac{V}{V} = \frac{7}{8}$$

$$\frac{\rho_{ice}}{\rho_{o.w.}} = 0.875$$

Specific gravity of the iceberg with respect to ocean water: S

$$S = \frac{\rho_{ice}}{\rho_{o.w.}} = \textbf{0.875}$$

Problem 4.11: A boat sails from sea water of specific weight 10.04 kN/m^3 to fresh water of specific weight 9.81 kN/m^3 and sinks by 35 mm. A man weighing 686 N (69.92 kgf) gets out and the boat rises to its original water level. Find the weight of the boat.

Solution: Given data:

Specific weight of sea water:
$$w_{sw} = 10.04 \text{ kN/m}^3$$

Specific weight of fresh water:
$$w_{FW} = 9.81 \text{ kN/ m}^3$$

Weight of a man: $W_{man} = 686 \text{ N} = 0.686 \text{ kN}$

Let W = weight of the boat and the man in kN.

                                    = weight of the sea water displaced by the boat

                                    $\forall$ = volume of sea water displaced by the boat

By definition of specific weight:

$$w_{sw} = \frac{W}{\forall}$$

$$10.04 = \frac{W}{\forall}$$

or $W = 10.04 \, \forall$ ... (1)

When the ship sails in fresh water:

Since the man has gone out of the boat, the weight of fresh water displaced weight of boat:

$$W_{boat} = (\rho \forall g)_{FW}$$

$$W_{boat} = w_{FW} \times \forall \quad \because \text{ Sp. wt. of fresh water: } w_{FW} = \rho g$$

$$W - 0.686 = 9.81 \times \forall$$

$$W - 0.686 = 9.81 \forall \qquad \qquad ...(2)$$

From Eqs. (1) and (2), we get

$$10.04 \, \forall - 0.686 = 9.81 \forall$$

$$10.04 \, \forall - 9.81 \forall = 0.686$$

$$0.23 \forall = 0.686$$

or $\forall = 2.98 \text{ m}^3$

Substituting the value of $\forall$ in Eq. (1), we get

$$W = 10.04 \times 2.98 = 29.919 \text{ kN}$$

also W = weight of boat + weight of man

$$W = W_{boat} + W_{man}$$

$$29.919 = W_{boat} + 0.686$$

or $W_{boat} = 29.919 - 0.686$

$$= \mathbf{29.233 \text{ kN}}$$

Problem 4.12: A solid cone of relative density 0.80 floats in water. What should be its minimum apex angle so that it may float its apex downwards in stable equilibrium.

Solution: Distance of centre of gravity G from the vertex O

$$OG = \frac{3}{4}H$$

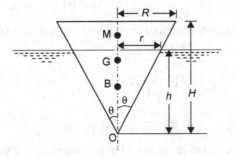

Fig. 4.16: Schematic for Problem 4.12

Distance of centre of buoyancy B from the vertex O,

$$OB = \frac{3}{4}H$$

Let M denotes the position of metacentre

Then $\qquad BM = \dfrac{I}{\forall}$

where $\qquad I = \dfrac{\pi}{4}r^4$, MOI

$\forall$ = volume of water displaced by solid cone

$$= \frac{1}{3}\pi r^2 h$$

$$BM = \frac{\frac{\pi}{4}r^4}{\frac{1}{3}\pi r^2 h} = \frac{3}{4}\frac{r^2}{h} \qquad \text{... (1)}$$

From triangle AOB, we get

$$\tan \theta = \frac{r}{h}$$

or $\qquad r = h \tan \theta$

Substituting $r = h \tan \theta$ in Eq. (1), we get

$$BM = \frac{3}{4} \times \frac{(h \tan \theta)^2}{h}$$

$$= \frac{3}{4} h \tan^2 \theta$$

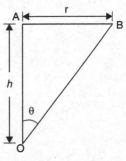

Fig. 4.17: Triangle AOB

$$OM = OB + BM$$

$$= \frac{3}{4} h + \frac{3}{4} h \tan^2 \theta$$

$$= \frac{3}{4} h (1 + \tan^2 \theta) = \frac{3}{4} h \sec^2 \theta$$

For stable equilibrium, M should be at a level higher than G

i.e., $$OM > OG$$

$$\frac{3}{4} h \sec^2 \theta > \frac{3}{4} H$$

or $$\sec^2 \theta > \frac{H}{h}$$ ... (2)

Specific gravity or relative density of a solid cone: $S = 0.80$ (given)

The depth of solid cone submerged in water:

$$h = S^{1/3} H$$

Substituting the value of h in Eq. (2), we get

$$\sec^2 \theta > \frac{1}{S^{1/3}}$$

or $$\cos^2 \theta < S^{1/3}$$

For equilibrium to be just stable

$$\cos^2 \theta = S^{1/3} \quad \text{or} \quad \cos \theta = S^{1/6}$$

Substituting the value of S = 0.8

$$\cos \theta = (0.8)^{1/6} = 0.96349$$

or $$\theta = \cos^{-1} (0.9634) = 15.53°$$

$\therefore$ Apex angle: $$2\theta = 2 \times 15.53° = \textbf{31.06°}$$

Problem 4.13: A wooden pole 150 mm × 150 mm × 6 m hangs vertically from a vertically string so that 4 m length of the pole is submerged in water. Find the tension in the string. Take the specific gravity of the wooden pole is 0.76.

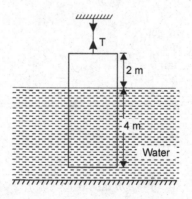

Fig. 4.18: Schematic for Problem 4.13

Solution: Given data:

Width of square cross-sectional area of a wooden pole: $b = 150$ mm $= 0.15$ m

∴   Cross-sectional area of a wooden pole: A

$$= b \times b = 0.15 \times 0.15 = 0.0225 \text{ m}^2$$

Length of pole: $l = 6$ m

Length of pole submerged in water $= 4$ m

Specific gravity of pole: $S = 0.76$

Density of pole: $\rho = 0.76 \times 1000 = 760$ kg/m^3

Let $T =$ tension in the string

Buoyant force + tension in the string = weight of the pole

$$(\rho \, g \, \forall)_{\text{water}} + T = (\rho \, gV)_{\text{pole}}$$

$$1000 \times 9.81 \times A \times 4 + T = 760 \times 9.81 \times 0.0225 \times 6$$

$$1000 \times 9.81 \times 0.0225 \times 4 + T = 760 \times 9.81 \times 0.0225 \times 6$$

$$882.9 + T = 1006.50$$

or $T = \textbf{123.6 N}$

Problem 4.14 : Two spheres weighing 12 kN and 50 kN are 2 m in diameter. They are connected by a short rope and placed in water. Find the tension in the connecting rope. Find also what portion of the lighter sphere will protrude above the water surface.

Solution: Given data:

Weight of lighter sphere: $W_1 = 12$ kN $= 12000$ N

Weight of heavier sphere: $W_2 = 50$ kN $= 50000$ N

Diameter of each sphere: $d = 2$ m

∴   Volume of each sphere: $V = \dfrac{\pi d^3}{6} = \dfrac{3.14(2)^3}{6}$

$$= 4.186 \text{ m}^3$$

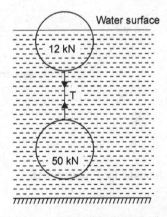
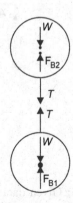

Fig. 3.19: Schematic for Problem 4.14

Consider the lower sphere:

For the equilibrium of the lower sphere:

Upward forces = downward forces

$$T + F_{B1} = W_1$$

$$T + (\rho \, g\forall_1)_{water} = 50000 \qquad \because \forall_1 = V$$

$$T + 41064.66 = 50000$$

or $$T = 8935.34 \; N$$

Now consider the upper sphere:

For the equilibrium of the upper sphere:

Upward forces = downward forces

$$F_{B2} = W_2 + T$$

$$(\rho \, g\forall_2)_{water} = 1200 + 8935.34$$

$$9.81 \times 1000 \times \forall_2 = 20935.34$$

or $$\forall_2 = 2.134 \; m^3$$

∴ Volume of the portion of the upper sphere protruding above the water surface:

$$= V - \forall_2 = 4.186 - 2.134 = \mathbf{2.052 \; m^3}$$

Percentage of the portion protruding above the water surface $= \dfrac{2.052}{4.186} \times 100$

$$= \mathbf{49.02\%}$$

Problem 4.15: The wooden beam shown in Fig. 4.20 is 150 mm × 150 mm and 6 m long. It is hinged at A and remains in equilibrium at α with the horizontal. Find the inclination α. Specific gravity of wood is 0.72.

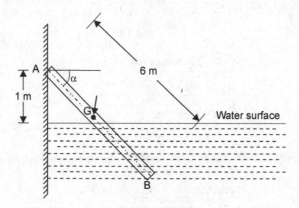

Fig. 4.20: Schematic for Problem 4.15

Solution: Given data:

The side of square cross-section of the beam:
$$b = 150 \text{ mm} = 0.15 \text{ m}$$

Cross-sectional area of the beam: $A = b \times b = 0.15 \times 0.15 = 0.0225 \text{ m}^2$

Length of beam: $l = 6 \text{ m}$

Specific gravity of the wooden beam: $S = 0.72$

$\therefore$ Density of the wooden beam: $\rho = S \times 1000 = 0.72 \times 1000 = 720 \text{ kg/m}^3$

Weight of the beam: $W = (\rho g \forall)_{\text{wood}} = 720 \times 9.81 \times A \times l$
$$= 720 \times 9.81 \times 0.0225 \times 6 = 953.53 \text{ N}$$

which acts at G, the centre of gravity of the beam.

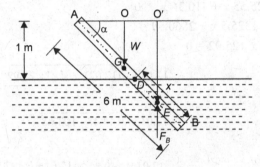

Fig. 4.21: Schematic 2 for Problem 4.15

Let x is length of the beam to be immersed under water.

The buoyant force (F_B) acts at E, the centre of buoyancy.
$$F_B = (\rho g \forall)_{\text{water}} = \rho g \times Ax$$
$$= 1000 \times 9.81 \times 0.225 \times x$$
$$= 220.725 x$$

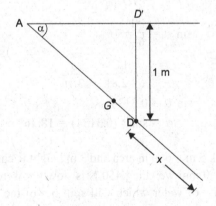

Fig. 4.22: Schematic 3 for Problem 4.15

and $AG = \dfrac{6}{2} = 3 \text{ m}$

$$AE = AB - BE$$

$$AE = 6 - \frac{x}{2}$$

Taking moments about the hinge A; $M_A = 0$

$$W \times AO - F_B \times AO' = 0$$

$$953.53 \times 3 \cos \alpha - 220.725 \, x \times \left(6 - \frac{x}{2}\right) \cos \alpha = 0$$

or $953.35 \times 3 - 220.925 \, x \times 6 + 220.75 \, x \times \dfrac{x}{2} = 0$

$2860.59 - 1325.55 \, x + 110.36 \, x^2 = 0$

or$110.36 \, x^2 - 1325.5 \, x + 2860.59 = 0$

or $x^2 - 12.01 \, x + 25.92 = 0$

$$x = \frac{12.01 \pm \sqrt{(12.02)^2 - 4 \times 1 \times 25.92}}{2 \times 1}$$

$$x = \frac{12.01 \pm \sqrt{144.48 - 103.68}}{2}$$

$$= \frac{12.01 \pm \sqrt{40.8}}{2} = \frac{12.01 \pm 6.387}{2}$$

Possible value of $\qquad x = \dfrac{12.01 - 6.387}{2} = 2.81 \text{ m}$

In $\angle AD'D$

$$\sin \alpha = \frac{D'D}{AD}$$

$$\sin \alpha = \frac{1}{6 - x}$$

$$= \frac{1}{6 - 2.81} = \frac{1}{3.16}$$

$$\sin \alpha = 0.3134$$

or $\qquad \alpha = \sin^{-1}(0.3134) = \textbf{18.26}°$

Problem 4.16: A tank 5 m × 6 m in area and 3 m height is completely full of water. If a solid cube of side 700 mm weighing 450 N is slowly lowered into the water until it floats, find the quantity of water which will spill out of the tank. Assume that no waves are formed in the above action.

Solution: Given data:

The side of cube: $\qquad b = 700 \text{ mm} = 0.7 \text{ m}$

∴   Cross-sectional area of cube: $A = b \times b = b^2 = (0.7)^2 = 0.49 \text{ m}^2$

Weight of cube: $\qquad W = 450 \text{ N}$

Let h = depth of cube sinks into the water

Now buoyant force on the cube = weight of the cube

$$F_B = W$$

$$(\rho g \forall)_{water} = 450$$

$$1000 \times 9.81 \times 0.49 \times h = 450$$

or $$h = 0.09361 \text{ m}$$

The quantity of water spill out = volume of water displaced by the cube

$$= \forall = Ah = 0.49 \times 0.09361 = 0.0486 \text{ m}^3 = \textbf{45.86 litres}$$

Problem 4.17: The wooden cone of specific gravity 0.70 is immersed in water as shown in Fig 4.20, by attaching of a 750 N weight at its lower end. Find the height of the cone.

Solution: Given data:

Specific gravity of cone: $S = 0.70$

∴ Density of cone: $\rho = 0.70 \times 1000 = 700 \text{ kg/m}^3$

Base diameter of cone: $d = 1 \text{ m}$

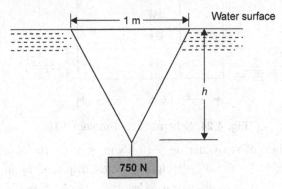

Fig. 4.23: Schematic for Problem 4.17

Volume of cone:
$$V = \frac{1}{3} \frac{\pi d^2}{4} \times h$$

$$= \frac{1}{12} \times 3.14 \, (1)^2 \times h = 0.2616 \, h$$

For the equilibrium of the cone:

Net downward forces = net upward forces

Weight of cone + 750 = buoyant force

$$(\rho g V)_{wood} + 750 = (\rho g \forall)_{water}$$

$$700 \times 9.81 \times 0.2616 \, h + 750 = 1000 \times 9.81 \times 0.2616 \, h$$

$$1796.40 \, h + 750 = 2566.29 \, h$$

or $$750 = 769.89 \, h$$

or $$h = \textbf{0.97416 m}$$

Problem 4.18: A cylindrical buoy, 1.6 m in diameter × 1.3 m in length and weighing 5 kN floats in seawater with its axis vertical. A 500 N load is placed centrally at the top of the buoy. If the buoy is to remain in stable equilibrium, find the maximum permissible height of the centre of gravity of the load above the top of the buoy. Specific weight of seawater is 10 kN/m³.   *(GGSIP University, Delhi Dec. 2001)*

Solution: Given data:

Diameter of the buoy: $d = 1.6$ m
Length of the buoy: $l = 1.3$ m
Weight of the buoy: $W_b = 5$ kN = 5000 N
Load placed centrally at the top of the buoy: $W = 500$ N
∴ Total weight of the arrangement: $W_T = W_b + W$
                                             $= 5000 + 500 = 5500$ N

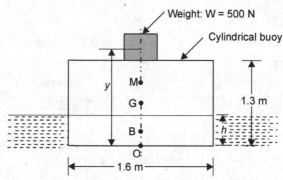

Fig. 4.24: Schematic for Problem 4.18

Specific weight of sea water: $w = 10$ kN/m³ $= 10 \times 10^3$ N/m³
Total weight: W_T = weight of water displaced by the arrangement
                       $W_T = \rho\, g\forall = w.\ \forall$ ∵ Specific weight: $w = \rho g$
or Volume of water displaced:

$$\forall = \frac{W_T}{w} = \frac{5500}{10 \times 10^3} = 0.55 \text{ m}^3$$

also $\forall$ = cross-sectional area of the buoy × depth of submerged buoy

$$0.55 = \frac{\pi}{4} d^2 \times h$$

$$0.55 = \frac{3.14}{4} \times (1.6)^2 \times h$$

or $h = 0.2736$ m

Height of centre of buoyancy (B) above base point O:

$$OB = \frac{h}{2} = 0.1368 \text{ m}$$

We know that the metacentric height:

$$BM = \frac{I}{\forall}$$

where

$$I = \frac{\pi}{4} d^4$$

$$= \frac{3.14}{64} \times (1.6)^4 = 0.3215 \text{ m}^4$$

$$\therefore \qquad BM = \frac{0.3215}{0.55} = 0.5845 \text{ m}$$

$$\therefore \qquad OM = OB + BM = 0.1368 + 0.5845 = 0.7213 \text{ m}$$

Let $\qquad y$ = height of CG of the load above the base.

To find the position of combined CG above the base point, taking moments about O, we get

$$W_b \times \frac{l}{2} + W \times y = W_T \times OG$$

$$5000 \times \frac{1.3}{2} + 500 \times y = 5500 \times OG$$

$$3250 + 500 \ y = 5500 \ OG$$

or $\qquad OG = \dfrac{3250 + 500 y}{5500}$

$$= 0.5909 + 0.0909 \ y$$

The equilibrium will be stable when $OM > OG$

i.e. $\qquad 0.7213 > (0.5909 + 0.0909 \ y)$

or $\qquad 0.1304 > 0.9090 \ y$

or $\qquad 0.0909 \ y < 0.1304$

or $\qquad y < 1.434 \text{ m}$

But the height of buoy = 1.3 m

Therefore, the height of CG of the load above the buoy should not be greater than $(1.435 - 1.3) = \textbf{0.135 m}$

Problem 4.19: A ship has a displacement of 6×10^7 N sea water. The moment of inertia of its plan area at water line is 10000 m^4. When a load of 250 kN is shifted in a transverse direction on the deck by 5.5 m the ship tilts over by an angle of 4°, calculate the metacentric height of the ship and the height of its CG above the centre of buoyancy. Specific gravity of sea water is 1.1

Solution: Given data:

Weight of ship: $\qquad W$ = weight of water displaced

$$= 6 \times 10^7 \text{ N}$$

Movable weight: $\qquad w = mg = 250 \text{ kN} = 250 \times 10^3 \text{ N}$

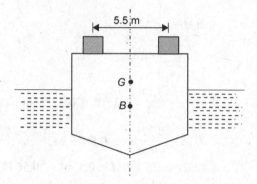

Fig. 4.25: Schematic for Problem 4.19

Distance moved by weight w: $x = 5.5$ m

Angle of heel: $\theta = 4°$

$$= 4 \times \frac{\pi}{180} \text{ rad} = \frac{4 \times 3.14}{180} \text{ rad} = 0.0697 \text{ rad}$$

MOI of the plan: $I = 10000 \text{ m}^4$

Density of sea water: $\rho = 1.1 \times 1000 = 1100 \text{ kg/m}^3$

∴ Metacentric height: $GM = \dfrac{mgx}{W.\theta}$

$$= \frac{205 \times 10^3 \times 5.5}{6 \times 10^7 \times 0.0647} = \mathbf{0.3287 \text{ m}}$$

Weight of water displaced by ship = $\rho g \forall$

∴ Volume of water displaced: $\forall = \dfrac{W}{\rho g} = \dfrac{6 \times 10^7}{1100 \times 9.81} = 5560.18 \text{ m}^3$

$$BM = \frac{I}{\forall} = \frac{10000}{5560.18} = 1.798 \text{ m}$$

The height of centre of gravity (G) above the centre of buoyancy (B):

$$BG = BM - GM = 1.798 - 0.328 = \mathbf{1.469 \text{ m}}$$

SUMMARY

1. If a body is fully or partly submerged in a fluid, the vertically upward force acts on the body is equal to the weight of the fluid displaced by the body. The upward force (F_B) is known as **force of buoyant** or **buoyancy** or **buoyant force**.

    Mathematically,

Contd...

Buoyant force: F_B = weight of liquid displaced by the body

$\qquad\qquad\qquad = \rho g \forall$

where ρ = density of the liquid

$\qquad\qquad\qquad \forall$ = volume of liquid displaced by the body

2. The point at which the force of buoyancy acts is known as **centre of buoyancy**. It is denoted by letter 'B'.

3. **Meta-centre** is the point about which a body starts oscillating when the body floating in a liquid is tilted by a small angle. It is denoted by letter 'M'.

4. **Meta-centric height:** The distance between the meta-centre (M) of a floating body and the centre of gravity (G) of the body is called meta-centric height. It is denoted by GM.

5. Condition of equilibrium of a floating and submerged body are:

Equilibrium	Floating body	Submerged body
(a) Stable equilibrium	M is above G	B is above G
(b) Unstable equilibrium	M is below G	B is below G
(c) Neutral equilibrium	M & G coincide	B & G coincide

where M = meta-centre,

$\qquad\qquad\qquad G$ = centre of gravity

$\qquad\qquad\qquad B$ = centre of buoyancy

6. Determination of Meta-centric height (GM)

   (*a*) **By Analytical Method**

$$GM = \frac{I}{\forall} - BG \qquad\qquad \text{when } B \text{ is above } G.$$

   and

$$GM = \frac{I}{\forall} + BG \qquad\qquad \text{when } B \text{ is below } G.$$

   where I = *MOI* of the horizontal section of the body at liquid level about its longitudinal axis (y-axis).

$\qquad\qquad\qquad \forall$ = volume of liquid displaced by body

$\qquad\qquad BG$ = distance between centre of gravity (G) and centre of buoyancy (B).

   (*b*) **By Experimental Method**

$$GM = \frac{mgx}{W\theta}$$

   where m = movable mass

$\qquad\qquad\qquad x$ = distance through which mass (m) is moved

$\qquad\qquad\qquad W$ = weight of the ship or floating body including movable weight (mg)

$\qquad\qquad\qquad \theta$ = angle of heel

ASSIGNMENT - 1

1. Define the following terms:

    (a) Buoyancy   (b) Centre of buoyancy

2. Explain the meta-center and meta-centric height.

                                        (GGSIP University, Delhi, Dec. 2005)

3. State and explain the principle of floatation.

4. Explain in brief the following types of equilibrium of floating bodies:

    (i) Stable equilibrium

    (ii) Unstable equilibrium and

    (iii) Neutral equilibrium   *(GGSIP University, Delhi, Dec. 2001)*

5. With neat sketches, explain the condition of equilibrium of a floating body and a submerged body.   *(GGSIP University, Delhi, Dec. 2002)*

6. Show that the distance between the meta-centre and centre of buoyancy is

    given by $BM = \dfrac{I}{\forall}$

    where I = MOI of the plan of the floating body at water surface about longitudinal axis.

                                $\forall$ = volume of the body submerged in liquid

7. Derive an expression for the metacentric height of a floating body experimentally. Explain with the help of neat sketch.

ASSIGNMENT - 2

1. Find the volume of the water displaced and position of centre of buoyancy for a wooden block of width 2 m and depth 2.5 m, when it floats horizontally in water. The density of wooden block is 720 kg/m^3 and its length 6 m.

                        Ans. 21.6 m^3, 0.9 from free surface or base.

2. A metal ball weighs 7000 N in air and 5000 N in water. What is its volume and specific gravity.   **Ans.** 0.20387 m^3, 3.5

3. Find the density of a metallic body which floats at the interface of mercury of sp. gravity 13.6 and water such that 30% of its volume is submerged in mercury and 70% in water as shown in Fig. 4.26.   **Ans.** 4780 kg/m^3

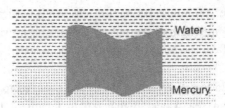

Fig. 4.26: Q-3

4. Find the density of a metallic body which floats at the interfaces of mercury of sp. gravity 13.6, oil of sp. gravity 0.9 and water such that 20% of its volume is submerged in mercury, 50% in water and 30% in oil as shown in Fig 4.27.

   Ans. 3490 kg/m^3

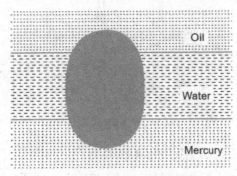

Fig. 4.27: Q-4

5. A solid cylinder of diameter 3 m has a height of 4 m. Find the metacentric height of the cylinder when it is floating in water with its axis vertically. The specific gravity of the cylinder is 0.7. State whether the equilibrium is stable or unstable.   **Ans.** – 0.1982 m, unstable equilibrium.

6. A rectangle pontoon 10 m long, 7 m broad and 2.5 m deep weighs 686.7 kN. It carries on its upper deck an empty boiler of 5 m diameter weighing 588.6 kN. The centre of gravity of the boiler and the position are at their respective centres along a vertically line. Find the metacentric heights, weight density of sea water is 10.104 kN/m^3.   **Ans.** 0.120 m

7. A body weight 4.50 N in water and 6 N in oil of specific gravity 0.80. Find the volume and weight of the body. Find also the density and specific gravity of the material of the body. **Ans.** 7.645 × 10^{-4} m^3, 12 N, 1600 kg/m^3,1.6

8. A iceberg floats in an ocean so that one-seventh of its volume is above the surface. Find the specific gravity of the iceberg with respect to ocean water.
   Ans. 0.8571

9. A boat sails from sea water weighing 10.06 kN/m^3 to fresh water weighing 9.81 kN/m^3 and sinks by 40 mm. A man weighs 580 N gets out and the boat rises to its original water level. Find the weight of the boat. **Ans.** 22.76 kN

10. A wooden pole 100 mm × 100 mm × 5 m hangs vertically from a vertically string so that 3 m length of the pole is submerged in water. Find the tension in the string. Take the specific gravity of the wooden pole is 0.65

Ans. 24.52 N

11. A wood pole 25 mm × 25 mm × 44 m and attached to a steel weight at the bottom floats in sea water so that 0.6 m length of the pole is exposed. Calculate the steel weight attached at bottom of the wood pole as shown in Fig. 4.28. Specific gravity of sea water, wood and steel may be taken as 1.025, 0.60 and 7.85 respectively.

Ans. 7.6515 N

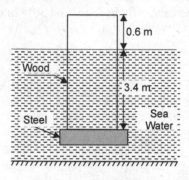

Fig. 4.28: Q-11

12. Two spheres weighing 10 kN and 25 kN are each 1.6 m in diameter. They are connected by a short rope and placed in water as shown in fig. 4.29. Find the tension in the connecting rope. Find also what portion of the lighter sphere will protrude above the free surface of water. **Ans.** $T = 3.958$ kN, 0.722 m^3

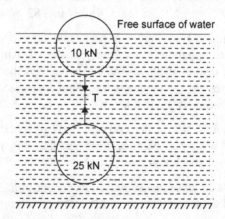

Fig. 4.29: Q-12

13. The wooden beam as shown in Fig. 4.30, is 200 mm × 200 mm and 5 m long. It is hinged at point A and remains in equilibrium at angle α with the horizontal. Find the inclination α. Take specific gravity of wood is 0.60.

Ans. $\alpha = 18.43°$

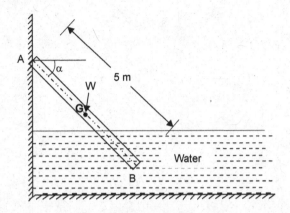

Fig. 4.30: Q-13

14. The wooden cone of the specific gravity 0.60 is immersed in water as shown in Fig. 4.31, by attaching a 650 N weight at its lower end. Find the height: h of the cone. **Ans.** $h = 0.633$ m

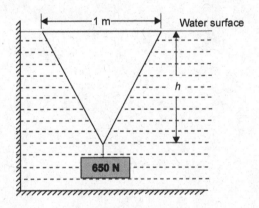

Fig. 4.31: Q-14

15. A tank 4 m × 6 m in area and 3 m height is completely full of water. If a solid cube of side 500 mm weighing 425 N is slowly lowered into the water until it floats, find the quantity of water which will spill out of the tank. Assume that no waves are formed in the above action neglect the effect of adhesion at the edge of the tank. **Ans.** 0.04325 m³

◻◻◻

5

Kinematics of Fluid Motion

5.1 INTRODUCTION

Kinematics of fluid motion deals with the flow of particles without considering forces causing the motion. This deals with the velocity and acceleration of fluid particles in motion. In this chapter, we will study

- the types of flow lines
- the types of fluid flow
- the methods of determining velocity and acceleration
- derivation of continuity equation for an incompressible or compressible flow
- description of a real flow in terms of a stream function.

5.2 METHODS OF DESCRIBING FLUID MOTION

The motion of fluid particles may be described by the following two methods:

(*a*) Langrangian method, and
(*b*) Eulerian method.

5.2.1 Langrangian Method

By this method we study the behaviour (*i.e.*, velocity, acceleration, density, *etc.*) of a single fluid particle at every instant of time as the particle moves to different positions.

By this method, study of the kinematics behaviour of a single fluid particle is a very complicated process and therefore this method is not generally adopted.

5.2.2 Eulerian Method

It deals with flow pattern of all the particles simultaneously in one section. In mechanics of fluids, this method is commonly used, because of its mathematical simplicity. Moreover, in mechanics of fluids the movement of an individual fluid particle is not of much importance.

S. Kumar, *Fluid Mechanics (Vol. 1)*,
https://doi.org/10.1007/978-3-030-99762-5_5

Due to its mathematical simplicity and easy application, this method is most commonly used in fluid mechanics.

5.3 TYPES OF FLOW LINES

The flow lines are classified as:

  (*a*) Stream line
  (*b*) Path line
  (*c*) Streak line.

  (*a*) **Stream Line:** A stream line is an imaginary line drawn in a flow field such that a tangent drawn at any point on it indicates the direction of the velocity vector at that point.

Considering a particle moving along a stream line for very short distance *ds* having components *dx*, *dy* and *dz* along the three mutually perpendicular coordinate axes. Let the components of the velocity vector V along *x*, *y* and *z*-directions be *u*, *v* and *w* respectively.

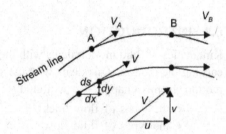

Fig. 5.1: Stream lines.

Let *t* is the time taken by a fluid particle to move a distance *ds* along the stream line with velocity V is

$$t = \frac{ds}{V}$$

which is the same as:

$$t = \frac{dx}{u} = \frac{dy}{v}$$

Hence the differential equation of the streamline may be written as:

$$\frac{dx}{u} = \frac{dy}{v}$$

Properties of stream lines:

  (*i*) There cannot be any movement of the fluid mass across the stream lines.

  (*ii*) A stream line neither intersects itself nor two stream lines can cross themselves.

  (*iii*) Stream line spacing varies inversely as the velocity; converging of stream line in any particular direction shows accelerated flow in that direction.

  (*b*) **Path Line:** A path line is a line or path traced by a single fluid particle during its motion over some time period.

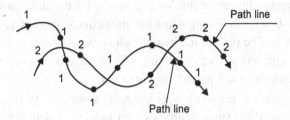

Fig. 5.2: Path lines.

(c) **Streak Line or Filament Line:** A streak line is a curve or line which gives an instantaneous picture of the location of the fluid particles, which has passed through a prescribed point in the flow. For example: Visible plume from a chimney represents such a locus because all the smoke particles have passed through the chimney's mouth.

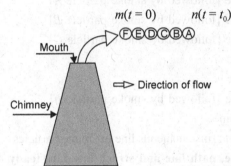

$m(t = 0)$ $n(t = t_0)$

Mouth

Direction of flow

Chimney

- line mn is called streak line at time $t = t_0$
- line mn is called stream line at time $t = t_0$
- line mn is called path line traced by smoke particle (A) at time $t = t_0$

Fig. 5.3: Stream line, path line and streak line of smoke particles flowing left to right, issued from the chimney mouth. [the flow is steady].

Let us suppose smoke particles used from chimney's mouth are in alphabetic from A to L. Consider the flow of air from west to east with uniform velocity V. It can be seen that stream line, as shown in Fig. 5.3 as defined above, is straight line from left to right.

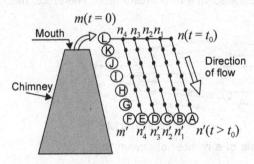

$m(t = 0)$

Mouth

$n_4 \, n_3 \, n_2 \, n_1$ $n(t = t_0)$

Direction of flow

Chimney

m' $n_4' \, n_3' \, n_2' \, n_1'$ $n'(t > t_0)$

Fig. 5.4: Representation of stream line, streak line and path lines.

- line mnn' is path line of smoke particle (A) at time $t > t_0$
- line mn_1n_1' is path line of smoke particle (B)
- line mn_2n_2' is path line of particle (C)
- line mn_3n_3' is path line of particle (D)
- line mn_4n_4' is path line of particle (E)
- line mm' is path of particle (F)
- line $mm'n'$ is streak line. (Instantaneous picture of the flow particles, all particles injected from the mouth of a chimney)
- line mm' is stream line.

If point *m* represents the mouth of a chimney, all the fluid particles that pass through point *m* will move in a straight line and streak line will coincide with the stream line through point *m*. The path of the particle which was at point *m* at line $t = 0$, will also be in the straight line $m - n$; coinciding with the stream line and the streak line, where *n* is the position of the particle at time $t = t_0$.

Now, let the wind direction changes suddenly at time $t = t_0$, it starts blowing in the southeast direction. All smoke particles start moving in that direction thereafter. The stream lines for any line $t > t_0$ will then be straight lines in the northwest to southeast direction as shown in fig. 5.4. It may be noted that a stream line represents an instantaneous picture of the flow field and does not depend upon the history of the fluid flow.

From Fig. 5.4.

line *m-n-n′* = path line [followed by smoke particle *A*]

line $m\text{-}n_1\text{-}n_1'$ = path line [followed by smoke particle *B*]

line $m\text{-}n_2\text{-}n_2'$ = path line [followed by smoke particle *C*]

line $m\text{-}n_3\text{-}n_3'$ = path line

line $m\text{-}n_4\text{-}n_4'$ = path line

line $m\text{-}m_1$ = path line [followed by smoke particle *F*]

= stream line

line *m-m′-n′* = streak line [instantaneous line of smoke particles]

There is no distinction among stream line, path line and streak line for steady flow.

5.4 STREAM TUBE

By definition of a stream line, it is clear that there can be no velocity normal to a steam line at any point on it. The stream tube is a collection of a number of stream lines forming an imaginary tube. There is no flow through the walls of a stream tube. The properties of the flow are constant across the section of a stream tube. Therefore, the flow in a stream tube is one dimensional.

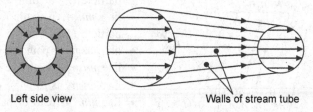

Left side view   Walls of stream tube

Fig. 5.5: Stream tube consists of a number of stream lines.

Flow can enter or leave a stream tube only through its end cross-sections, and not its side surface. The stream tube can have regular or irregular shape.

5.5 TYPES OF FLUID FLOW

The fluid flow is classified as:

    (*a*) Steady and unsteady flow.

    (*b*) Uniform and non-uniform flow.

    (*c*) Laminar and turbulent flow.

    (*d*) Compressible and incompressible flow.

    (*e*) Rotational and irrotational flow, and

    (*f*) One, two and three-dimensional flows.

5.5.1 Steady and Unsteady Flow

A flow is considered to be steady if fluid flow parameters such as velocity, pressure, density, temperature *etc.* at any point do not change with time. If any one of these parameters changes with time, the flow is said to be unsteady.

    Mathematically,

$$\frac{\partial V}{\partial t} = 0 \qquad \frac{\partial p}{\partial t} = 0, \quad \frac{\partial \rho}{\partial t} = 0 \quad \text{for steady flow}$$

$$\frac{\partial V}{\partial t} \neq 0 \qquad \frac{\partial p}{\partial t} \neq 0, \quad \frac{\partial \rho}{\partial t} \neq 0 \quad \text{for unsteady flow}$$

5.5.2 Uniform and Non-uniform Flow

A type of flow in which velocity: V, pressure: p, density: ρ, temperature: T, *etc.* at any given time do not change with respect to space (*i.e.*, length of direction of the flow) is called uniform flow.

    Mathematically,

$$\frac{\partial V}{\partial s} = 0, \quad \frac{\partial p}{\partial s} = 0 \quad \text{for uniform flow}$$

    In case of non-uniform flow, velocity, pressure, density *etc.* at given time change with respect to space.

    Mathematically,

$$\frac{\partial V}{\partial s} = 0, \quad \frac{\partial p}{\partial s} \neq 0 \quad \text{for non-uniform flow}$$

    We use the terms uniform and non-uniform flow often in connection with open channels. In a channel where the section of the channel is uniform and the depth of flow is uniform, the flow will be uniform as the velocity will be the same at the sections. But if the sectional dimensions of the channel are different at different sections, the depths of flow will be different at different sections. Obviously, the velocity will be different at different sections and the flow will be non-uniform. Whether the flow is uniform or non-uniform, if the rate of flow is constant the flow is steady and if the rate of flow changes with time the flow is unsteady. Thus, we may come across steady or unsteady or uniform or non-uniform flow. Any type of flow can exist independently of the other.

    A combination of two types of flow is also possible. Some combinations are:

    (*i*) Steady uniform flow.   (*ii*) Steady non-uniform flow.

  (*iii*) Unsteady uniform flow.   (*iv*) Unsteady non-uniform flow.

(*i*) **Steady Uniform Flow:** The flow characteristics of this type of flow do not vary with respect to both time and space. For example, Flow through a prismatic pipe or channel at constant discharge.

(*ii*) **Steady Non-uniform Flow:** If the flow characteristics do not change with respect to time but vary from location to location, the flow is called steady non-uniform flow: For example:

- Flow through a nozzle.
- Flow approaching an overfall in a channel.

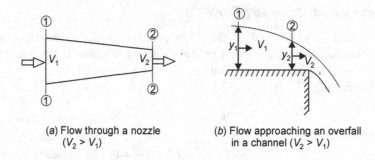

(*a*) Flow through a nozzle
($V_2 > V_1$)

(*b*) Flow approaching an overfall
in a channel ($V_2 > V_1$)

Fig. 5.6: Steady non-uniform flow.

(*iii*) **Unsteady Uniform Flow:** If the flow characteristics at any point changes with time but do not vary with space, this type of flow is called unsteady uniform flow. Unsteady uniform flow is impracticable.

(*iv*) **Unsteady Non-uniform Flow:** If the flow characteristics change with respect to both time and space, the flow is called unsteady non-uniform flow. The unsteady flow in practice is generally understood as unsteady non-uniform flow. For example:

- Flow through a pipe with a valve in operation in unsteady flow may be treated as example of unsteady non-uniform.

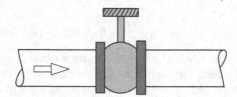

Fig. 5.7: Flow through a pipe with a valve in operation.

5.5.3 Laminar and Turbulent Flow

Laminar Flow

In this type of flow, the particles of fluid move in orderly and well-defined paths. Laminar flow is also called viscous flow or stream line flow. This type of flow is only possible at slow speed and in viscous fluid. For example, flow through a capillary tube, flow of blood in veins and arteries.

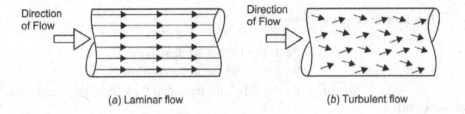

(a) Laminar flow   *(b)* Turbulent flow

Fig. 5.8: Laminar and turbulent flow

Turbulent Flow

In this type of flow, the particles of fluid move irregularly and disorderly *i.e.*, fluid particles move in a zig-zag way. The zig-zag movement of fluid particles is responsible for high energy loss.

The flow either laminar or turbulent is determined by using of a non-dimensional number *i.e.*, Reynolds number (Re).

Reynolds Number (Re)

The Reynolds number is the ratio of forces due to inertia and viscosity. It is denoted by Re.

Mathematically:

Reynolds number: $\quad \text{Re} = \dfrac{\text{Inertia force}}{\text{Viscous force}} = \dfrac{F_i}{F_v}$

where   Inertia force: F_i = mass × acceleration = $ma = \rho Va = \rho L^3 . \dfrac{V^2}{L} = \rho L^2 V^2$

and   Viscous force: F_v = shear stress × area $= \mu \dfrac{du}{dy} . L^2 = \mu \dfrac{V}{L} L^2 = \mu L V$

$\therefore \qquad\qquad \text{Re} = \dfrac{\rho V^2 L^2}{\mu V L} = \dfrac{\rho V L}{\mu} = \dfrac{V L}{\nu}$

where kinematic viscosity:

$$\nu = \dfrac{\text{Dynamic viscosity}(\mu)}{\text{Density of fluid }(\rho)}$$

$$\nu = \dfrac{\mu}{\rho}$$

L = a characteristic length.

$\text{Re} = \dfrac{VL}{\nu}$ this equation is used when fluid flows over plate.

$\text{Re} = \dfrac{VD}{\nu}$ this equation is used when fluid flow through pipe.

where D = diameter of pipe.

For flow over plate: $\left[\mathrm{Re} = \dfrac{VL}{v} \right]$

$\mathrm{Re} \leq 3 \times 10^5$ for laminar flow

$\mathrm{Re} \geq 5 \times 10^5$ for turbulent flow

$3 \times 10^5 < \mathrm{Re} < 5 \times 10^5$, for transition flow [flow may be laminar or turbulent]

For flow through pipes or tubes: $\left[\mathrm{Re} = \dfrac{VD}{v} \right]$

$\mathrm{Re} \leq 2000$ for laminar flow

$\mathrm{Re} \geq 4000$ for turbulent flow

$2000 < \mathrm{Re} < 4000$ for transition flow [flow may be laminar or turbulent]

The value of the Reynolds number in a flow gives an idea about its nature. For example, at higher Reynolds number the magnitude of viscous forces is small compared to the inertia forces. And at lower Reynolds number the magnitude of viscous forces is large compared to the inertia forces.

5.5.4 Compressible and Incompressible Flow

Compressible Flow

If the density of the fluid changes from point to point in the fluid flow, it is referred to as compressible flow.

Mathematically,

Density: $\rho \neq c$

For example: Gases (like air, carbon dioxide *etc.*) are compressible.

Incompressible Flow

If the density of the fluid remains constant at every point in the fluid flow, it is referred to as an incompressible flow.

Mathematically,

Density: $\rho = c$

For example: Liquids are generally incompressible in nature.

Compressible or incompressible flows are also defined on basis of Mach number (M). It is index of the ratio of inertia and elastic forces.

Mathematically,

$$M^2 = \frac{\text{Inertia force}}{\text{Elastic force}} = \frac{\rho L^2 V^2}{K L^2}$$

$$M^2 = \frac{\rho V^2}{K} \quad \text{where } K = \text{bulk modulus of the fluid.}$$

$$M^2 = \frac{V^2}{K/\rho}$$

where $\dfrac{K}{\rho} = a^2$ where $a = $ the speed of sound in the flowing medium.

$$\therefore M^2 = \frac{V^2}{a^2}$$

$$\text{or } \boldsymbol{M} = \frac{V}{a}$$

This relation gives another important definition of the Mach number (M) as the ratio of the fluid velocity to the velocity of sound in the flowing medium.

If the Mach number, $M > 0.2$, the fluid is considered to be compressible.

If the Mach number, $M < 0.2$, the fluid assumed to be incompressible.

Mach number is used to determine of compressibility in a particular gas flow and the flow is classified into different regimes:

    (*i*) Transonic flow

    (*ii*) Subsonic flow

    (*iii*) Sonic flow

    (*iv*) Supersonic flow

    (*v*) Hypersonic flow

    (*vi*) High hypersonic flow

    (*i*) **Transonic flow:** A flow is called transonic flow if the Mach number (M) lies between 0.9 and 1.1 (*i.e.,* the Mach number varies between the values slightly less than one to values slightly greater than one)

    Mathematically,

    For transonic flow, $0.9 < M < 1.1$

    (*ii*) **Subsonic flow:** A flow is called subsonic flow if the Mach number (M) is less than one. Therefore for subsonic flow, the velocity of flow is less than the velocity of sound.

    Mathematically,

$$M < 1, \quad V < a \quad \text{for subsonic flow}$$

    (*iii*) **Sonic flow:** A flow is called sonic flow if the Mach number (M) is equal to one. Therefore for sonic flow, the velocity of flow is equal to the velocity of sound.

    Mathematically,

    Mach number: $M = 1, \quad V = a \quad$ for sonic flow.

    (*iv*) **Supersonic flow:** A flow is called supersonic flow if the Mach number (M) is greater than one and less than seven.

    Mathematically,

$$1 < M < 7 \quad \text{for supersonic flow}$$

    (*v*) **Hypersonic flow:** A flow is called hypersonic flow if the Mach number is equal to and greater than 7 and less than 10.

    Mathematically,

$$7 \leq M < 10 \quad \text{for hypersonic flow.}$$

    (*vi*) **High hypersonic flow:** A flow is called high hypersonic flow if the Mach number is greater and equal than 10.

    Mach number: $M \geq 10 \quad$ for high hypersonic flow.

Table 5.1

Regime	Transonic Flow	Subsonic Flow	Sonic Flow	Supersonic Flow	Hypersonic Flow	High Hypersonic Flow
Mach Number: M	$0.9 < M < 1$	$M < 1$	$M = 1$	$1 < M < 7$	$7 \leq M < 10$	$M \geq 10$

5.5.5 Rotational and Irrotational Flow

(*i*) **Rotational Flow:** The rotational flow is that type of flow in which the fluid particles while flowing along stream lines, also rotate about their own axis. For example, if we assumed the earth is a fluid particle, the flow of the earth is rotational flow because earth is moving around the run but also rotates about its own axis.

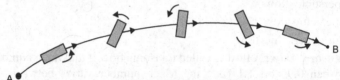

Fig. 5.9: Rotational flow.

(*ii*) **Irrotational Flow:** The irrotational flow is that type of flow in which the fluid particles while flowing along stream lines, do not rotate about their own axis.

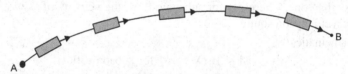

Fig. 5.10: Irrotational flow.

5.5.6 One, Two and Three-dimensional Flows

(*i*) **One-Dimensional Flow:** For a steady one-dimensional flow, the velocity is a function of one space coordinate only. The variation of velocities in other mutually perpendicular direction is assumed negligible.

Mathematically,

$u = f(x)$, $v = 0$ and $w = 0$

(*ii*) **Two-Dimensional Flow:** A type of flow in which the velocity is a function of time and two rectangular space coordinates say x and y. For a steady two-dimensional flow the velocity is a function of two space coordinates only. The variation in the IIIrd direction is negligible.

Mathematically,

$u = f_1(x, y)$, $v = f_2(x, y)$ and $w = 0$

(*iii*) **Three-Dimensional Flow:** In this type of flow, the velocity is a function of time and three mutually perpendicular directions. But for a steady three-dimensional flow, the fluid parameters are functions of three space coordinate (*x*, *y* and *z*) only.

Mathematically,

$$u = f_1(x, y, z), \ v = f_2(x, y, z) \text{ and } w = f_3(x, y, z).$$

5.6 RATE OF FLOW

The quantity of the fluid flowing per unit time through a section of a pipe or a open channel is called **rate of flow**. The rate of flow is expressed as the weight of the fluid flowing per second across the section is called **weight flow rate**. The mass of the fluid flowing across the section per second is called **mass flow rate**. The volume of the fluid flowing across the section per second is called **volume flow rate** or **discharge**.

Units

(*i*) N/s (*ii*) kg/s (*iii*) m³/s or litre/s.

In the case of incompressible fluids, the rate of flow is usually expressed as the volume of the fluid flowing per second.

In the case of compressible fluids, the rate of flow is usually expressed as the weight of the fluid flowing per second.

Consider a fluid flowing through a pipe in which

A = cross-sectional area of pipe

V = average velocity of fluid across the section

v = specific volume $= \dfrac{1}{\rho}$

Then, Discharge: $Q = AV = mv$

or $AV = mv$

or $m = \dfrac{AV}{v}$

Mass flow rate: $m = \rho AV$ kg/s

On multiplying g both sides, we get

$$mg = \rho g AV$$

or $W = \rho g AV$ N/s

where $\dot{W}$ = weight slow rate; N/s

ρ = density of fluid; kg/m³

g = 9.81 m/s²

A = cross-sectional area; m²

V = average velocity of fluid; m/s

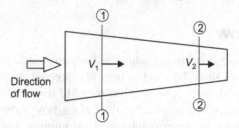

5.7 CONTINUITY EQUATION

The equation based on the **law of conservation of mass** is called **continuity equation**. It means that mass of fluid is neither created nor destroyed. The quantity of fluid entering the stream tube at one end per unit time should be equal to the quantity of the fluid leaving the stream tube at the other end per unit of time: consider two cross-sections of a converging duct as shown in Fig. 5.11.

Fig. 5.11: Continuity equation

Let V_1 = average velocity of section (1)-(1)

          V_2 = average velocity at section (2)-(2)

          A_1 = cross-sectional area at section (1)-(1)

          A_2 = cross-sectional area at section (2)-(2)

          ρ_1 = density at section (1)-(1)

          ρ_2 = density at section (2)-(2)

According to law of conservation of mass,

Rate of flow at section (1)-(1) = rate of flow at section (2)-(2)

$$m_1 = m_2 = m \qquad \because m = \rho A V = \text{constant}$$

$$\rho_1 A_1 V_1 = \rho_2 A_2 V_2 = m = \textbf{constant} \qquad ...(i)$$

Equation. (*i*) is called **continuity equation.** [It states that mass flow is the product of density, cross sectional area and velocity at any section in the flow field is constant], applicable to the compressible as well as incompressible fluid.

If the fluid is incompressible, then $\rho_1 = \rho_2$, the Eq. (*i*), is reduced to

$$A_1 V_1 = A_2 V_2 \qquad\qquad ...(ii)$$

Equation. (*ii*) is also called **continuity equation,** is used for **incompressible flow.**

Significance

The continuity equation tells us whether fluid flow is possible or not. If fluid flow satisfies continuity equation, then fluid flow is possible. If fluid flow does not satisfy continuity equation then fluid flow is impossible.

Problem 5.1: The diameters of the pipe at sections 1-1 and 2-2 are 100 mm and 250 mm respectively. If the discharge through the pipe is 0.06 m³/s, find the average velocities at the two sections.

Solution: Given data:

     At section (1)-(1)

Diameter: $D_1 = 100$ mm $= 0.1$ m

At section (2)-(2)

Diameter: $D_2 = 250$ mm $= 0.25$ m

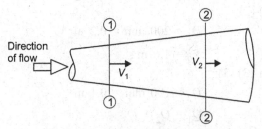

Fig. 5.12: Schematic for Problem 5.1

Let V_1 and V_2 are average velocities at the section (1)-(1) and (2)-(2) respectively. Discharge passing through the pipe: $Q = 0.06$ m³/s.

$$Q = A_1 V_1$$

$$0.06 = \frac{\pi}{4} D_1^2 \times V_1$$

$$0.06 = \frac{3.14}{4} \times (0.1)^2 \times V_1$$

$$V_1 = \textbf{7.64 m/s}$$

also

$$Q = A_2 V_2$$

$$0.06 = \frac{\pi}{4} D_2^2 \times V_2$$

$$0.06 = \frac{3.14}{4} \times (0.25)^2 \times V_2$$

or

$$V_2 = \textbf{1.22 m/s}$$

Problem 5.2: A 300 mm diameter pipe conveying water branches into two pipes of diameter 200 mm and 100 mm respectively. If the average velocities in the 300 mm diameter pipe and 200 mm diameter pipe are 2.5 m/s and 1.6 m/s respectively, determine the velocity in the 100 mm diameter pipe.

Solution:

Let Q_1, Q_2 and Q_3 be the discharge passing through sections (1)-(1), (2)-(2) and (3)-(3) respectively.

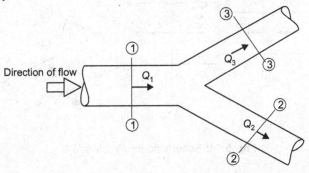

Fig. 5.13: Schematic for Problem 5.2

Diameter at section (1)-(1):

$$D_1 = 300 \text{ mm} = 0.3 \text{ m}$$

and

$$V_1 = 2.5 \text{ m/s}$$

At section (2)-(2):

$$D_2 = 200 \text{ mm} = 0.2 \text{ m}$$

and

$$V_2 = 1.6 \text{ m/s}$$

and at section (3)-(3):

$$D_3 = 100 \text{ mm} = 0.1 \text{ m}$$

∴

$$Q_1 = Q_2 + Q_3$$

∵ Discharge: $Q = AV$

$$A_1 V_1 = A_2 V_2 + A_3 V_3$$

$$\frac{\pi}{4} D_1^2 \times V_1 = \frac{\pi}{4} D_2^2 \times V_2 + \frac{\pi}{4} D_3^2 \times V_3$$

$$D_1^2 V_1 = D_2^2 V_2 + D_3^2 V_3$$

$$(0.3)^2 \times 2.5 = (0.2)^2 \times 1.6 + (0.1)^2 \times V_3$$

$$0.225 = 0.064 + 0.01 \times V_3$$

or

$$0.01 \, V_3 = 0.161$$

$$V_3 = \textbf{16.1 m/s}$$

Problem 5.3:   Water flows through a pipe AB 1.2 m in diameter at 3 m/s and then passes through a pipe BC 1.5 m in diameter. At C, the pipe branches. Branch CD is 0.8 m in diameter and carries one-third of the flow in AB. The flow velocity in branch CE is 2.5 m/s. Find the volume rate of flow in AB, the velocity in BC, the velocity in CD and diameter of CE.

Solution: Ist step to draw the pipe layout of given problem:

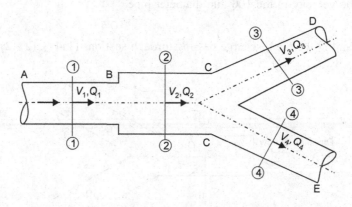

Fig. 5.14: Schematic for Problem 5.3

Given data:

Diameter of pipe AB : d_1 = 1.25 m

$\therefore$ Cross-sectional area at section (1)-(1) in pipe AB:

$$A_1 = \frac{\pi}{4}d_1^2 = \frac{3.14}{4} \times (1.2)^2 = 1.13 \text{ m}^2$$

Velocity at section (1)-(1) in pipe AB:

$$V_1 = 3 \text{ m/s}$$

Diameter of pipe BC: d_2 = 1.5 m

$\therefore$ Cross-sectional area at section (2)-(2) in pipe BC:

$$A_2 = \frac{\pi}{4}d_2^2 = \frac{3.14}{4} \times (1.5)^2 = 1.766 \text{ m}^2$$

Diameter of pipe CD: d_3 = 0.8 m

$\therefore$ Cross-sectional area at section (3)-(3) in pipe CD:

$$A_3 = \frac{\pi}{4}d_3^2 = \frac{3.14}{4} \times (0.8)^2 = 0.502 \text{ m}^2$$

Discharge at section (3)-(3): $Q_3 = \dfrac{Q_1}{3}$

Velocity at section (4)-(4) in pipe CE:

$$V_4 = 2.5 \text{ m/s.}$$

Find:

(i)   Volume rate of flow in AB: Q_1 = ?

(ii)   Velocity in BC: V_2 = ?

(iii) Velocity in CD: V_3 = ?

(iv) Diameter of CE: d_4 = ?

(i)   Volume rate of flow in AB: Q_1

Applying continuity equation at section (1)-(1), we get

$$Q_1 = A_1 V_1$$
$$Q_1 = 1.13 \times 3 = \mathbf{3.39 \text{ m}^3/s}$$

(ii) Velocity in BC: V_2

Discharge: $Q_2 = Q_1 = 3.39 \text{ m}^3/s$

Applying continuity equation at section (2)-(2), we get

$$Q_2 = A_2 V_2$$
$$3.39 = 1.766 \times V_2$$

or $V_2 = \mathbf{1.919 \text{ m/s}}$

(iii) Velocity in CD: V_3

Discharge: $Q_3 = \dfrac{Q_1}{3} = \dfrac{3.39}{3} = 1.13 \text{ m}^3/s$

Applying continuity equation at section (3)-(3), we get

$$Q_3 = A_3 V_3$$

or
$$V_3 = \frac{Q_3}{A_3} = \frac{1.13}{0.502} = \textbf{2.25 m/s}$$

(iv) Diameter of CE: d_4

Discharge: $Q_2 = Q_3 + Q_4$

or $Q_4 = Q_2 - Q_3 = 3.39 - 1.13 = 2.26 \text{ m}^3/\text{s}$

Applying continuity equation at section (4)-(4), we get

$$Q_4 = A_4 V_4$$

or
$$A_4 = \frac{Q_4}{V_4} = \frac{2.26}{2.5} = 0.904 \text{ m}^2$$

also
$$A_4 = \frac{\pi}{4} d_4^2$$

$$0.904 = \frac{3.14}{4} \times d_4^2$$

or
$$d_4^2 = 1.1515$$

$$d_4 = \textbf{1.075 m}$$

Problem 5.4: A jet of water from a 25 mm diameter nozzle is directed vertically upwards. Assuming that the jet remains circular and neglecting any loss of energy, what will be the diameter at a point 5 m above the nozzle, if the velocity with which the jet leaves the nozzle is 12 m/s.

Solution: Given data:

Diameter of nozzle: $d_1 = 25 \text{ mm} = 0.025 \text{ m}$

Velocity of jet leaving the nozzle:

$$V_1 = 12 \text{ m/s}$$

Height of jet at section (2):

$$s = 5 \text{ m}$$

let $V_2 = $ Velocity of jet at section (2).

Initial velocity of jet at sesction (1):

$$u = V_1 = 12 \text{ m/s}$$

Final velocity of jet at section (2):

$$v = V_2$$

Substituting the value of u, V, a and s in the relation:

$$v^2 - u^2 = 2 as$$

$$v^2 - (12)^2 = 2 \times (-9.81) \times 5 \quad \because a = -g$$

$$= -9.81 \text{ m/s}^2$$

The –ve shows that jet moved upward

$$v^2 - 144 = -98.1$$

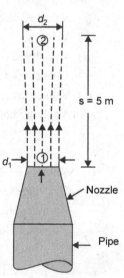

Fig. 5.15: Schematic for Problem 5.4

or $\qquad v^2 = -98.1 + 144 = 45.9$

or $\qquad v = \sqrt{45.9} = 6.77 \text{ m/s} = V_2$

Applying continuity equation at sections (1) and (2), we get

$$A_1 V_1 = A_2 V_2$$

$$\frac{\pi}{4} d_1^2 \times V_1 = \frac{\pi}{4} d_2^2 \times V_2$$

or $\qquad d_1^2 \times V_1 = d_2^2 \times V_2$

$$(0.025)^2 \times 12 = d_2^2 \times 6.77$$

or $\qquad d_2^2 = 0.001107$

or $\qquad d_2 = 0.03327 \text{ m} = \mathbf{33.27 \text{ mm}}$

5.8 CONTINUITY EQUATION IN THREE-DIMENSIONS IN CARTESIAN COORDINATES (x, y, z)

Consider a rectangular fluid element of length dx, dy and dz in direction of x, y and z respectively as shown in Fig. 5.16

Let u, v and w are the inlet velocity components in the directions of x, y, and z respectively.

Mass of fluid entering the face $abcd$, along x-direction:

$\qquad$ = density × area of face $abcd$ × velocity in x-direction

$\qquad$ = $\rho.dy.dz\ u = \rho.u.dy.dz$

Mass of fluid leaving the face $a'b'c'd'$, along x-direction

$$= \rho.u.dy.dz + \frac{\partial}{\partial x}(\rho u\, dy\, dz)dx$$

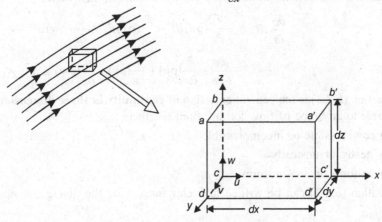

Fig. 5.16: Rectangular fluid element

∴ Gain of mass per unit time in small fluid element in x-direction = mass of fluid entering in face $abcd$ – mass of fluid leaving the face $a'b'c'd'$

$$= \rho\, u\, dy\, dz - [\rho\, u\, dy\, dz + \frac{\partial}{\partial x}(\rho u\, dy\, dz)dx\,]$$

$$= -\frac{\partial}{\partial x}(\rho u \, dy \, dz \, dx)$$

$$= -\frac{\partial}{\partial x}(\rho u) \, dx \, dy \, dz \quad \bigg| \quad \because dx. \, dy. \, dz = dV, \text{ volume of small fluid element}$$

$$= -\frac{\partial}{\partial x}(\rho u) \, dV$$

Similar expression may be obtained considering the faces normal to y- and z-directions.

Gain of mass per unit time in small fluid element in y-direction,

$$= -\frac{\partial}{\partial y}(\rho v) \, dV$$

and gain of mass per unit time in small fluid element in z-direction,

$$= -\frac{\partial}{\partial z}(\rho w) \, dV$$

Therefore, net gain of mass per unit time within the element,

$$= -\frac{\partial}{\partial x}(\rho u) \, dV - \frac{\partial}{\partial y}(\rho v) \, dV - \frac{\partial}{\partial z}(\rho w) \, dV \quad \ldots(5.8.1)$$

Considering the time rate of variation of density, the change in the mass of the element per unit time

$$= \frac{\partial \rho}{\partial t} \, dx. dy. dz = \frac{\partial \rho}{\partial t}. dV \qquad \ldots(5.8.2)$$

Equations (5.8.1) and (5.8.2) represent the same quantity, hence

$$\frac{\partial \rho}{\partial t}. dV = -\frac{\partial}{\partial x}(\rho u) \, dV - \frac{\partial}{\partial y}(\rho v) \, dV - \frac{\partial}{\partial z}(\rho w) \, dV$$

or

$$\frac{\partial \rho}{\partial t} + \frac{\partial}{\partial x}(\rho u) + \frac{\partial}{\partial y}(\rho v) + \frac{\partial}{\partial z}(\rho w) = 0 \quad \ldots(5.8.3)$$

Equation (5.8.3) is the general **equation of continuity in three dimensions** and is applicable to any type of flow for any fluid whether:

- compressible or incompressible
- steady or unsteady
- uniform or non-uniform

Equation (5.8.3) can be written in vector form using the divergence operator

$$\frac{\partial \rho}{\partial t} + \nabla.(\rho \vec{V}) = 0$$

For steady flow of incompressible fluids,

$$\frac{\partial \rho}{\partial t} = 0, \, \rho = \text{constant}$$

Equation (5.8.3) is reduced to

$$\frac{\partial u}{\partial x}+\frac{\partial v}{\partial y}+\frac{\partial w}{\partial z} = 0 \qquad\qquad ...(5.8.4)$$

Equation (5.8.4) is the continuity equation in three-dimensions and is applicable only for steady state and incompressible fluid flow:

If $w = 0$, Eq. (5.8.4) reduced to

$$\frac{\partial u}{\partial x}+\frac{\partial v}{\partial y} = 0 \qquad\qquad ...(5.8.5)$$

Equation (5.8.5) is the continuity equation in two dimensions.

If $w = v = 0$, Eq. (5.8.4) reduced to

$$\frac{\partial u}{\partial x} = 0 \qquad\qquad ...(5.8.6)$$

Equation (5.8.6) is the continuity equation in one dimension.

$$\partial u = 0$$

or $\qquad\qquad u = $ constant

If A is the cross-sectional area perpendicular to the velocity u, then the rate of flow is:

$$Q = Au = \text{constant for steady flow.}$$

This equation shows that, if area of flow A is constant, then the velocity of flow u must also be constant.

5.9 CONTINUITY EQUATION: INTEGRAL FORM

The continuity equation can be expressed in an integral form by an application of the statement of conservation of mass to a finite control volume of any shape and size.

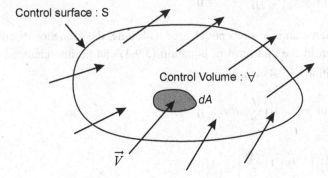

Fig. 5.17: Control Volume

Let $\forall = $ control volume is bounded by the control surface S.

The efflux of mass across the control surface S is given by

$$= \iint\limits_{S} \rho \vec{V}.d\vec{A} \qquad\qquad ...(5.9.1)$$

where $\vec{V}$ = velocity vector at an elemental area $d\vec{A}$ which is treated as a vector by considering its positive direction along the normal drawn outward from the surface.

The rate of mass accumulation within the control volume becomes

$$= \frac{\partial}{\partial t} \iiint\limits_{\forall} \rho.d\forall \qquad \qquad ...(5.9.2)$$

where $\qquad\qquad d\forall$ = elemental volume,

ρ = density of fluid

$\forall$ = total volume bounded by the control surface S.

According to continuity equation:

Net rate of mass efflux from control volume + rate of accumulation of mass in the control volume = 0

$$\boxed{\iint\limits_{S} \rho\vec{V}.d\vec{A} + \frac{\partial}{\partial t} \iiint\limits_{\forall} \rho.d\forall = 0} \qquad \qquad ...(5.9.3)$$

Equation (5.9.3) is integral form of the continuity equation.

The first term of the equation (5.9.3) can be converted into a volume integral by the use of the Gauss divergence theorem as:

$$\iint\limits_{S} \rho\vec{V}.d\vec{A} = \iiint\limits_{\forall} \nabla.(\rho\vec{V})d\forall$$

$\therefore$ Equation (5.9.3) becomes

$$\iiint\limits_{\forall} \nabla.(\rho\vec{V})d\forall + \frac{\partial}{\partial t} \iiint\limits_{\forall} \rho.d\forall = 0 \qquad \qquad ...(5.9.4)$$

Since the volume $\forall$ does not change with time, the sequence of differentiation and integration in the 2nd term of equation (5.9.4) can be interchanged.

$\therefore$ Equation (5.9.4) can be written as

$$\iiint\limits_{\forall} \nabla.(\rho\vec{V})d\forall + \iiint\limits_{\forall} \frac{d}{dt}\rho.d\forall = 0$$

$$\iiint\limits_{\forall} \left[\nabla.(\rho\vec{V}) + \frac{\partial\rho}{\partial t} \right]d\forall = 0 \qquad \qquad ...(5.9.5)$$

Equation (5.9.5) is valid for any arbitrary control volume irrespective of its shape and size. So we can write

$$\boxed{\nabla.(\rho\vec{V}) + \frac{\partial\rho}{\partial t} = 0} \qquad \qquad ...(5.9.6)$$

Equation (5.9.6) is differential form of the continuity equation.

5.10 CONTINUITY EQUATION IN CYLINDRICAL COORDINATES (r, θ, z)

Let us consider an elementary volume: $dV = rd\theta.dr.dz$

V_r, V_θ, V_z are inlet velocities along r-direction, θ-direction and z-direction respectively.

Mass flow rate of fluid entering in radial (r) direction:

$$= \rho \times \text{velocity in } r\text{-direction} \times \text{area}$$
$$= \rho \, V_r \, .rd\theta.dz$$

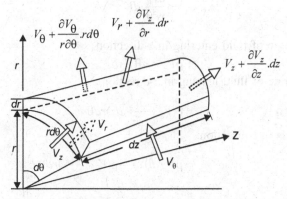

Fig. 5.18: Continuity equation in cylindrical volume element.

Mass flow rate of fluid leaving in radial (r) direction:

$$= \rho(V_r + \frac{\partial V_r}{\partial r}.dr)(r + dr)d\theta dz$$

$$= \rho[r.V_r + r\frac{\partial V_r}{\partial r}.dr + V_r.dr + \frac{\partial V_r}{\partial r}.dr^2]d\theta dz$$

$$= \rho[r.V_r + r\frac{\partial V_r}{\partial r}.dr + V_r.dr]d\theta dz$$

[square of small term (dr^2) can be neglected]

Gain of mass in radial (r) direction:

$$= \rho V_r.r \, d\theta \, dz - \rho r V_r \, d\theta \, dz - \rho r\frac{\partial V_r}{\partial r}.dr \, d\theta dz - \rho V_r \, dr \, d\theta \, dz$$

$$= -\rho r\frac{\partial V_r}{\partial r}.dr \, d\theta dz - \rho V_r \, dr \, d\theta \, dz$$

$$= -\rho\left[\frac{\partial V_r}{\partial r} + \frac{V_r}{r}\right]r V_r \, d\theta \, dz \quad = -\rho\left[\frac{\partial V_r}{\partial r} + \frac{V_r}{r}\right]dV \quad \because \, dV = rd\theta.dr.dz$$

Now consider mass flow rate in θ-direction:

Mass flow rate of fluid entering in θ-direction:

$$= \rho \times \text{velocity in } \theta\text{-direction} \times \text{area}$$
$$= \rho V_\theta \, dr \, dz$$

Mass flow rate of fluid leaving in θ-direction:

$$= -\rho\left(V_\theta + \frac{\partial V_\theta}{r\partial \theta}.rd\theta\right)dr.dz$$

Gain of mass in θ-direction

$$= \rho V_\theta \, dr \, dz - \rho \left(V_\theta + \frac{\partial V_\theta}{r \partial \theta} . r d\theta \right) dr.dz$$

$$= \rho V_\theta \, dr \, dz - \rho V_\theta \, dr \, dz - \rho \frac{\partial V_\theta}{r \partial \theta} . r d\theta . dr.dz$$

$$= -\rho \frac{\partial V_\theta}{r \partial \theta} . r d\theta . dr.dz$$

$$= -\rho \frac{\partial V_\theta}{r \partial \theta} . dV$$

Mass flow rate of fluid entering in z-direction
$$= \rho V_z \, r \, d\theta \, . \, dr$$
Mass flow rate of fluid leaving in z-direction

$$= \rho (V_z + \frac{\partial V_z}{\partial z} . dz) r d\theta dr$$

Gain of mass in z-direction

$$= \rho V_z r d\theta . dr - \rho (V_z + \frac{\partial V_z}{\partial z}) r d\theta dz$$

$$= \rho V_z r d\theta . dr - \rho V_z \, r d\theta \, dz - \rho + \frac{\partial V_z}{\partial z} . dz \, r d\theta dr$$

$$= -\rho r \frac{\partial V_z}{\partial_z} . r \, d\theta dz \, dr$$

$$= -\rho \frac{\partial V_z}{\partial z} dV$$

$\therefore$ Net gain in mass flow rate in cylindrical volume element:

$$= -\rho \left[\frac{\partial V_r}{\partial r} + \frac{V_r}{r} \right] dV - \rho \frac{\partial V_\theta}{\partial \theta} dV - \rho \frac{\partial V_z}{\partial z} dV \quad ...(5.10.1)$$

But mass of fluid element $= \rho \times$ volume of fluid element
$$= \rho \, dV$$
Rate of increase of fluid mass in the element with time

$$= \frac{\partial (\rho dV)}{\partial t} = \frac{\partial \rho}{\partial t} . dV \quad ...(5.10.2) \quad \because \, dV = \text{constant}$$

According to principle of continuity, the mass is neither created nor destroyed in the fluid element, hence net gain of mass flow rate in the fluid element is equal to the rate of increase of mass of fluid in element.

Hence equating Eqs. (5.10.1) and (5.10.2) we get

$$-\rho \left[\frac{\partial V_r}{\partial r} + \frac{V_r}{r} \right] dV - \rho \frac{1}{r} \frac{\partial V_\theta}{\partial \theta} - \rho \frac{\partial V_z}{\partial z} dV \quad = \frac{\partial \rho}{\partial t} dV$$

or

$$-\rho \frac{1}{r} \frac{\partial}{\partial r} (r V_r) - \rho \frac{1}{r} \frac{\partial V_\theta}{\partial \theta} - \rho \frac{\partial V_z}{\partial z} \quad = \frac{\partial \rho}{\partial t}$$

or
$$\boxed{\frac{\partial \rho}{\partial t} + \frac{1}{r}\frac{\partial}{\partial r}(\rho r V_r) + \frac{1}{r}\frac{\partial(\rho V_\theta)}{\partial \theta} + \frac{\partial(\rho V_z)}{\partial z} = 0}$$ ...(5.10.3)

Equation (5.10.3) is three-dimensional cylindrical coordinate (r, θ, z) is applicable for:
(a) steady or unsteady flow   (b) compressible or incompressible flow

For steady for $\dfrac{\partial \rho}{\partial t} = 0$

Incompressible flow $\rho = c$

Equation (5.10.3) becomes

$$\frac{\rho}{r}\frac{\partial(rV_r)}{\partial r} + \frac{\rho}{r}\frac{\partial V_\theta}{\partial \theta} + \frac{\partial(\rho V_z)}{\partial z} = 0$$

$$\frac{1}{r}\frac{\partial(rV_r)}{\partial r} + \frac{1}{r}\frac{\partial V_\theta}{\partial \theta} + \frac{\partial V_z}{\partial z} = 0$$ ...(5.10.4)

Equation (5.10.4) is continuity equation in cylindrical coordinate for 3-D steady and incompressible flow.

The continuity equation in cylindrical coordinate in 2-D (r, θ) flow.

Equation (5.10.4) becomes

$$\frac{1}{r}\frac{\partial(rV_r)}{\partial r} + \frac{1}{r}\frac{\partial V_\theta}{\partial \theta} = 0$$

or $\dfrac{\partial(rV_r)}{\partial r} + \dfrac{\partial V_\theta}{\partial \theta} = 0$ ...(5.10.5)

Equation (5.10.5) is continuity equation in cylindrical for 2-D steady and incompressible flow.

5.11 VELOCITY AND ACCELERATION

Let, u, v and w are the velocity components of the resultant velocity (V) along x, y, and z-directions respectively. The velocity components u, v and w are functions of space coordinate and time. Mathematically, the velocity components are given as:

$$u = f_1 (x, y, z, t)$$
$$v = f_2 (x, y, z, t)$$
and $w = f_3 (x, y, z, t)$

Resultant velocity vector:

$$V = ui + vj + wk$$
$$V = \sqrt{u^2 + v^2 + w^2}$$

Let a_x, a_y and a_z are the acceleration components in x, y and z-directions respectively. Then by the chain rule of differentiation, we get

$$a_x = \frac{du}{dt} = \frac{\partial u}{\partial x}\cdot\frac{dx}{dt} + \frac{\partial u}{\partial y}\cdot\frac{dy}{dt} + \frac{\partial u}{\partial z}\cdot\frac{dz}{dt} + \frac{\partial u}{\partial t}$$

But $\dfrac{\partial x}{\partial t} = u, \dfrac{dy}{dt} = v$ and $\dfrac{dz}{dt} = w$

$$\therefore \qquad a_x = \frac{du}{dt} = u\frac{\partial u}{\partial x} + v\frac{\partial u}{\partial y} + w\frac{\partial u}{\partial z} + \frac{\partial u}{\partial t}$$

Similarly, $\qquad a_y = \frac{dv}{dt} = u\frac{\partial v}{\partial x} + v\frac{\partial v}{\partial y} + w\frac{\partial v}{\partial z} + \frac{\partial v}{\partial t} \qquad$ ...(5.11.1)

and $\qquad a_z = \frac{dw}{dt} = u\frac{\partial w}{\partial x} + v\frac{\partial w}{\partial y} + w\frac{\partial w}{\partial z} + \frac{\partial w}{\partial t}$

For steady flow, $\dfrac{dV}{dt} = 0$, where V is resultant velocity

or $\qquad \dfrac{\partial u}{\partial t} = \dfrac{\partial v}{\partial t} = \dfrac{\partial w}{\partial t} = 0$

Hence the acceleration components in x, y and z-directions become

$$a_x = \frac{\partial u}{\partial t} = u\frac{\partial u}{\partial x} + v\frac{du}{dy} + w\frac{\partial u}{\partial z}$$

$$a_y = \frac{\partial v}{\partial t} = u\frac{\partial v}{\partial x} + v\frac{dv}{dy} + w\frac{\partial v}{\partial z} \qquad \text{...(5.11.2)}$$

and $\qquad a_z = \dfrac{\partial w}{\partial t} = u\dfrac{\partial w}{\partial x} + v\dfrac{dw}{dy} + w\dfrac{\partial w}{\partial z}$

$\therefore$ Acceleration vector: $\vec{a} = a_x i + a_y\, j + a_z k$

$$= \sqrt{a_x^2 + a_y^2 + a_x^2}$$

Local Acceleration

It is defined as the rate of increase of velocity with respect to time at a definite or fixed point in the flow field. In the Eq. (5.11.1) given by the expressions $\dfrac{\partial u}{\partial t}, \dfrac{\partial v}{\partial t}$ and $\dfrac{\partial w}{\partial t}$ are known as **local acceleration.**

Convective Acceleration

It is defined as the rate of change of velocity due to change of position of fluid particles in a fluid flow. The expressions in Eq. (5.11.2) are known as **convective acceleration.**

Problem 5.5: A nozzle is a channel of varying cross-sectional area. Its cross-sectional area converges linearly along its length. The inside diameters at inlet and outlet are 750 mm and 250 mm respectively, and the axial length is 300 mm. Determine the convective acceleration at a section half way along the axis of the nozzle if the discharge is constant at 0.015 m³/s.

Solution: Given data:

Diameter at inlet: $\qquad d_1 = 750$ mm = 0.75 m

$\therefore$ Cross-sectional area: $A_1 = \dfrac{\pi}{4}d_1^2 = \dfrac{3.14}{4} \times (0.75)^2 = 0.4415\,\text{m}^2$

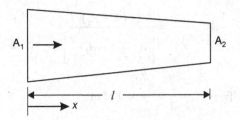

Fig. 5.19: Schematic for Problem 5.5

Diameter at outlet: $\quad d_2 = 250$ mm $= 0.25$ m

$\therefore$ Cross-sectional area: $A_2 = \dfrac{\pi}{4}d_2^2 = \dfrac{3.14}{4}\times(0.25)^2 = 0.049\,\text{m}^2$

Length of nozzle: $\quad l = 300$ mm $= 0.3$ m

Discharge: $\quad Q = 0.015$ m³/s

The cross-sectional area at any section at a distance x from the inlet:

$$A = A_1 - (A_1 - A_2)\frac{x}{l}$$

Velocity at any section: $\quad u = \dfrac{\text{Discharge}}{\text{Cross-sectional area}} = \dfrac{Q}{A_1 - \dfrac{(A_1 - A_2)x}{l}}$

$$\frac{du}{dx} = \frac{d}{dx}\left[\frac{Q}{A_1 - (A_1 - A_2)\dfrac{x}{l}}\right] = \frac{d}{dx}Q\left[A_1 - (A_1 - A_2)\frac{x}{l}\right]^{-1}$$

$$= Q\frac{d}{dx}\left[A_1 - (A_1 - A_2)\frac{x}{l}\right]^{-1}$$

$$= Q\frac{\left[A_1 - (A_1 - A_2)\dfrac{x}{l}\right]^{-2}}{(-1)}\times\left[-\frac{(A_1 - A_2)}{l}\right]$$

$$= \frac{Q(A_1 - A_2)/l}{\left[A_1 - (A_1 - A_2)\dfrac{x}{l}\right]^{-2}}$$

In one-dimensional flow, the convective acceleration is given by $u\dfrac{du}{dx}$

$\therefore \qquad u\dfrac{du}{dx} = \dfrac{Q}{A_1 - (A_1 - A_2)\dfrac{x}{l}}\times\dfrac{Q(A_1 - A_2)/l}{\left[A_1 - (A_1 - A_2)\dfrac{x}{l}\right]^2}$

$$= \frac{Q^2(A_1 - A_2)/l}{\left[A_1 - (A_1 - A_2)\dfrac{x}{l}\right]^3}$$

At a section halfway, $x = \dfrac{l}{2} = \dfrac{0.3}{2} = 0.15\,\text{m}$

$\therefore$ $u\dfrac{du}{dx} = \dfrac{(0.015)^2(0.4415 - 0.047)/0.3}{\left[0.4415 - (0.4415 - 0.047)\times\dfrac{0.15}{0.3}\right]^3}$

$$= \frac{2.958 \times 10^{-4}}{0.1457} = \textbf{0.023 m/s}^2$$

Problem 5.6: The velocity vector for a steady in compressible flow is given by
$$V = (6xt + yz^2)\,i + (3t + xy^2)j + (xy - 2xyz - 6tz)k$$

Verify whether the continuity equation is satisfied.

Solution: Given data:

Velocity vector: $V = (6xt + yz^2)\,i + (3t + xy^2)j + (xy - 2xyz - 6tz)k$ ...(i)

In general form velocity vector:

$$V = ui + vj + wk \qquad\qquad ...(ii)$$

From Eqs. (i) and (ii), we get

$$u = 6xt + yz^2$$
$$v = 3t + xy^2$$

and $w = xy - 2xyz - 6tz$ are velocity components along
x, y and z-directions respectively.

The continuity equation for three-dimensions incompressible steady flow.

$$\frac{\partial u}{\partial x} + \frac{\partial v}{\partial y} + \frac{\partial w}{\partial z} = 0$$

Now $\dfrac{\partial u}{\partial x} = \dfrac{\partial}{\partial x}(6xt + yz^2) = 6t$

$$\frac{\partial v}{\partial y} = \frac{\partial}{\partial y}(3t + xy^2) = 2xy$$

and $\dfrac{\partial w}{\partial z} = \dfrac{\partial}{\partial z}(xy - 2xyz - 6tz) = -2xy - 6t$

$\therefore$ $\dfrac{\partial u}{\partial x} + \dfrac{\partial v}{\partial y} + \dfrac{\partial w}{\partial z} = 6t + 2xy - 2xy - 6t = 0$

Hence, the continuity equation is satisfied.

Problem 5.7: Consider a two-dimensional incompressible steady flow field with velocity components in rectangular coordinates given by:

$$u(x, y) = \frac{k(x^2 - y^2)}{(x^2 + y^2)^2} \text{ and } v(x, y) = \frac{2kxy}{(x^2 + y^2)^2}$$

where k is an arbitrary non-zero constant. Is the equation of continuity satisfied.

Solution: Given data:

The velocity components along x and y-directions:

$$u(x, y) = \frac{k(x^2 - y^2)}{(x^2 + y^2)^2}, \quad v(x, y) = \frac{2kxy}{(x^2 + y^2)^2}$$

The continuity equation for two-dimensional incompressible steady flow field

$$\frac{\partial u}{\partial x} + \frac{\partial v}{\partial y} = 0$$

Now
$$\frac{\partial u}{\partial x} = \frac{\partial}{\partial x} \frac{k(x^2 - y^2)}{(x^2 + y^2)^2} = k \frac{\partial}{\partial x} (x^2 - y^2)(x^2 + y^2)^{-2}$$

$$= k[(x^2 - y^2)(-2)(x^2 + y^2)^{-3} \times 2x + (x^2 + y^2)^{-2} \times 2x]$$

$$= k \left[\frac{-4x(x^2 - y^2)}{(x^2 + y^2)^3} + \frac{2x}{(x^2 + y^2)^2} \right]$$

$$= 2kx \left[\frac{-2(x^2 - y^2)}{(x^2 + y^2)^3} + \frac{1}{(x^2 + y^2)^2} \right]$$

and
$$\frac{\partial v}{\partial y} = \frac{\partial}{\partial y} \frac{2kxy}{(x^2 + y^2)^2} = 2kx \frac{\partial}{\partial y} y(x^2 + y^2)^{-2}$$

$$= 2kx \left[y(-2)(x^2 + y^2)^{-3} \times 2y + (x^2 + y^2)^{-2} \times 1 \right]$$

$$= 2kx \left[\frac{-4y^2}{(x^2 + y^2)^3} + \frac{1}{(x^2 + y^2)^2} \right]$$

$$\therefore \quad \frac{\partial u}{\partial x} + \frac{\partial v}{\partial y} = 2kx \left[\frac{(-2x^2 + 2y^2)}{(x^2 + y^2)^3} + \frac{1}{(x^2 + y^2)^2} - \frac{4y^2}{(x^2 + y^2)^3} + \frac{1}{(x^2 + y^2)^2} \right]$$

$$= 2kx \left[\frac{(-2x^2 + 2y^2)}{(x^2 + y^2)^3} + \frac{2}{(x^2 + y^2)^2} - \frac{4y^2}{(x^2 + y^2)^3} \right]$$

$$= 2kx \left[\frac{-2x^2 + 2y^2 + 2(x^2 + y^2) - 4y^2}{(x^2 + y^2)^3} \right]$$

$$= 2kx \left[\frac{-2x^2 + 2y^2 + 2x^2 + 2y^2 - 4y^2}{(x^2 + y^2)^3} \right]$$

$$= 2kx \left[\frac{0}{(x^2 + y^2)^3} \right] = 0$$

Hence, the continuity equation is satisfied.

Problem 5.8: Given the velocity field:

$$V = (6 + 2xy + t^2)i - (xy^2 + 10t)j + 25k$$

What is the acceleration of a particle at (3, 0, 2) at time $t = 1.5$?

Solution: Velocity field: $V = (6 + 2xy + t^2)i - (xy^2 + 10t)j + 25k$

The velocity components u, v and w are:

$$u = 6 + 2xy + t^2$$
$$v = -(xy^2 + 10t)$$
$$w = 25$$

The acceleration components in x, y and z directions are given by:

$$a_x = \frac{du}{dt} = u\frac{\partial u}{\partial x} + v\frac{\partial u}{\partial y} + w\frac{\partial u}{\partial z} + \frac{\partial u}{\partial t}$$

$$a_y = \frac{dv}{dt} = u\frac{\partial v}{\partial x} + v\frac{\partial v}{\partial y} + w\frac{\partial v}{\partial z} + \frac{\partial v}{\partial t}$$

$$a_z = \frac{dw}{dt} = u\frac{\partial w}{\partial x} + v\frac{\partial w}{\partial y} + w\frac{\partial w}{\partial z} + \frac{\partial w}{\partial t}$$

From the velocity components,

$$\frac{\partial u}{\partial x} = 2y, \frac{\partial u}{\partial y} = 2x, \frac{\partial u}{\partial z} = 0, \frac{\partial u}{\partial t} = 2t$$

$$\frac{\partial v}{\partial x} = -y^2, \frac{\partial v}{\partial y} = -2xy, \frac{\partial v}{\partial z} = 0, \frac{\partial v}{\partial t} = -10$$

$$\frac{\partial w}{\partial x} = 0, \frac{\partial w}{\partial y} = 0, \frac{\partial w}{\partial z} = 0, \frac{\partial w}{\partial t} = 0$$

Substituting these values; the acceleration components at (3, 0, 2) at time $t = 1.5$

$$a_x = (6 + 2xy + t^2)2y - (xy^2 + 10t)2x + 25 \times 0 + 2t$$

$$= (6 + 2xy + t^2)2y - (xy^2 + 10t)2x + 2t$$

$$= 0 - 90 + 3 = -87 \text{ units}$$

$$a_y = (6 + 2xy + t^2)(-y^2) - (xy^2 + 10t)(-2xy) + 25 \times 0 - 10 = -10 \text{ units}$$

and $$a_z = (6 + 2xy + t^2) \times 0 - (xy^2 + 10t) \times 0 + 25 \times 0 + 0 = 0$$

$\therefore$ Acceleration: $a = a_x i + a_y j + a_z k = -87i - 10j + 0 = -87i - 10j$

Resultant acceleration: $a = \sqrt{(-87)^2 + (-10)^2}$ = **87.57 units**

Problem 5.9: What is the irrotational velocity field associated with the potential
$$\phi = 3x^2 - 3x + 3y^2 + 16t^2 + 12zt.$$
Does the flow field satisfy the incompressible continuity equation?

Solution: Given:

Velocity potential: $\phi = 3x^2 - 3x + 3y^2 + 16t^2 + 12zt$

The velocity components (u, v, w) in terms of ϕ is given by

$$\frac{\partial \phi}{\partial x} = -u$$

$$6x - 3 = -u$$

or $u = -6x + 3$

Differentiating w.r.t. x, we get

$$\boxed{\frac{\partial u}{\partial x} = -6}$$

Similarly $\frac{\partial \phi}{\partial y} = -v$

$$6y = -v$$

or $v = -6y$

Differentiating w.r.t y, we get

$$\boxed{\frac{\partial v}{\partial y} = -6}$$

and $\frac{\partial \phi}{\partial z} = -w$

$$12t = -w$$

or $w = -12t$

Differentiating w.r.t z, we get

$$\boxed{\frac{\partial w}{\partial z} = 0}$$

This incompressible continuity equation for a three-dimensional flow:

$$\frac{\partial u}{\partial x} + \frac{\partial v}{\partial y} + \frac{\partial w}{\partial z} = 0$$

$$-6 - 6 + 0 = 0$$

$$-12 \neq 0 \text{ which is not true.}$$

The flow field does not satisfy the equation of continuity and so it cannot exist.

5.12 STREAM FUNCTION (ψ)

It is defined as the scalar function of space and time, such that its partial derivative with respect to any direction gives the velocity component at right angles to that direction. It is denoted by ψ (*psi*). It is applicable to two-dimensional flow.

Mathematically, it is defined as

$$\left.\begin{array}{l}\dfrac{\partial \psi}{\partial x} = v \\[2ex] \dfrac{\partial \psi}{\partial y} = -u\end{array}\right\} ...(5.12.1)$$

The continuity equation for two-dimensional flow:

$$\frac{\partial u}{\partial x} + \frac{\partial v}{\partial y} = 0$$

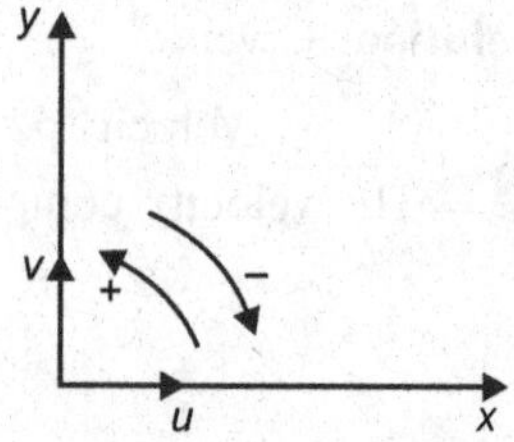

Fig. 5.20: Sign convention.

Substituting the values of $u = -\dfrac{\partial \psi}{\partial y}$ and $v = \dfrac{\partial \psi}{\partial x}$ in above equation, we get,

$$\frac{\partial}{\partial x}\left(-\frac{\partial \psi}{\partial y}\right) + \frac{\partial}{\partial y}\left(\frac{\partial \psi}{\partial x}\right) = 0$$

$$-\frac{\partial^2 \psi}{\partial x \partial y} + \frac{\partial^2 \psi}{\partial y \partial x} = 0$$

Hence existence of stream function (ψ) means a possible case of fluid flow. The flow may be rotational or irrotational

The rotational component ω_z is given by

$$\omega_z = \frac{1}{2}\left(\frac{\partial v}{\partial x} - \frac{\partial u}{\partial y}\right)$$

Substituting the value of $u = -\dfrac{\partial \psi}{\partial y}$ and $v = \dfrac{\partial \psi}{\partial x}$ in above equation, we get

$$\omega_z = \frac{1}{2}\left[\frac{\partial}{\partial x}\left(\frac{\partial \psi}{\partial x}\right) - \frac{\partial}{\partial y}\left(-\frac{\partial \psi}{\partial y}\right)\right]$$

$$= \frac{1}{2}\left[\frac{\partial^2 \psi}{\partial x^2} + \frac{\partial^2 \psi}{\partial y^2}\right]$$

For irrotational flow, $\omega_z = 0$. Hence above equation becomes as

$$\frac{\partial^2 \psi}{\partial x^2} + \frac{\partial^2 \psi}{\partial y^2} = 0$$

This equation is called **Laplace equation.**

The properties of stream function (ψ) are:

(*a*) If stream function (ψ) exists, it is a possible case of fluid flow which may be rotational or irrotational flow.

(*b*) If stream function (ψ) satisfies the Laplace equation $\left[\dfrac{\partial^2 \psi}{\partial x^2} + \dfrac{\partial^2 \psi}{\partial y^2} = 0\right]$, it is

a possible case of an irrotational flow.

As we know, no fluid flows across a stream line. Such that the flow between two stream lines must remain unchanged. For example, consider two streams 1 and 2 respectively. The mass flows between stream 1 and 2 across A_1A_2 must be the same as across B_1B_2 or A_1B_2. Density of fluid being constant, volumetric flow between them must also be constant. Consider two adjacent stream 1 and 2 as shown in Fig. 5.21.

Let ψ be the stream function of 1 and $(\psi + d\psi)$ of stream 2. The flow between origin O and stream 1 and 2 is due to increase in stream function ψ. Similarly, the flow between stream 1 and 2 is due to increase in stream function $d\psi$. So

$$\psi = f(x, y) \text{ for steady flow}$$

$$d\psi = \frac{\partial \psi}{\partial x}.dx + \frac{\partial \psi}{\partial y}dy \qquad \text{...(5.12.2)}$$

Flow across   1-2 = flow across 1-3 + flow across 3-2

$Vds = -vdx + udy$ per unit thickness of flow

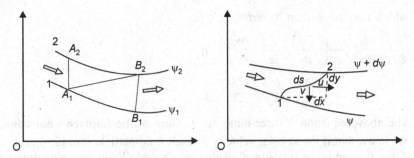

Fig. 5.21: Stream function

The minus sign indicates that the velocity v is acting in the downward direction.

$$Vds = \frac{\partial \psi}{\partial x}.dx + \frac{\partial \psi}{\partial y}.dy = d\psi$$

i.e., $dq = d\psi$

The stream function also defined as the flow rate per unit thickness of flow between two stream lines. The unit of ψ is m²/s; *i.e.*, discharge per unit thickness of flow.

From Eqs. (5.12.1) and (5.12.2), we get

$$d\psi = -vdx + udy$$

$-vdx$ is the flow through 1-3 per unit thickness of flow, $-ve$ sign show that velocity v is acting in the downward direction.

5.13 VELOCITY POTENTIAL (ϕ)

It is defined as a scalar quantity dependent upon space and time. Its negative derivative with respect to any direction gives the velocity in that direction. It is represented by ϕ (phi).

Mathematically, it is defined as

$$u = -\frac{\partial \phi}{\partial x}, \ v = -\frac{\partial \phi}{\partial y}, \ w = -\frac{\partial \phi}{\partial z}$$

where u, v and w are the components of velocity in x, y and z-directions.

In cylindrical coordinates (r, θ, z), the velocity components are:

$$V_r = -\frac{\partial \phi}{\partial x}, V_\theta = -\frac{1}{r}\frac{\partial \phi}{\partial \theta}, V_z = -\frac{\partial \phi}{\partial z}$$

The –ve sign means that the velocity potential decreases in the direction of flow. The velocity potential is not a physical quantity which could be directly measured and therefore, its zero position may be arbitrarily chosen.

The continuity equation for three-dimensional steady flow of incompressible fluids:

$$\frac{\partial u}{\partial x}+\frac{\partial v}{\partial y}+\frac{\partial w}{\partial z} = 0$$

which may be written in terms of ϕ as.

$$\frac{\partial}{\partial x}\left(-\frac{\partial \phi}{\partial x}\right)+\frac{\partial}{\partial y}\left(-\frac{\partial \phi}{\partial y}\right)+\frac{\partial}{\partial z}\left(-\frac{\partial \phi}{\partial z}\right) = 0$$

$$\frac{\partial^2 \phi}{\partial x^2}+\frac{\partial^2 \phi}{\partial y^2}+\frac{\partial^2 \phi}{\partial z^2} = 0$$

The above equation is three-dimensional form of the **Laplace's equation.**

We know that if the angular velocity of each fluid particle about its mass centre is zero, the flow must be irrotational. This sets up a condition for velocity potential to exit in a flow field. The existence of velocity potential in a flow field ensures that the flow must be irrotational. It is for this reason that an irrotational flow is often called as potential flow. Lines drawn in the flow field along which velocity potential (ϕ) are constant are called as equipotential lines.

If velocity potential (ϕ) exists and satisfies the continuity equation and must also satisfy the Laplace's equation, it represents the possible steady, incompressible irrotational flow.

The angular velocities of a fluid element about its mass centre are:

$$\omega_x = \frac{1}{2}\left(\frac{\partial w}{\partial y}-\frac{\partial v}{\partial z}\right)$$

$$\omega_y = \frac{1}{2}\left(\frac{\partial u}{\partial z}-\frac{\partial w}{\partial x}\right)$$

and

$$\omega_z = \frac{1}{2}\left(\frac{\partial v}{\partial x}-\frac{\partial u}{\partial y}\right)$$

Substituting the values of $u = -\frac{\partial \phi}{\partial x}$, $v = -\frac{\partial \phi}{\partial y}$ and $w = -\frac{\partial \phi}{\partial z}$ in the above angular velocities, we get

$$\omega_x = \frac{1}{2}\left[\frac{\partial}{\partial y}\left(-\frac{\partial \phi}{\partial z}\right) - \frac{\partial}{\partial z}\left(-\frac{\partial \phi}{\partial y}\right)\right] = \frac{1}{2}\left[-\frac{\partial^2 \phi}{\partial y \partial z} + \frac{\partial^2 \phi}{\partial z \partial y}\right]$$

$$\omega_y = \frac{1}{2}\left[\frac{\partial}{\partial z}\left(-\frac{\partial \phi}{\partial x}\right) - \frac{\partial}{\partial x}\left(-\frac{\partial \phi}{\partial z}\right)\right] = \frac{1}{2}\left[-\frac{\partial^2 \phi}{\partial z \partial x} + \frac{\partial^2 \phi}{\partial x \partial z}\right]$$

and
$$\omega_z = \frac{1}{2}\left[\frac{\partial}{\partial z}\left(-\frac{\partial \phi}{\partial y}\right) - \frac{\partial}{\partial y}\left(-\frac{\partial \phi}{\partial x}\right)\right] = \frac{1}{2}\left[-\frac{\partial^2 \phi}{\partial x \partial y} + \frac{\partial^2 \phi}{\partial y \partial x}\right]$$

If ϕ is a continuous function, then

$$\frac{\partial^2 \phi}{\partial y \partial z} = \frac{\partial^2 \phi}{\partial z \partial y}; \frac{\partial^2 \phi}{\partial z \partial x} = \frac{\partial^2 \phi}{\partial x \partial y}; \text{ etc.}$$

$\therefore$
$$\omega_x = \omega_y = \omega_z = 0$$

The resultant angular velocity:

$$\omega = \sqrt{\omega_x^2 + \omega_y^2 + \omega_z^2} = 0$$

Problem 5.10: A stream function is given by $\psi = 5x - 6y$. Find the velocity components and also magnitude and direction of the resultant velocity at any point.

Solution: Given data:

Stream function: $\qquad \psi = 5x - 6y$

$\therefore \qquad\qquad \dfrac{\partial \psi}{\partial x} = 5 \qquad$ and $\qquad \dfrac{\partial \psi}{\partial y} = -6$

we know

$$\frac{\partial \psi}{\partial x} = v \qquad \text{and} \qquad \frac{\partial \psi}{\partial y} = -u$$

or $\qquad\qquad v = \dfrac{\partial \psi}{\partial x} \qquad$ and $\qquad u = -\dfrac{\partial \psi}{\partial x} = -(-6)$

$\qquad\qquad\qquad = \textbf{5 units/s} \qquad\qquad\qquad = \textbf{6 units/s.}$

Resulting velocity: $\qquad V = \sqrt{u^2 + v^2} = \sqrt{(6)^2 + (5)^2}$

$$= \sqrt{36 + 25} = \sqrt{61} = \textbf{7.81 units/s}$$

Direction of the resultant velocity is given by:

$$\tan \theta = \frac{v}{u} = \frac{5}{6}$$

$$\tan \theta = 0.8333$$

or $\qquad\qquad \theta = \tan^{-1}(0.8333) = \textbf{39.80°}$

Problem 5.11: In a two-dimensional flow $\psi = 3xy$. Prove that the flow is irrotational. Also determine the corresponding velocity potential.

Solution: Given data:

Stream function: $\psi = 3xy$

The Laplace equation

$$\frac{\partial^2 \psi}{\partial x^2} + \frac{\partial^2 \psi}{\partial y^2} = 0$$

Stream function: $\psi = 3xy$

$$\frac{\partial \psi}{\partial x} = 3y \qquad\qquad \frac{\partial \psi}{\partial y} = 3x$$

$$\frac{\partial^2 \psi}{\partial x^2} = 0 \qquad\qquad \frac{\partial^2 \psi}{\partial y^2} = 0$$

$\therefore$ The Laplace equation

$$0 + 0 = 0$$
$$0 = 0$$

The stream function satisfies the Laplace equation. Hence, the flow is irrotational. Now the velocity components u and v in terms of ψ are:

$$u = -\frac{\partial \psi}{\partial y} = -3x$$

$$v = \frac{\partial \psi}{\partial x} = 3y$$

and velocity potential: ϕ

we know $\dfrac{\partial \phi}{\partial x} = -u = -(-3x) = 3x$

and $\dfrac{\partial \phi}{\partial y} = -v = -(3y) = -3y$

$$\frac{\partial \phi}{\partial y} = -3y$$

$$d\phi = -3ydy$$

On integrating, we get

$$\int d\phi = \int -3ydy$$

$$\phi = -3\frac{y^2}{2} + c \qquad\qquad\qquad ...(i)$$

where c is a constant which is independent of y but can be function of x. Differentiating above Eq. (i) w.r. t. x, we get

$$\frac{\partial \phi}{\partial x} = 0 + \frac{\partial c}{\partial x}$$

$$3x = \frac{\partial c}{\partial x}$$

or

$$dc = 3x \, dx$$

On integrating, we get

$$\int dc = \int 3x \, dx$$

$$c = \frac{3x^2}{2}$$

Substituting the value of c in Eq. (*i*), we get.

Potential function: $\phi = -\frac{3}{2}y^2 + \frac{3}{2}x^2$.

Problem 5.12: The stream function for a two-dimensional flow in given by $\psi = 2xy$. Calculate the velocity at the point $P(2, 3)$. Find the value of velocity potential.

(GGSIP University, Delhi, Dec. 2002)

Solution: Given data:

Stream function: $\psi = 2xy$

The velocity components in the direction of x and y are:

$$u = -\frac{\partial \psi}{\partial y} = -\frac{\partial}{\partial y}(2xy) = -2x$$

and

$$v = \frac{\partial \psi}{\partial x} = -\frac{\partial}{\partial x}(2xy) = 2y$$

$\therefore$ The velocity at point $P(2, 3)$:

$$v = ui + vj = -2xi + 2yj$$
$$= -2 \times 2i + 2 \times 3k = -4i + 6j$$

The resultant velocity at point P:

$$V = \sqrt{(-4)^2 + (6)^2} = \sqrt{16 + 36} = \sqrt{52} = 7.21 \text{ units/s}$$

The velocity potential: ϕ

we know

$$\frac{\partial \phi}{\partial x} = -u = -(-2x) = 2x$$

and

$$\frac{\partial \phi}{\partial y} = -v = -2y$$

$$\frac{\partial \phi}{\partial y} = -2y$$

$$d\phi = -2y \, dy \qquad \qquad \dots(i)$$

On integrating Eq. (*i*), we get

$$-\int d\phi = \int -2y \, dy$$

$$\phi = -2\frac{y^2}{2} + c$$

$$\phi = -y^2 + c \qquad \qquad \dots(ii)$$

where c = constant of integration which is independent of y but can be function of x only.

Differentiating Eq. (ii) w.r.t. x, we get

$$\frac{\partial \phi}{\partial x} = \frac{\partial c}{\partial x}$$

or

$$\frac{\partial c}{\partial x} = \frac{\partial \phi}{\partial x}$$

$$\frac{\partial c}{\partial x} = 2x \qquad\qquad \because \frac{\partial \phi}{\partial x} = 2x$$

or $dc = 2x\, dx$

On integrating above equation, we get

$$\int dc = \int 2x\,dx$$

$$c = \frac{2x^2}{2}$$

$$c = x^2$$

Substituting the value of $c = x^2$ in Eq. (ii), we get

$$\phi = -y^2 + x^2$$

Velocity potential: $\phi = x^2 - y^2$

$\therefore$ Velocity potential (ϕ) at point P (2, 3)

$$= (2)^2 - (3)^2 = 4 - 9 = -\mathbf{5\ units}$$

Problem 5.13: Calculate the velocity components if the velocity potential function follows the law:

$$\phi = \log_e \frac{x}{y}$$

Solution: Given data:

Velocity potential function: $\phi = \log_e \dfrac{x}{y}$

We know that the velocity components of the velocity potential function are:

$$\frac{\partial \phi}{\partial x} = -u$$

and

$$\frac{\partial \phi}{\partial y} = -v$$

Now

$$\frac{\partial \phi}{\partial x} = -u$$

$$\frac{\partial}{\partial x}\left(\log_e \frac{x}{y}\right) = -u$$

$$\frac{1}{x/y} \cdot \frac{1}{y} = -u$$

or

$$u = -\frac{1}{x}$$

$\therefore$ The velocity along x-direction: $u = -\frac{1}{x}$.

and

$$\frac{\partial \phi}{\partial y} = -v$$

$$\frac{\partial}{\partial y}\left(\log_e \frac{x}{y}\right) = -v$$

$$\frac{1}{x/y} \cdot \frac{d}{dy}(xy^{-1}) = -v$$

$$\frac{1}{x/y}(-1)x.y^{-2} = -v$$

or

$$v = \frac{1}{y}$$

The velocity components of velocity potential function $\phi = \log_e \dfrac{x}{y}$ are $\left(-\dfrac{1}{x}, \dfrac{1}{y}\right)$.

Problem 5.14: The velocity potential for a two-dimensional potential flow is given by $\phi = x(2y - 1)$. Determine:

(*i*) The velocity at the point $P(4, 5)$ and

(*ii*) The value of stream function ψ at the point P.

(GGSIP University, Delhi, Dec. 2004)

Solution: Given data:

Velocity potential: $\qquad \phi = x(2y - 1)$

(*i*) The velocity components in the direction of x and y are:

$$u = -\frac{\partial \phi}{\partial x} = -\frac{\partial}{\partial x}[x(2y-1)] = -(2y-1)$$

and

$$v = -\frac{\partial \phi}{\partial y} = -\frac{\partial}{\partial y}[x(2y-1)] = -\frac{\partial}{\partial y}[2xy - x] = -2x$$

At the point $P(4, 5)$, *i.e.*, at $x = 4$, $y = 5$

$\therefore \qquad u = -(2y - 1) = -(2{\times}5{-}1) = -9$ units/s

and $\qquad v = -2x = -2 \times 4 = -8$ units/s

$\therefore$ The velocity at the point P:

$$V = ui + vj = -9i - 8j$$

or The resultant velocity at point P:

$$V = \sqrt{u^2 + v^2} = \sqrt{(-9)^2 + (-8)^2} = 12.04 \ \textbf{units/s}$$

(ii) The value of stream function ψ at the point *P*.

We know that

$$\frac{\partial \psi}{\partial y} = -u = -[-(2y - 1)] = 2y - 1$$

and

$$\frac{\partial \psi}{\partial x} = v$$

$$\frac{\partial \psi}{\partial x} = -2x \qquad \qquad \qquad \dots(i)$$

On integration above Eq. (*i*) w.r.t. *x*, we get

$$\int d\psi = \int -2x dx$$

$$\psi = \frac{2x^2}{2} + c$$

$$\psi = -x^2 + c \qquad \qquad \qquad \dots(ii)$$

where *c* = constant of integration, it is independent on *x* but it can be a function of *y*.

Differentiating the above equation w.r.t. *y*, we get

$$\frac{\partial \psi}{\partial y} = \frac{\partial c}{\partial y}$$

$$2y - 1 = \frac{\partial c}{\partial y} \qquad \qquad \because \frac{\partial \psi}{\partial y} = 2y - 1$$

or $\qquad \qquad \qquad dc = (2y - 1) \ dy$

On integrating above equation, we get

$$\int dc = \int (2y - 1) dy$$

$$c = \frac{2y^2}{2} - y$$

$$c = y^2 - y$$

Substituting the value of $c = y^2 - y$ in Eq. (*ii*), we get

$$\psi = -x^2 + y^2 - y$$

or $\qquad \qquad \qquad \psi = y^2 - x^2 - y$

∴ Stream function ψ at point *P*(4, 5)

$$= 5^2 - 4^2 - 5 = 25 - 16 - 5 = \textbf{4 units}$$

Problem 5.15: In a 2-dimensional incompressible flow, the fluid velocity components are given by $u = x - 4y$ and $v = -y - 4x$. Show that velocity potential exists and find its form as well as stream function.

Solution: Given

Velocity components:

$$u = x - 4y,$$
$$v = -y - 4x$$

$$\frac{du}{dx} = 1, \quad \frac{dv}{dy} = -1$$

We know that continuity equation

$$\frac{\partial u}{\partial x} + \frac{\partial v}{\partial y} = 0$$

$$1 - 1 = 0$$

$$0 = 0$$

The continuity equation is satisfied and the flow is possible and irrotational. Hence, the velocity potential (ϕ) exists.

$$u = -\frac{\partial \phi}{\partial x} \text{ and } v = \frac{\partial \phi}{\partial y}$$

$$x - 4y = -\frac{d\phi}{dx}$$

or
$$d\phi = (4y - x)dx$$

On integrating above equation, we get

$$\phi = 4yx - \frac{x^2}{2} + c \qquad \qquad \dots (i)$$

where c = constant of integrating, independent on x and dependent on y. Differentiating Eq. (i) w.r.t. y, we get

$$\frac{d\phi}{dy} = 4x + \frac{\partial c}{\partial y} \qquad \qquad \because \frac{d\phi}{\partial y} = -v$$

$$-v = 4x + \frac{dc}{dy}$$

$$-(-y - 4x) = 4x + \frac{dc}{dy}$$

$$y + 4x = 4x + \frac{dc}{dy}$$

or
$$\frac{dc}{dy} = y$$

$$dc = y \, dy$$

On integrating, we get

$$c = \frac{y^2}{2}$$

Substituting the value of $c = \frac{y^2}{2}$ in Eq. (i), we get

$$\phi = 4xy - \frac{x^2}{2} + \frac{y^2}{2}$$

$$\phi = \frac{1}{2}(y^2 - x^2) + 4xy$$

We know that the stream function in velocity components.

$$v = \frac{\partial \psi}{\partial x}, \; -u = \frac{\partial \psi}{\partial y}$$

$$-y - 4x = \frac{\partial \psi}{\partial x}$$

or $\partial \psi = (-y - 4x) \, dx$

On integrating, we get

$$\psi = -xy - \frac{4x^2}{2}$$

$$\psi = -xy - 2x^2 + c \qquad\qquad\qquad ...(ii)$$

where c = constant of integrating, independent on x and dependent only on y. Differentiating Eq. (ii) w.r.t. y, we get

$$\frac{d\psi}{dy} = -x + \frac{\partial c}{\partial y}$$

$$-u = -x + \frac{\partial c}{\partial y}$$

$$-(x - 4y) = -x + \frac{\partial c}{\partial y}$$

$$-x + 4y = -x + \frac{\partial c}{\partial y}$$

or $\dfrac{dc}{dy} = 4y$

$$dc = 4y \, dy$$

On integrating above equation, we get

$$c = \frac{4y^2}{2} = 2y^2$$

Substituting the value of $c = 2y^2$ in Eq. (ii), we get

$$\psi = -xy - 2x^2 + 2y^2$$

$$\psi = 2(y^2 - x^2) - xy.$$

Problem 5.16: If potential function is $3xy$, find x and y components of velocity at $(1, 3)$ and $(3, 3)$. Determine the discharge passing between stream lines through these points.

Solution: Given:

Potential function: $\phi = 3xy$

We know that the velocity components in terms of ϕ are given by:

$$u = -\frac{\partial \phi}{\partial x} \quad \text{and} \quad v = -\frac{\partial \phi}{\partial y}$$

$$u = -\frac{\partial}{\partial x}(3xy) \qquad v = -\frac{\partial}{\partial y}(3xy)$$

$$u = -3y, \qquad\qquad v = -3x$$

The velocity components at (1, 3) are:

$$u = -3 \times 3 = -9 \text{ units/s}$$

and $v = -3 \times 1 = -3 \text{ units/s}$

The velocity components at (3, 3) are:

$$u = -3 \times 3 = -9 \text{ units/s}$$

and $v = -3 \times 3 = -9 \text{ units/s}$

The velocity components in terms of ψ are given by:

$$\frac{\partial \psi}{\partial x} = v, \quad \frac{\partial \psi}{\partial y} = -u$$

$$\frac{\partial \psi}{\partial x} = -3x, \quad \frac{\partial \psi}{\partial y} = -(-3y) = 3y$$

$$d\psi = -3x dx$$

On integrating above equation, we get

$$\psi = -\frac{3x^2}{2} + c \qquad\qquad\qquad ...(i)$$

where c = constant of integrating, it is independent on x but it can be a function of y.
Differentiating Eq. (i) w.r. t. y, we get

$$\frac{\partial \psi}{\partial y} = \frac{\partial c}{\partial y}$$

$$3y = \frac{\partial c}{\partial y}$$

or $\dfrac{\partial c}{\partial y} = 3y$

$$dc = 3y \, dy$$

On integrating above equation, we get

$$c = \frac{3y^2}{2}$$

Substituting the value of c in Eq. (i), we get

Stream function: $\psi = -\dfrac{3x^2}{2} + \dfrac{3y^2}{2}$

Discharge between the streamlines passing through (1, 3) and (3, 3)

$$= \Psi_{(1,3)} - \Psi_{(3,3)}$$

$$= -\frac{3}{2} \times (1)^2 + \frac{3 \times (3)^2}{2} - \frac{3 \times (3)^2}{2} + 3 \times \frac{(3)^2}{2}$$

$$= \frac{-3}{2} + \frac{27}{2} = \frac{-3 + 27}{2} = \frac{24}{2} = 12 \text{ units}$$

Problem 5.17: The velocity components in a two-dimensional flow field for an incompressible fluid are as follows:

$$u = \frac{y^3}{3} + 2x - x^2 y$$

and $$v = xy^2 - 2y - \frac{x^3}{3}.$$

Obtain an expression for the stream function Ψ.

Solution: Given relation: $u = \dfrac{y^3}{3} + 2x - x^2 y$,

and $$v = xy^2 - 2y - \frac{x^3}{3}$$

The velocity components in terms of stream function (Ψ) are:

$$\frac{\partial \Psi}{\partial y} = -u \quad \bigg| \quad \frac{\partial \Psi}{\partial x} = v$$

$$\frac{\partial \Psi}{\partial y} = -\frac{y^3}{3} - 2x + x^2 y \qquad \frac{\partial \Psi}{\partial x} = xy^2 - 2y - \frac{x^3}{3} \quad ...(i)$$

$$d\Psi = \left(-\frac{y^3}{3} - 2x + x^2 y \right) dy$$

On integration, we get $$\Psi = \frac{-y^4}{3 \times 4} - 2xy + \frac{x^2 y^2}{2} + C$$

$$\Psi = \frac{-y^4}{12} - 2xy + \frac{x^2 y^2}{2} + C \qquad ...(ii)$$

where C is a constant which is independent of y but can be function of x. Differentiating Eq. (ii) w.r.t. x, we get

$$\frac{\partial \Psi}{\partial x} = 0 - 2y + \frac{2xy^2}{2} + \frac{dC}{dx}$$

$$\frac{\partial \Psi}{\partial x} = -2y + xy^2 + \frac{dC}{dx} \qquad ...(iii)$$

Equating Eq. (*i*) and Eq. (*iii*), we have

$$xy^2 - 2y - \frac{x^3}{3} = -2y + xy^2 + \frac{dC}{dx}$$

or

$$\frac{dC}{dx} = \frac{-x^3}{3}$$

$$dC = \frac{-x^3}{3} dx$$

On integrating, we get $\quad C = -\dfrac{x^4}{3 \times 4}$

$$C = -\frac{x^4}{12}$$

Substituting the value of C in Eq. (*ii*), we get

$$\Psi = -\frac{y^4}{12} - 2xy + \frac{x^2 y^2}{2} - \frac{x^4}{12}$$

$$\boxed{\Psi = -\frac{x^4}{12} - \frac{y^4}{12} + \frac{x^2 y^2}{2} - 2xy}$$

Problem 5.18: The velocity components in a two-dimensional flow field for an incompressible fluid are $u = \dfrac{y^3}{3} + 2x - x^2 y$, $v = xy^2 - 2y - \dfrac{x^3}{3}$. Show that these functions represent a possible case of an irrotational flow. Also get an expression for the stream function.

Solution: Given velocity components:

$$u = \frac{y^3}{3} + 2x - x^2 y$$

$$v = xy^2 - 2y - \frac{x^3}{3}$$

$$\frac{du}{dx} = 0 + 2 - 2xy \qquad\qquad \frac{dv}{dy} = 2xy - 2 - 0$$

$$= 2 - 2xy \qquad\qquad\qquad = 2xy - 2$$

We know that the continuity equation

$$\frac{du}{dx} + \frac{du}{dy} = 0$$

$$2 - 2xy + 2xy - 2 = 0$$

$$0 = 0$$

The continuity is satisfied and the flow is possible and irrotational.

We know that the stream function in velocity components:

$$v = \frac{\partial \psi}{\partial x}, \quad -u = \frac{\partial \psi}{\partial y}$$

$$xy^2 - 2y - \frac{x^3}{3} = \frac{\partial \psi}{\partial x}$$

or

$$d\psi = \left(xy^2 - 2y - \frac{x^3}{3} \right) dx$$

On integrating, we get $\quad \psi = \dfrac{x^2 y^2}{2} - 2xy - \dfrac{x^4}{3 \times 4}$

$$\psi = \frac{x^2 y^2}{2} - 2xy - \frac{x^4}{12} + C \qquad \qquad ...(i)$$

where $\qquad \qquad C =$ constant of integrating, independent on x and dependent only on y.

Differentiating Eq. (*i*) *w. r. t. y*, we get

$$\frac{d\psi}{dy} = \frac{2x^2 y}{2} - 2x + \frac{dC}{dy}$$

$$-u = x^2 y - 2x + \frac{dC}{dy}$$

$$\frac{-y^3}{3} - 2x + x^2 y = x^2 y - 2x + \frac{dC}{dy}$$

or

$$\frac{dC}{dy} = \frac{-y^3}{3}$$

or

$$dC = \frac{-y^3}{3} dy$$

On integrating, we get $\quad C = \dfrac{-y^4}{3 \times 4} = \dfrac{-y^4}{12}$

Substituting the value of C in Eq. (*i*), we get

$$\psi = \frac{x^2 y^2}{2} - 2xy - \frac{x^4}{12} - \frac{y^4}{12}$$

$$\psi = -\frac{x^4}{12} - \frac{y^4}{12} + \frac{x^2 y^2}{12} - 2xy$$

5.14 TYPES OF MOTION OR DEFORMATION OF FLUID ELEMENTS

Fluid particles can undergo the following four types of displacements:

- (i) Pure (or linear) translation,
- (ii) Linear deformation,
- (iii) Angular (or shear) deformation, and
- (iv) Pure rotation.

5.14.1 Pure (or linear) Translation

Pure translation takes place when the fluid element moves only in linear direction without rotating and deformation. It means angular velocity of the fluid element is zero and length of fluid element is unchanged *i.e., ab = a'b'* as shown in Fig. 5.22. This type of motion normally takes place when liquid flows in a pipe of uniform cross-sectional area.

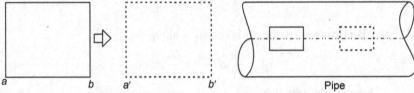

Fig. 5.22: Pure translation; lengths ab, *a'b'* are equal and parallel to each other.

5.14.2 Linear Deformation

Linear deformation takes place when the fluid element moves only in linear direction with deformation and without rotating. In this type of motion, length of fluid element increase and width of fluid element decreases but both length and width remain parallel to each other as shown in Fig. 5.23. The linear deformation normally takes place when fluid element moves on centre line in a nozzle.

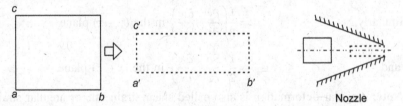

Fig. 5.23: Linear deformation (or Elongation); length and width of fluid element parallel to each other but changed (length increases and width decreases)

5.14.3 Angular (or shear) Deformation

The angular deformation at a point (O) in fluid element is defined as the average of rates of angular changes for two mutually perpendicular linear fluid elements. This type of motion normally takes place in free vortex such as liquid flowing down in sink.

Mathematically,

Angular deformation: $\epsilon_{xy} = \dfrac{1}{2}[\alpha + \beta]/dt$

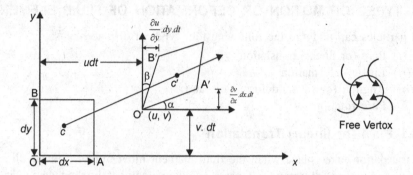

Fig. 5.24: Fluid element translated, angular distortion, no rotation.

where α is caused by the variation of v along the x-axis

$$\therefore \qquad \alpha = \frac{\partial v}{\partial x}.dx.\frac{dt}{dx} = \frac{\partial v}{\partial x}.dt$$

and β is caused by the variation of u along the y-axis,

$$\therefore \qquad \beta = \frac{\partial u}{\partial y}.dy.\frac{dt}{dy} = \frac{\partial u}{\partial y}.dt \text{ m}$$

$\therefore$ Angular deformation:

$$\in_{xy} = \frac{1}{2}\left(\frac{\partial v}{\partial x}.dt.+\frac{\partial u}{\partial y}.dt\right)/dt$$

$$= \frac{1}{2}\left(\frac{\partial v}{\partial x}+\frac{\partial u}{\partial y}\right) \text{ in the } (x-y) \text{ plane}$$

$$= \frac{1}{2}\left(\frac{\partial u}{\partial y}+\frac{\partial v}{\partial x}\right)$$

Similarly, $\qquad \in_{yz} = \frac{1}{2}\left(\frac{\partial v}{\partial z}+\frac{\partial w}{\partial y}\right) \text{ in the } (y-z) \text{ plane}$

and $\qquad \in_{xz} = \frac{1}{2}\left(\frac{\partial u}{\partial z}+\frac{\partial w}{\partial x}\right) \text{ in the } (x-z) \text{ plane.}$

Note: Angular deformation is also called shear strain rate or angular strain.

5.14.4 Pure Rotation

It takes place when fluid element rotates about its axis which is perpendicular to the plane of motion.

Let two-dimensional flow of a fluid element in x-y plane. OA and OB two mutually perpendicular axes of fluid element along x and y axes respectively.

Let u = velocity at point O along x-axis

$$u+\frac{\partial u}{\partial y}.dy = \text{velocity at point B along } x\text{-axis}$$

$$v = \text{velocity at point O along } y\text{-axis}$$

$$v + \frac{\partial v}{\partial x}.dx = \text{velocity at point A along } y\text{-axis}$$

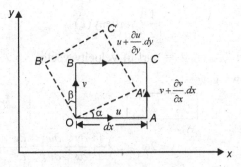

Fig. 5.25: Pure rotation

These different velocities will cause angular velocities of linear elements OB and OA. During time interval dt the elements OB and OA would move relative to point O, the fluid element gets displaced and occupies the dotted position $OA'C'B'$.

Since velocity at O and at A in y-axis are different, (*i.e.*, velocity at point A is more than velocity at point O) fluid element OA will rotate in anticlockwise (+ve) direction, the angular velocity of element OA about the z-axis is

$$\omega_{OA} = \text{velocities at points } O \text{ and } A \text{ per unit length}$$

$$= \frac{\left(v + \frac{\partial v}{\partial x}.dx - v \right)}{dx} = \frac{\partial v}{\partial x}$$

and the angular velocity of element OB about z-axis is

$$\omega_{OB} = \text{velocities at points } O \text{ and } B \text{ per unit length}$$

$$= \frac{-\left(u + \frac{\partial u}{\partial y}.dy \right) + u}{dy} = -\frac{\partial u}{\partial y}$$

Average of the angular velocities of elements OA and OB gives the rotation about z-axis.

$$\therefore \qquad \omega_z = \frac{1}{2}\left(\frac{\partial v}{\partial x} - \frac{\partial u}{\partial y} \right) \qquad \text{for } (x - y) \text{ plane}$$

Similarly,
$$\omega_x = \frac{1}{2}\left(\frac{\partial w}{\partial y} - \frac{\partial v}{\partial z} \right) \qquad \text{for } (y - z) \text{ plane}$$

and
$$\omega_y = \frac{1}{2}\left(\frac{\partial u}{\partial z} - \frac{\partial w}{\partial x} \right) \qquad \text{for } (z - x) \text{ plane}$$

The angular velocity is written in vector form:

$$\vec{\omega} = \frac{1}{2}[\omega_x i + \omega_y j + \omega_z k]$$

$$= \frac{1}{2}\left(\vec{\nabla} \times \vec{V}\right) = \frac{1}{2} \, \text{curl} \, \vec{V}$$

$$= \frac{1}{2} \times \text{vorticity}(\vec{\xi}) \qquad \qquad \qquad \text{...(1)}$$

where vorticity: $\zeta = \vec{\nabla} \times \vec{V}$ = curl of velocity vector

From Eq. (1), we get

Vorticity $\boxed{\vec{\xi} = 2\vec{\omega}}$ $\qquad$ | $\qquad$ ζ = Greek letter zeta

$$\zeta_z = 2\omega_z = 2 \times \frac{1}{2}\left(\frac{\partial v}{\partial x} - \frac{\partial u}{\partial y}\right) = \frac{\partial v}{\partial x} - \frac{\partial u}{\partial y}$$

and

$$\zeta_x = 2\omega_x = 2 \times \frac{1}{2}\left(\frac{\partial w}{\partial y} - \frac{\partial v}{\partial z}\right) = \frac{\partial w}{\partial y} - \frac{\partial v}{\partial z}$$

$$\zeta_y = 2\omega_y = 2 \times \frac{1}{2}\left(\frac{\partial u}{\partial z} - \frac{\partial w}{\partial x}\right) = \frac{\partial u}{\partial z} - \frac{\partial w}{\partial x}$$

5.15 VORTICITY

It is defined as twice the angular velocity of a fluid particle. Thus vorticity is used to measure of rotation of a fluid particle. It is denoted by ζ (Greek letter: zeta).

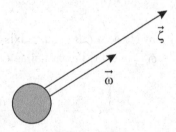

Fig. 5.26: The vorticity vector is equal to twice the angular velocity vector of a rotating fluid particle.

Mathematically,

Vorticity: $\qquad \qquad \zeta = 2\omega$

If the vorticity at a point in a flow is non-zero, the fluid particle that happens to occupy that point in space is rotating, the flow in the region is called rotational.

If the vorticity in a region of the flow is zero, fluid particles there are not rotating; the flow in that region is called irrotational.

For example: Fluid particles within the viscous boundary layer near a solid wall are rotational (*i.e.*, non-zero vorticity), while fluid particles outside the boundary layer are irrotational (*i.e.*, vorticity is zero)

The rotation of fluid elements is associated with wakes, boundary layer, flow through turbo-machinery (fans, blower, turbines, pumps, compressors *etc*), and flow with heat transfer.

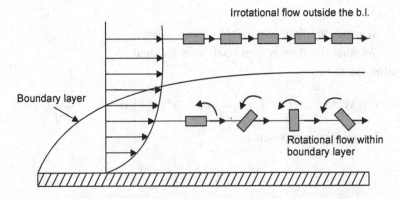

Fig. 5.27: The difference between rotational and irrotational flow.

Problem 5.19: The velocity vector of fluid flow is given by $\vec{V} = 8x^3 i - 10x^2 yj$. Find the shear strain rate and whether the flow is rotational or irrotational.

Solution: Given

Velocity vector: $\qquad \vec{V} = 8x^3 i - 10x^2 yj$

$\therefore \qquad\qquad u = 8x^3, \quad v = -10x^2 y$

The shear strain rate

$$= \frac{1}{2}\left(\frac{\partial v}{\partial x} + \frac{\partial u}{\partial y}\right)$$

$$= \frac{1}{2}\left(\frac{\partial(-10x^2 y)}{\partial x} + \frac{\partial(8x^3)}{\partial y}\right)$$

$$= \frac{1}{2}[-10 \times 2xy + 0] = -10\ xy$$

The rotation in x-y plane:

$$\omega_z = \frac{1}{2}\left(\frac{\partial v}{\partial x} - \frac{\partial u}{\partial y}\right)$$

$$\omega_z = \frac{1}{2}\left(\frac{\partial(-10x^2 y)}{\partial x} - \frac{\partial(8x^3)}{\partial y}\right)$$

$$\omega_z = \frac{1}{2}[-10 \times 2xy + 0]$$

$$\omega_z = -10xy$$

$$\omega_z \neq 0$$

Hence, the fluid flow is rotational.

Problem 5.20: The velocity vector for two-dimensional fluid flow is given by:

$$\vec{V} = \left(\frac{y^3}{3} + 2x - x^2y\right)i + \left(xy^2 - 2y - \frac{x^3}{3}\right)j$$

Find:

(a) Whether the flow is possible or impossible.

(b) Whether the flow is rotational or irrotational.

Solution: Given

Velocity vector: $\vec{V} = \left(\dfrac{y^3}{3} + 2x - x^2y\right)i + \left(xy^2 - 2y - \dfrac{x^3}{3}\right)j$

∴ The velocity components:

$$u = \frac{y^3}{3} + 2x - x^2y \text{ and}$$

$$v = xy^2 - 2y - \frac{x^3}{3}$$

(a) Whether the flow is possible or impossible:

The continuity equation for 2-D flow;

$$\frac{\partial u}{\partial x} + \frac{\partial v}{\partial y} = 0$$

$$\frac{\partial}{\partial x}\left(\frac{y^3}{3} + 2x - x^2y\right) + \frac{\partial}{\partial y}\left(xy^2 - 2y - \frac{x^3}{3}\right) = 0$$

$$0 + 2 - 2xy + 2xy - 2 + 0 = 0$$

$$0 = 0, \text{ to satisfy the continuity equation.}$$

Hence, the flow is possible.

(b) Whether the flow is rotational or irrotational:

We know that the rotation in x-y plane, about z-axis.

$$\omega_z = \frac{1}{2}\left(\frac{\partial v}{\partial x} - \frac{\partial u}{\partial y}\right)$$

$$= \frac{1}{2}\left[\frac{\partial}{\partial x}\left(xy^2 - 2y - \frac{x^3}{3}\right) - \frac{\partial}{\partial y}\left(\frac{y^3}{3} + 2x - x^2y\right)\right]$$

$$= \frac{1}{2}\left[\left(y^2 - 0 - \frac{3x^2}{3}\right) - \left(\frac{3y^2}{3} + 0 - x^2\right)\right]$$

$$= \frac{1}{2}[y^2 - x^2 - y^2 + x^2] = 0$$

$$\omega_z = 0$$

The rotation is zero, which means the flows is irrotational.

Problem 5.21: (*a*) If the expression for the stream function is described by $\psi = x^3 - 3xy^2$, determine whether the flow is (*i*) rotational, (*ii*) irrotational.

(*b*) If the flow is irrotational, then indicate the correct value of the velocity potential: (*i*) $\phi = y^3 - 3x^2y$, (*ii*) $\phi = -3x^2y$.

Solution: Given

(*a*) Stream function: $\qquad\qquad \psi = x^3 - 3xy^2$

We know that the vorticity vector:

$$\zeta_z = \frac{\partial v}{\partial x} - \frac{\partial u}{\partial y} \qquad\qquad \dots(1)$$

The velocity components in terms of ψ are given by

$$v = \frac{\partial \psi}{\partial x}, \qquad\qquad u = -\frac{\partial \psi}{\partial y}$$

$$v = \frac{\partial}{\partial x}(x^3 - 3xy^2), \qquad u = -\frac{\partial}{\partial y}(x^3 - 3xy^2)$$

$$v = 3x^2 - 3y^2, \qquad\qquad u = 6xy$$

and $\qquad\qquad \dfrac{\partial v}{\partial x} = 6x, \qquad\qquad \dfrac{\partial u}{\partial y} = 6x$

Substituting the above values in Eq. (1), we get

Vorticity vector: $\qquad\qquad \zeta_z = 6x - 6x$

$$\zeta_z = 0$$

Hence, the flow is irrotational.

(*b*) For an irrotational flow, the Laplace equation must be satisfied, *i.e.*,

$$\frac{\partial^2 \phi}{\partial x^2} + \frac{\partial^2 \phi}{\partial y^2} = 0 \qquad\qquad \dots(2)$$

Let us check the validity for each expression for ϕ:

(*i*) $\qquad\qquad\qquad \phi = y^3 - 3x^2y \qquad\qquad \dots(3)$

Differentiate w.r. t. x, we get

$$\frac{\partial \phi}{\partial x} = -6xy$$

Again $\qquad\qquad \boxed{\dfrac{\partial^2 \phi}{\partial x^2} = -6y}$

Differentiate Eq. (3) w.r.t. y, we get

$$\frac{\partial \phi}{\partial y} = 3y^2 - 3x^2$$

Again $\qquad\qquad \boxed{\dfrac{\partial^2 \phi}{\partial y^2} = 6y}$

$\therefore$ From Eq. (2), we get

$$-6y + 6y = 0$$
$$0 = 0$$

(*ii*) $\phi = -3x^2y$ ...(4)

Differentiate w.r.t. x, we get

$$\frac{\partial\phi}{\partial x} = -6xy$$

Again $\boxed{\dfrac{\partial^2\phi}{\partial x^2} = -6y}$

Differentiate Eq. (4) w.r.t. y, we get

$$\frac{\partial\phi}{\partial y} = -3x^2$$

Again $\dfrac{\partial^2\phi}{\partial y^2} = 0$

$\therefore$ From Eq. (2), we get

$$-6y + 0 \neq 0$$

Hence, the correct value of the velocity potential:

$$\phi = y^2 - 3x^2y$$

5.16   VORTEX FLOW OR WHIRLING FLOW

If the fluid continuously flows along a curved path about a fixed axis, the flow is called **vortex** or **whirling flow.** In other words, a fluid flow in which the stream lines are concentric circles is known as the vortex flow. The vortex flow is classified into two categories as:

    (*a*) Forced vortex flow, and   (*b*) Free vortex flow.

5.16.1 Forced Vortex Flow

In this type of vortex flow, the fluid masses are made to move in a curved path under the action of an external force or torque.

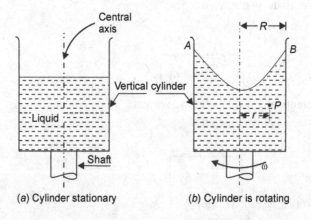

Fig. 5.28: Demonstration of forced vortex flow.

Let a vertical cylinder of radius R containing a liquid be rotating about its axis at an angular velocity ω. Consider a point P at a radial distance r from the axis of rotation.

The velocity at point P: $v = \omega.r$

where r = radius of fluid particle from the axis of rotation.

$\therefore$ The angular velocity: $\omega = \dfrac{v}{r}$ = constant

or $\qquad\qquad\qquad\qquad \dfrac{v}{r}$ = **constant = c** $\qquad\qquad\qquad$ ...(5.15.1)

Equation (5.15.1) shows that in forced vortex flow, the tangential velocity of the fluid particles at any point is directly proportional to the radial distance from the axis of rotation.

Examples of forced vortex are:

(i) Rotation of water in a washing machine.

(ii) Rotation of liquid in a vertical cylinder

(iii) Rotation of liquid inside the impeller of a centrifugal pump

(iv) Rotation of water through the runner of a reaction turbine.

5.16.2 Free Vortex Flow

When no external force or torque is required to rotate the fluid mass, that type of flow is known as **free vortex flow**. The liquid is rotating in free vortex due to fluid pressure itself or due to the gravity (i.e., weight of liquid) or due to rotation previously imparted. The free vortex motion is also known as potential vortex or irrotational vortex.

Example of free vortex flow are:

(i) A whirlpool in a river.

(ii) Flow of liquid in a centrifugal pump casing after it has left the impeller.

(iii) Flow of liquid through a hole/outlet provided at the bottom of a shallow vessel (e.g. wash basin, bath tub etc.).

Let us consider a fluid particle of mass m at a radius distance r from the axis of rotation, having a tangential velocity v, then

$$\text{Angular momentum} = \text{mass} \times \text{velocity} = mv$$
$$\text{Moment of momentum} = mvr$$

$\therefore$ Rate of change in moment of momentum $= \dfrac{\partial}{\partial t}(mvr)$ = Torque

But for free vortex flow, external torque is zero

$$\dfrac{\partial}{\partial t}(mvr) = 0$$

On integrating, we get

$$mvr = \text{constant}$$

Since mass m is constant, then

$$vr = c$$

where c is called strength of vortex.

$$v = \frac{c}{r}$$

$$v \propto \frac{1}{r}$$

Thus, tangential velocity (v) of fluid is inversely proportional to distance r from the axis of rotation.

Equation of Motion for Vortex Flow

Let us consider a cylindrical fluid element whose axis is in a radial direction, with respect to the stream line.

Let dA = area of cross-sectional of fluid element

dr = length of small fluid element

r = radius of the element from O

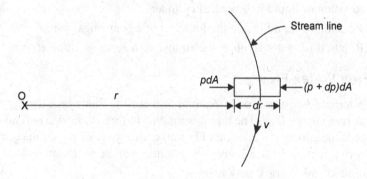

Fig. 5.29: Forces on a cylindrical fluid element

Let v be the velocity along the stream line.

The force acts radially towards the centre due to pressure difference corresponding to the element is called contripetal force. And the force acts radially outward away from the centre 'O' due to radial acceleration is called centrifugal force.

Resolving the force radially,

Centripetal force = centifugal force

$(p + dp).dA - p.dA$ = mass of the element × radial acceleration

$$= m \, a_r$$

$$pdA + dp.dA - p.dA = \rho.dA.dr.\frac{v^2}{r}$$

$$dpdA = \rho dA.dr.\frac{v^2}{r}$$

$$\frac{dp}{dr} = \rho\frac{v^2}{r} \qquad\qquad ...(5.15.2)$$

The Eq. (5.15.2) gives the pressure variation along the radial direction for a force. The expression $\dfrac{dp}{dr}$ is called pressure gradient in the radial direction, $\dfrac{dp}{dr}$ is +ve, hence pressure increases with increase of radius r.

The pressure variation in the vertical plane is given by the hydrostatic law.

$$\frac{dp}{dz} = -\rho g \qquad \qquad \ldots(5.15.3)$$

In Eq. (5.15.3) z is measured vertically in the upward direction.

The pressure p varies w.r.t. r and z or p is function of both r and z

$\therefore \qquad\qquad\qquad p = f(r, z)$

$$dp = \frac{\partial p}{\partial r}.dr + \frac{\partial p}{\partial z}.dz \qquad \qquad \ldots(5.15.4)$$

Substituting the value of $\dfrac{\partial p}{\partial r}$ and $\dfrac{\partial p}{\partial z}$ from Eqs. (5.15.2) and (5.15.3) in Eq. (5.15.4), we get

$$dp = \frac{\rho v^2}{r}.dr - \rho g dz \qquad \qquad \ldots(5.15.5)$$

Eq. (5.15.5) gives the variation of pressure of a rotating fluid in any direction.

Equation of forced vortex flow:

For forced vortex flow

$$v = \omega r$$

where angular velocity, ω = constant

Substituting the value of v in Eq. (5.15.4), we get

$$dp = \frac{\rho \omega^2 r^2}{r}.dr - \rho g dz$$

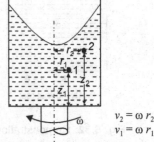

Consider two points 1 and 2 in the fluid having forced vortex flow as shown in Fig. 5.30.

Integrating the above equation for points 1 and 2; we get

$v_2 = \omega\, r_2$
$v_1 = \omega\, r_1$

Fig. 5.30: Forced vortex flow

$$\int_1^2 dp = \int_1^2 \rho \omega^2 r.dr - \int_1^2 \rho g dz$$

$$p_2 - p_1 = \frac{\rho \omega^2}{2}[r_2^2 - r_1^2] - \rho g(z_2 - z_1)$$

$$= \frac{\rho}{2}[\omega^2 r_2^2 - \omega^2 r_1^2] - \rho g(z_2 - z_1)$$

$$p_2 - p_1 = \frac{\rho}{2}(v_2^2 - v_1^2) - \rho g(z_2 - z_1)$$

If the points 1 and 2 lie on the free surface of the liquid, then $p_1 = p_2$ Hence above equation becomes:

$$0 = \frac{\rho}{2}(v_2^2 - v_1^2) - \rho g(z_2 - z_1)$$

$$\rho g(z_2 - z_1) = \frac{\rho}{2}(v_2^2 - v_1^2)$$

$$(z_2 - z_1) = \frac{1}{2g}(v_2^2 - v_1^2)$$

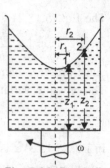

If point 1 lies on the axis of rotation, then

$$v_1 = \omega r_1 = 0 \qquad \because r_1 = 0$$

$$\therefore \qquad z_2 - z_1 = \frac{1}{2g}v_2^2$$

Let $z_2 - z_1 = z$

Fig. 5.31: Parabolic free surface

$$\therefore \qquad z = \frac{v_2^2}{2g} = \frac{(\omega r_2)^2}{2g} \qquad ...(5.15.6)$$

Thus z varies with the square of r, hence Eq. (5.15.6) is a parabola. This means the free surface of the liquid is a paraboloid.

Case 1: If a Cylindrical Vessel is Closed at the Top: As shown in Fig 5.31 (*a*) the initial stage of the cylinder, when it is at rest.

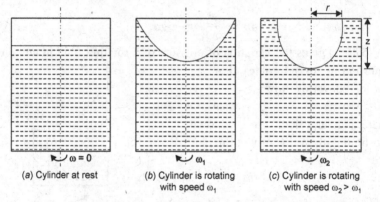

(*a*) Cylinder at rest   (*b*) Cylinder is rotating with speed ω_1 (*c*) Cylinder is rotating with speed $\omega_2 > \omega_1$

Fig. 5.32: Demonstration of forced vortex flow in a closed cylindrical vessel.

Fig. 5.32 (*b*) shows the shape as the paraboloid formed when the speed of rotation is ω_1. If the speed is increased further, say ω_2, the shape of paraboloid formed will be as shown in Fig. 5.32 (*c*). In this case the radius of the parabola at the top of the vessel is unknown. Also the height of the paraboloid formed corresponding to angular speed ω_2 is unknown. Thus to solve the two unknowns, we should have two equations. One equation is

$$z = \frac{(\omega_2 r_2)^2}{2g}$$

The second equation is obtained from the fact that for closed vessel, volume of air before rotation is equal to the volume of air after rotation.

∴ Volume of air before rotation

= volume of closed vessel – volume of liquid in vessel

and

Volume of air after rotation = volume of paraboloid formed

$$= \frac{\pi r^2 z}{2}.$$

Problem 5.22: An open circular cylinder of 140 mm diameter and 900 mm length contains water upto a height of 600 mm. Find the speed at which the cylinder is to be rotated about its vertical axis, so that the axial depth becomes zero.

Solution: Given data:

Diameter of cylinder: $D = 140$ mm $= 0.14$ m

$\therefore$ Radius of cylinder: $R = \dfrac{D}{2} = \dfrac{0.14}{2} = 0.07$ m

Length of cylinder: $l = 900$ mm $= 0.09$ m
Initial height of water: $h = 600$ mm $= 0.6$ m
When axial depth is zero.
The depth of paraboloid: $z = l = 0.9$ m

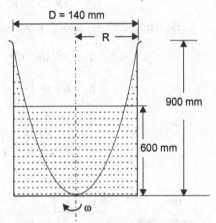

we know, $z = \dfrac{(\omega R)^2}{2g}$

$0.9 = \dfrac{\omega^2 \times (0.07)^2}{2 \times 9.81}$

or $\omega^2 = 3603.67$

$\omega = 60.03$ rad/s

also $\omega = \dfrac{2\pi N}{60}$

$60.03 = \dfrac{2 \times 3.14 \times N}{60}$

or Speed of the rotating cylinder:
$N = \textbf{573.53 rpm}$

Fig. 5.33: Schematic for Problem 5.22

Problem 5.23:   Prove that in case of forced vortex, the rise of liquid level at the ends is equal to the fall of liquid level at the axis of rotation.

Solution: Let

$R =$ radius of the cylinder.
$O - O =$ initial level of liquid in cylinder when the cylinder is at rest.
$\therefore$ Initial height of liquid in cylinder $= h$ as shown in Fig. 5.34 (a)

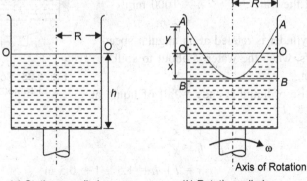

(a) Stationary cylinder   (b) Rotating cylinder
Fig. 5.34: Schematic for Problem 5.23

∴ Volume of liquid in cylinder = $\pi R^2 h$ ...(i)

Let the cylinder is rotated at constant velocity ω. The liquid will rise at the ends and will fall at the centre as shown in Fig. 5.34 (b).

Let $\qquad y$ = rise of liquid at the ends from O - O

$\qquad\qquad x$ = fall of liquid at the centre from O - O.

Then, volume of liquid in cylinder

$$= \text{volume of cylinder upto level } A \text{ - } A - \text{ volume of paraboloid}$$

$$= \pi R^2 (h+y) - \pi R^2 \times \frac{\text{height of paraboloid}}{2}$$

$$= \pi R^2 (h+y) - \frac{\pi R^2 (x+y)}{2} \qquad\qquad ...(ii)$$

Equating Eqs. (i) and (ii), we get

$$\pi R^2 h = \pi R^2 (h+y) - \frac{\pi R^2 (x+y)}{2}$$

or $\qquad\qquad h = h + y - \left(\frac{x+y}{2}\right)$

or $\qquad\qquad 0 = y - \frac{x}{2} - \frac{y}{2}$

$$0 = \frac{y}{2} - \frac{x}{2}$$

$$0 = y - x$$

or $\qquad\qquad \boxed{x = y} \qquad\qquad$ hence, proved.

Problem 5.24: An open cylinder of 200 mm diameter and 1500 mm length contains water upto a height of 1000 mm. Find the maximum speed at which the cylinder is to be rotated about its vertical axis so that no water spills.

Solution: Given data:

Diameter of the cylinder: $\quad D$ = 200 mm

$\qquad\qquad\qquad\qquad\qquad = 0.2$ m

∴ Radius of the cylinder: $\quad R$ = 0.1 m

Length of the cylinder: $\qquad l$ = 1500 mm

$\qquad\qquad\qquad\qquad\qquad = 1.5$ m

Height of the water: $\qquad h$ = 1000 mm

$\qquad\qquad\qquad\qquad\qquad = 1$ m

Let the cylinder is rotated at an angular speed of ω rad/s, when the water is about to spill.

Then using,

$\qquad$ Rise of liquid at ends = fall of liquid at centre

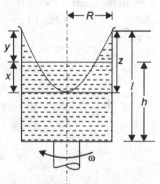

Fig. 5.35: Schematic for Problem 5.24

$$y = x$$

$$l - h = x$$

or $\qquad\qquad x = l - h = 1.5 - 1 = 0.5$ m

∴ Height of parabola: $\qquad z = x + y \qquad\qquad\qquad\qquad \because x = y$

$$= 2x = 2 \times 0.5 = 1 \text{ m}$$

We know,
$$z = \frac{(\omega R)^2}{2g}$$

$$1 = \frac{\omega^2 \times (0.1)^2}{2 \times 9.81}$$

or
$$\omega^2 = 1962$$

or
$$\omega = 44.29 \text{ rad/s}$$

also
$$\omega = \frac{2\pi N}{60}$$

$$44.29 = \frac{2 \times 3.14 \times N}{60}$$

or Maximum speed of the rotating cylinder:

$$N = \textbf{423.15 rpm}$$

Problem 5.25: An open cylindrical vessel, 45 cm in diameter and 60 cm deep is filled to brim with water. If the vessel is rotated about its axis at 300 rpm, how much water is spilled and what is the depth of water at the axis? Also estimate the speed when water surface just touches the bottom of the tank.

(*GGSIP University, Delhi, Dec. 2008*)

Solution: Given data:

Diameter of vessel: $D = 45$ cm $= 0.45$ m

∴ Radius of vessel: $R = \dfrac{D}{2} = \dfrac{0.45}{2} = 0.225$ m

Height of vessel: $h = 60$ cm $= 0.6$ m

Speed of vessel: $N = 300$ rpm

∴ Angular speed: $\omega = \dfrac{2\pi N}{60}$

$$= \frac{2 \times 3.14 \times 300}{60}$$

$$= 31.4 \text{ rad/s}$$

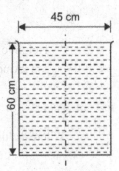

Fig. 5.36: Schematic 1 for Problem 5.25

Initial volume of water in the tank

$$= \text{cross-sectional area} \times \text{initial height of water}$$

$$= \frac{\pi}{4} D^2 \times h$$

$$= \frac{3.14}{4} \times (0.45)^2 \times 0.6 = 0.09537 \text{ m}^3$$

Height of parabola: $\qquad z = \dfrac{\omega^2 R^2}{2g} = \dfrac{(31.4)^2 \times (0.225)^2}{2 \times 9.81} = 2.54$ m

As the height of parabola is more than the height of vessel, the shape of imaginary parabola will be shown in Fig. 5.37.

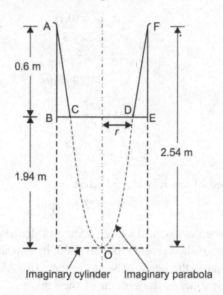

Fig. 5.37: Schematic 2 for Problem 5.25

Let $\qquad\qquad\qquad\qquad\qquad\qquad r$ = radius of the parabola at the bottom of the vessel.

Height of imaginary parabola = 2.54 − 0.6 = 1.94 m

Amount of water spilled = volume of water in portion ACDF

$\qquad\qquad\qquad\qquad\qquad$ = volume of parabola AOF − volume of parabola COD

Where volume of parabola AOF $= \dfrac{\pi}{4} D^2 \times \dfrac{\text{height of parabola}}{2}$

$$= \dfrac{3.14}{4} \times (0.45)^2 \times \dfrac{2.54}{2} = 0.2018 \text{ m}^3$$

For the imaginary parabola COD:

$$\omega = 31.4 \text{ rad/s}$$
$$r = \text{radius at the bottom of vessel}$$
$$z = 1.94 \text{ m}$$

also $\qquad\qquad\qquad\qquad\qquad z = \dfrac{\omega^2 r^2}{2g}$

$$1.94 = \dfrac{(31.4)^2 \times r^2}{2 \times 9.81}$$

or $\qquad\qquad\qquad\qquad\qquad r^2 = 0.03860$

$$r = 0.1964 \text{ m}$$

∴ Volume of imaginary parabola COD:

$$= \frac{1}{2} \times \text{ area of the top of the imaginary}$$

parabola × height of parabola

$$= \frac{1}{2} \times \pi r^2 \times 1.94$$

$$= \frac{1}{2} \times 3.14 \times (0.1964)^2 \times 1.94 = 0.1174 \text{ m}^3$$

∴   Amount of water spilled = 0.2018 − 0.1174 = **0.0844 m³**

Case-II: When water surface just touches the bottom of the tank, then height of parabola is equal to height of tank.

$$z = 0.6 = \frac{\omega^2 R^2}{2g}$$

$$0.6 = \frac{\omega^2 \times (0.225)^2}{2 \times 9.81}$$

or $\omega^2 = 232.53$

$\omega = 15.248$ rad/s

also $\omega = \frac{2\pi N}{60}$

$$15.248 = \frac{2 \times 3.14 \times N}{60}$$

or $N = 145.68$ rpm ≈ **146 rpm**

Problem 5.26: A vessel cylindrical in shape and closed at the top and bottom, contains water upto a height of 80 cm. The diameter of the vessel is 20 cm and length of vessel is 120 cm. The vessel is rotated about its vertical axis. Find the speed of rotation, when axial depth of water at the centre is zero.

(*GGSIP University, Delhi, Dec. 2004*)

Solution: Given data:

Initial height of water: $h = 80$ cm = 0.80 m

Diameter of vessel: $D = 20$ cm = 0.20 m

∴ Radius of vessel: $R = \frac{D}{2} = \frac{0.20}{2} = 0.1$ m

Length of vessel: $l = 120$ cm = 1.2 m

Axial depth of water at centre is zero.

i.e., $z = l = 1.2$ m

where z = height of paraboloid formed when the vessel is rotated.

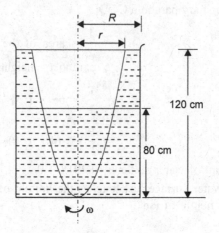

Fig. 5.38: Schematic for Problem 5.26

we know,
$$z = \frac{(\omega r)^2}{2g}$$

where $\qquad r$ = radius of paraboloid at the top of the vessel

$$1.2 = \frac{(\omega r)^2}{2 \times 9.81}$$

or $\qquad\qquad (\omega r)^2 = 23.544$

$$\omega r = 4.85 \text{ m/s} \qquad ...(i)$$

In case of closed vessel; we have

Volume of air before rotation $\qquad$ = volume of air after rotation

$$\frac{\pi}{4}D^2 . l - \frac{\pi}{4}D^2 \times h = \pi r^2 \times \frac{z}{2}$$

$$\frac{\pi}{4} \times (0.2)^2 \times 1.2 - \frac{\pi}{4} \times (0.2)^2 \times 0.80 = \pi r^2 \times \frac{1.2}{2}$$

$$0.012\pi - 0.008\,\pi = 0.6\ r^2\pi$$

$$0.004\,\pi = 0.6\ r^2\pi$$

or $\qquad\qquad 0.6\ r^2 = 0.004$

or $\qquad\qquad r^2 = 0.00666$

$$r = 0.0816 \text{ m}$$

Substituting the value of r in Eq. (i), we have

$$\omega \times 0.0816 = 4.85$$

or $\qquad\qquad \boldsymbol{\omega = 59.43 \text{ rad/s}}$

we have $\qquad\qquad \omega = \frac{2\pi N}{60}$

$$59.43 = \frac{2 \times 3.14 \times N}{60}$$

The speed of rotation: $\qquad N = \boldsymbol{567.80 \text{ rpm}}$

Problem 5.27: An open cylinder 0.3 m in diameter and 0.6 m high, two-third filled with oil of specific gravity 0.8 is rotated about its vertical axis. Determine the speed of rotation when

    (*i*) the oil starts spilling over brim.

    (*ii*) the point of the centre of the base is just exposed.

Solution: Given data:

Diameter of cylinder: $D = 0.3$ m

∴ Radius of cylinder: $R = \dfrac{D}{2} = \dfrac{0.3}{2} = 0.15$ m

Height of cylinder: $H = 0.6$ m

Initial height of oil: $h = \dfrac{2}{3}H = \dfrac{2}{3} \times 0.6 = 0.4$ m

Specific gravity of oil: $S = 0.8$

∴ Density of oil: $\rho = S \times \rho_{water} = 0.8 \times 1000 = 800$ kg/m^3

(*i*) The oil starts spilling over brim.

    In this case, we know that the condition rise of liquid at ends = fall of liquid at centre

$$y = x$$

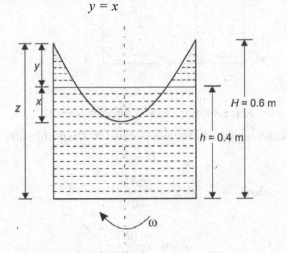

Fig. 5.39: Schematic 1 for Problem 5.27

$$H - h = x$$
$$0.6 - 0.4 = x$$

or $x = 0.2$ m

∴ Height of parabola, $z = x + y = 2x$ ∵ $x = y$

        $= 2 \times 0.2 = 0.4$ m

We know that, $z = \dfrac{\omega^2 R^2}{2g}$

$$0.4 = \frac{\omega^2 (0.15)^2}{2 \times 9.81}$$

or $\omega^2 = 348.8$

or $\omega = 18.67 \text{ rad/s}$

also $\omega = \frac{2\pi N}{60}$

$$18.67 = \frac{2 \times 3.14 \times N}{60}$$

or $N = \textbf{178.37 rpm}$

(*ii*) The point of the centre of the base is just exposed.

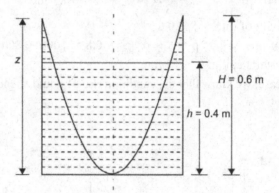

Fig. 5.40: Schematic 2 for Problem 5.27

In this case, height of parabola is equal to height of tank.

$$z = H = 0.6 \text{ m}$$

We know that, $z = \frac{\omega^2 R^2}{2g}$

$\therefore$ $0.6 = \frac{\omega^2 (0.15)^2}{2 \times 9.81}$

or $\omega^2 = 523.2$

or $\omega = 22.87 \text{ rad/s}$

also $\omega = \frac{2\pi N}{60}$

$\therefore$ $22.87 = \frac{2 \times 3.14 \times N}{60}$

or $N = \textbf{218.50 rpm}$

5.17 STREAM LINE

A line along which the stream function (ψ) is constant, is called stream line.

Mathematically,

$$\psi = \text{constant} \qquad \text{for stream line}$$

$\therefore$ The change in stream function on the stream line is zero

i.e., $$d\psi = 0$$

But $$\psi = f(x, y)$$

The total derivative for the stream function (ψ) may be written as:

$$d\psi = \frac{\partial \psi}{\partial x} dx + \frac{\partial \psi}{\partial y} dy$$

$$d\psi = v\, dx - u\, dy \qquad \qquad ...(5.17.1)$$

$$\because \quad \frac{\partial \psi}{\partial x} = v \text{ and } \frac{\partial \psi}{\partial y} = -u$$

For stream line, $\psi = \text{constant}$

$\therefore$ $d\psi = 0$

From Eq. (5.17.1), we get for stream line

$$vdx - udy = 0$$

or $$\boxed{\frac{dy}{dx} = \frac{v}{u}}$$ $...(5.17.2)$

where $\dfrac{dy}{dx} = $ slope of stream line.

5.17.1 Equipotential Line

A line along which the potential function (ϕ) is constant, is called equipotential line.

Mathematically,

$$\phi = \text{constant} \qquad \text{for equipotential line}$$

$\therefore$ The change in potential function on the equipotential line is zero.

i.e., $$d\phi = 0$$

but $$\phi = f(x, y)$$

The total derivative for the potential function (ϕ) may be written as:

$$d\phi = \frac{\partial \phi}{\partial x} dx + \frac{\partial \phi}{\partial y} dy$$

$$d\phi = -udx - vdy \qquad \qquad ...(5.17.3)$$

$$\because \quad \frac{\partial \phi}{\partial x} = -u \text{ and } \frac{\partial \phi}{\partial y} = -v$$

For potential line, $\phi = \text{constant}$ and therefore, $d\phi = 0$. From Eq. (5.17.3), we get for stream line

$$-udx - vdy = 0$$

or
$$\frac{dy}{dx} = -\frac{v}{u}$$ ...(5.17.4)

where $\dfrac{dy}{dx}$ = slope of equipotential line.

From Eqs. (5.17.2) and (5.17.4) it is seen that the product of the slope of the stream line (ψ = constant) and the slope of the equipotential line (ϕ = constant) at the point of intersection is equal to –1 (*i.e.*, $\dfrac{v}{u} \times \left(-\dfrac{u}{v}\right) = -1$). Thus, the stream lines are normal to the equipotential lines. The streamlines and equipotential lines form a net of mutually perpendicular lines, which is called a flownet. The flownet, which utilises the concepts of steam function and potential function, is an important tool in analysing two-dimensional, irrotational flow problems.

5.18   FLOW NET

The flow net is graphical representation of two-dimensional irrotational flow and consists of stream lines intersecting orthogonally of equipotential lines (they intersect at right angles) and in the process forming small curvilinear squares.

Let ψ, $\psi + d\psi$ and $\psi + 2d\psi$ are stream lines intersected by a series of equipotential lines ϕ, $\phi - d\phi$ and $\phi - 2d\phi$. It is to be noted that potential function (ϕ) decreases in the direction of flow.

If the distance between the stream lines at some small region of flow is dn, and the distance between equipotential lines is ds, then for a flow net: $ds = dn$, and the smaller these dimensions, the better will be the representation of flow. The condition may be obtained using following expressions:

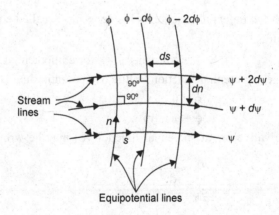

Fig. 5.41 Flow net.

$$-\frac{\partial \phi}{\partial x} = u, \quad -\frac{\partial \psi}{\partial y} = u$$

$$-\frac{\partial \phi}{\partial x} = -\frac{\partial \psi}{\partial y} = u$$

$$\frac{\partial \phi}{\partial n} = \frac{\partial \psi}{\partial s} = V_n$$

If $\qquad d\phi = d\psi$ then $dn = ds$

5.19 USES OF FLOW NET

The flow net helps us in showing and analysing the behaviour of certain two-dimensional, irrotational fluid flow problems which cannot be easily analysed directly by mathematical relations. Since the local velocities are inversely proportional to the spacing between the stream lines and potential lines, the flow net used to estimate the velocity at any point in the flow field if the velocity at a particular point is known. With the application of continuity equation, the velocity at any point in terms of reference velocity can be obtained. The pressure distribution is then obtained by application of the Bernoulli's equation.

Condition for drawing flow net:

(1) The flow must be steady
(2) The fluid should be ideal (*i.e.,* viscosity is zero).
   However, for rapidly accelerating flows, even for fluids of low viscosity the flow nets give good results.
(3) The flow should be two-dimensional, incompressible and irrotational.

5.20 METHODS OF DRAWING THE FLOW NET

The following methods can be used to draw the flow net:

1. Analytical method,
2. Graphical method.

5.20.1 Analytical (or Mathematical) Method

The flow net can be drawn by finding the expression stream function (ψ) and velocity function (ϕ) in terms of x and y and plotting them. The plot of stream function (ψ) gives the stream lines while that of velocity (or potential) function (ϕ) yields equipotential lines. This method can be applied for simple cases of parallel flows and ideal boundary conditions.

5.20.2 Graphical Method

This method consists of drawing stream lines and equipotential lines such that they cut orthogonally and form curvilinear squares. This method consumes lot of time and requires lot of erasing in order to achieve the condition of orthogonal intersection combined with square shaped net. However, in a region where the boundaries converge, diverge or bend, the flow net does not contain squares.

Now consider a flow net when the stream lines are diverging as shown in Fig. 5.42.

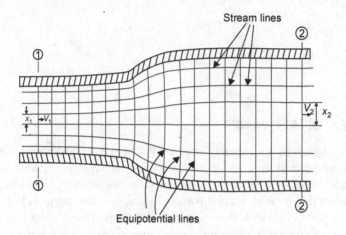

Fig. 5.42: Flow nets for gradually increasing boundary.

Let two section of the flow nets (1) and (2)

and x_1 = spacing between two stream lines at section (1),

v_1 = velocity of liquid particle at section (1),

x_2, v_2 = corresponding values at section (2).

Since there can be no flow across the stream lines, therefore discharge per unit width, between two adjacent stream lines will be equal. Therefore, discharge per unit width

$$q = x_1 v_1 = x_2 v_2$$

Thus, the velocity of liquid particles varies inversely with spacing between the stream lines. (*i.e.,* the velocity decreases with the increase in spacing and vice versa).

Also the flow nets shown in Fig. 5.43 for gradually decreasing boundary.

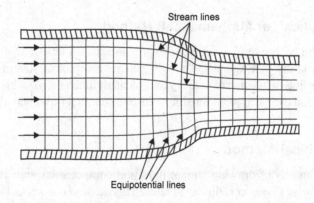

Fig. 5.43: Flow nets for gradually decreasing boundary.

5.21   SOURCE AND SINK FLOWS

5.21.1 Source Flow

The flow coming from a point (source : S) moving out radially in all directions of a plane at uniform rate, is called source flow.

A source flow is shown in Fig. 5.44.

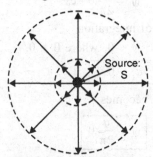

Fig. 5.44: Source flow.

S is the source from which the fluid moves radially outward as shown by the lines carrying arrow heads. The strength of a source is defined as the volume of fluid flow rate per unit depth (m^2/s). The strength of a source is denoted by letter q.

Let u_r = radial velocity of flow at a radius r from the source (S).

q = strength of a source in m^2/s

r = radius of the circle at radial velocity u_r.

The radial velocity u_r at radius r is given by:

$$u_r = \frac{q}{2\pi r} \qquad \text{...(5.21.1)}$$

From Eq. (5.21.1), it is observed that the radial velocity is a measure of source strength, both being directly proportional to each other.

But the radial velocity decreases with the increase of r and at a large distance away from the source, the velocity will be nearly equal to zero. The flow is in radial direction, hence the tangetial velocity (u_θ) is zero.

Equation of Stream Function

We know that the stream function (ψ) in radial and tangential velocities components.

$$\frac{1}{r}\frac{\partial \psi}{\partial \theta} = u_r : \text{radial velocity}$$

and $$\frac{\partial \psi}{\partial r} = -u_\theta : \text{tangential velocity}$$

From Eq. (5.21.1), the radial velocity for source

$$u_r = \frac{q}{2\pi r}$$

$$\therefore \qquad \frac{1}{r}\frac{\partial \psi}{\partial \theta} = \frac{q}{2\pi r}$$

or
$$d\psi = \frac{q}{2\pi}.d\theta$$

On integrating the above equation w.r.t. θ, we get

$$\psi = \frac{q}{2\pi}.\theta + c_1 \qquad\qquad\qquad ...(5.21.2)$$

where c_1 is a constant of integration

Let $\qquad\qquad \psi = 0$, where $\theta = 0$,

Then $\qquad\qquad \boxed{c_1 = 0}$

Hence the Eq. (5.21.2) becomes

$$\boxed{\psi = \frac{q}{2\pi}.\theta} \qquad\qquad\qquad ...(5.21.3)$$

In Equation (5.21.3) q is constant and the stream function is a function of θ. [*i.e.*, $\psi = f(\theta)$]. For a given value of θ, the stream function (ψ) will be constant and this will be a radial line. The stream lines can be plotted by having different values of θ. Here θ is taken in radians.

Plotting of Streamlines

When $\qquad\qquad\qquad \theta = 0,\ \psi = 0$

$$\theta = 45° = \frac{\pi}{4}\ \text{radians},\ \psi = \frac{q}{2\pi}\times\frac{\pi}{4} = \frac{q}{8}$$

$$\theta = 90° = \frac{\pi}{4}\ \text{radians},\ \psi = \frac{q}{2\pi}\times\frac{\pi}{2} = \frac{q}{4}$$

$$\theta = 135° = \frac{3}{4}\pi\ \text{radians},\ \psi = \frac{q}{2\pi}\times\frac{3}{4}\pi = \frac{3q}{8}$$

$$\theta = 180° = \pi\ \text{radians},\ \psi = \frac{q}{2\pi}\times\pi = \frac{q}{2}$$

$$\theta = 225° = \frac{5}{4}\pi\ \text{radians},\ \psi = \frac{q}{2\pi}\times\frac{5}{4}\pi = \frac{5}{8}q$$

$$\theta = 270° = \frac{3}{2}\pi\ \text{radians},\ \psi = \frac{q}{2\pi}\times\frac{3}{2}\pi = \frac{3}{4}q$$

$$\theta = 315° = \frac{7}{4}\pi\ \text{radians},\ \psi = \frac{q}{2\pi}\times\frac{7}{4}\pi = \frac{7}{8}q$$

The streamlines will be radial lines as shown in Fig. 5.43

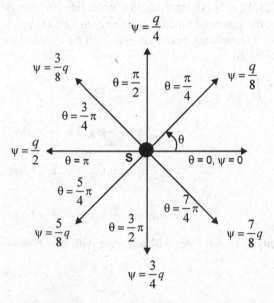

Fig. 5.45: Streamlines for source flow.

Equation of Potential Function

We know that the potential function (ϕ) in radial and tangential velocities components

$$u_r = \frac{\partial\phi}{\partial r}, \quad u_v = \frac{1}{r}\frac{\partial\phi}{\partial\theta}$$

also $\qquad\qquad u_r = \dfrac{q}{2\pi r}$ from Eq. (5.21.1)

Equating the two values of u_r, we get

$$\frac{\partial\phi}{\partial r} = \frac{q}{2\pi r}$$

or $\qquad\qquad d\phi = \dfrac{q}{2\pi}\dfrac{dr}{r}.$

On integrating the above equation w.r.t. r, we get

$$\int d\phi = \int \frac{q}{2\pi}\frac{dr}{r}$$

$$\phi = \frac{q}{2\pi}\int \frac{dr}{r} = \frac{q}{2\pi}\log_e r$$

$$\boxed{\phi = \frac{q}{2\pi}\log_e r} \qquad\qquad ...(5.21.4)$$

In Eq. (5.21.4), q is constant and the potential function is function of r. For a given value of r, the potential function will be constant. Hence, it will be a circle with at the source (S). The velocity potential lines will be circles with origin at the source as shown in Fig. 5.46.

At
$$r = 1, \; \phi = \frac{q}{2\pi} \log_e 1 = 0$$

$$r = 2, \; \phi = \frac{q}{2\pi} \log_e 2 = \frac{q}{2\pi} \times 0.6931 = \frac{q}{2.88\pi}$$

$$r = 3, \; \phi = \frac{q}{2\pi} \log_e 3 = \frac{q}{2\pi} \times 1.098 = \frac{q}{1.82\pi}$$

$$r = 4, \; \phi = \frac{q}{2\pi} \log_e 4 = \frac{q}{2\pi} \times 1.386 = \frac{q}{1.44\pi}$$

The velocity potential lines will be circle with origin at the source as shown in Fig. 5.46.

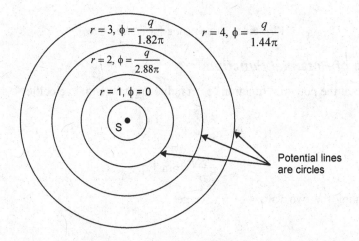

Fig. 5.46: Potential line for source flow.

Pressure distribution in a plane source flow:

Let $\qquad p$ = pressure at point 1 from the source (S)

$\qquad r$ = radius at point 1 from the source (S)

$\qquad u_r$ = radial velocity at point 1

$\qquad p_0$ = pressure at point 2, which is large distance away from the source.

The velocity will be zero at point 2.

Applying Bernoulli's equation at points 1 and 2, we get

$$\frac{p}{\rho g} + \frac{u_r^2}{2g} + z_1 = \frac{p_0}{\rho g} + \frac{u_{r2}^2}{2g} + z_1 \quad \because z_1 = z_1 : \text{plane of flow is horizontal. } u_{r2} = 0$$

$$\frac{p}{\rho g} + \frac{u_r^2}{2g} = \frac{p_0}{\rho g}$$

$$\frac{p}{\rho g} - \frac{p_0}{\rho g} = -\frac{u_r^2}{2g}$$

$$p - p_0 = -\frac{\rho u_r^2}{2}$$

$$p - p_0 = -\frac{\rho}{2}\left(\frac{q}{2\pi r}\right)^2 \qquad \because \ u_r = = \frac{q}{2\pi r} \text{ for source flow}$$

$$p - p_0 = -\frac{\rho q^2}{8\pi^2 r^2} \qquad\qquad\qquad ...(5.21.5)$$

$$\because \ \rho \text{ and q are constant}$$

Eq. (5.20.5) shows that the pressure difference is inversely proportional to the square of the radius from the source (S).

5.21.2 Sink Flow

The sink flow is just opposite to the source flow. In sink flow, the flow is radially inward to a common point where it disappears at constant rate. The pattern of stream lines and equipotential lines of a sink flow is the same as that of a source flow. All the equations derived for a source flow shall hold good for sink flow except that in sink flow equation, q is to be taken as $-q$.

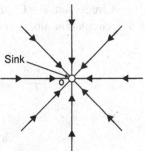

Fig. 5.47: Sink flow.

5.22 CIRCULATION

The circulation is defined as the line integral of the tangential component of velocity around a closed contour in the flow field. It is denoted by Greek letter Γ (gamma)

Mathematically,

Circulation: $\qquad \Gamma = \oint V_t \, ds = \oint \vec{V}.\vec{ds}$

$$= \oint (udx + vdy + wdz)$$

For a two dimensional flow.

$$\Gamma = \oint (udx + udy)$$

or $\qquad\qquad = \oint V \cos \alpha \, ds.$

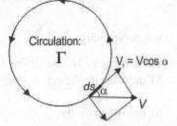

Fig. 5.48: Circulation in a flow field.

The integral sign $\oint$ indicates summation all around the contour. By sign convention, circulation is positive in the anticlockwise direction.

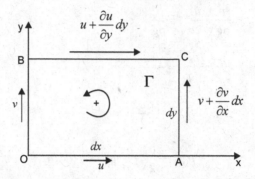

Fig. 5.49: Circulation around a rectangular element.

Consider a rectangular element $OACB$ with sides dx and dy parallel to x and y-directions respectively, the tangential velocities are indicated in Fig. 5.49. The circulation for rectangular element is obtained by summation of line integrals around the element in an anticlockwise direction.

Circulation: Γ = Circulation along OA + circulation along AB + circulation along CB + circulation along BO

$$= \Gamma_{OA} + \Gamma_{AB} + \Gamma_{CB} + \Gamma_{BO}$$

$$\Gamma = udx + \left(v + \frac{\partial v}{\partial x} dx \right) dy - \left(u + \frac{\partial u}{\partial y} dy \right) dx - vdy$$

$$= \frac{\partial v}{\partial x} dxdy - \frac{\partial u}{\partial y} dxdy = \left(\frac{\partial v}{\partial x} - \frac{\partial u}{\partial y} \right) dx.dy \quad \because dA = dxdy, \text{ area of plane.}$$

$$\Gamma = \left(\frac{\partial v}{\partial x} - \frac{\partial u}{\partial y} \right) dA$$

or $\qquad \dfrac{\Gamma}{dA} = \dfrac{\partial v}{\partial x} - \dfrac{\partial u}{\partial y}$

$$\frac{\Gamma}{dA} = \zeta : \text{vorticity}$$

where vorticity: $\zeta = \dfrac{\partial v}{\partial x} - \dfrac{\partial u}{\partial y}$

Thus vorticity is also defined as the circulation per unit area.

SI unit of circulation: Γ = m²/s

SI unit of vorticity: $\zeta = \dfrac{1}{s}$ or s^{-1}

5.23 DOUBLET

A doublet is formed by combining a source and a sink of equal strength in a special way, which are allowed to approach each other in such a manner that the product of their strength and the distance between them remains a constant.

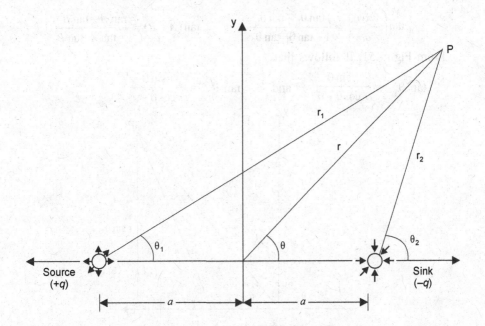

Fig. 5.50: The combination of a source and a sink of equal strength located along the x-axis.

Consider an irrotational and incompressible region of flow composed of a line source of strength q at location $(-a, 0)$ and a line sink of the same strength (but opposite sign) at $(a, 0)$, as shown in Fig. 5.50.

For line source $(-a, 0)$;

Stream function: $\psi_1 = \dfrac{q}{2\pi} \theta_1$ $\hspace{4cm}$ (i)

Similarly for line sink $(a, 0)$,

Stream function: $\psi_2 = \dfrac{q}{2\pi} \theta_2$ $\hspace{4cm}$ (ii)

Superposition enables us to simply add the two stream functions, Eq (i) and Eq (ii), to obtain the combined stream function for the pair is

$$\psi = \psi_1 + \psi_2$$

$$= \frac{q}{2\pi}\theta_1 - \frac{q}{2\pi}\theta_2$$

$$\psi = \frac{q}{2\pi}(\theta_1 - \theta_2)$$

$$\frac{2\pi\psi}{q} = \theta_1 - \theta_2$$

or

Taking the tangent of both sides, we get

$$\tan\left(\frac{2\pi\psi}{q}\right) = \tan(\theta_1 - \theta_2)$$

$$\tan\left(\frac{2\pi\psi}{q}\right) = \frac{\tan\theta_1 - \tan\theta_2}{1 + \tan\theta_1 \ \tan\theta_2} \qquad \therefore \tan(A+B) = \frac{\tan A - \tan B}{1 + \tan A \ \tan B}$$

From Fig. 5.51, if follows that

$$\tan\theta_1 = \frac{r\sin\theta}{r\cos\theta + a} \qquad \text{and} \qquad \tan\theta_2 = \frac{r\sin\theta}{r\cos\theta - a}$$

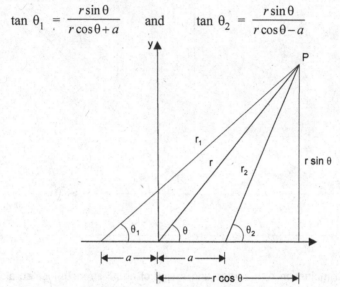

Fig. 5.51: Demonstrating law of triangle

$$\tan\left(\frac{2\pi\psi}{q}\right) = \frac{\dfrac{r\sin\theta}{r\cos\theta + a} - \dfrac{r\sin\theta}{r\cos\theta - a}}{1 + \dfrac{r\sin\theta}{(r\cos\theta + a)} \times \dfrac{r\sin\theta}{(r\cos\theta - a)}} = \frac{\left[\dfrac{1}{r\cos\theta + a} - \dfrac{1}{r\cos\theta - a}\right]r\sin\theta}{1 + \dfrac{r^2\sin^2\theta}{r^2\cos^2\theta - a^2}}$$

$$= \frac{\left[\dfrac{r\cos\theta - a - r\cos\theta - a}{r^2\cos^2\theta - a^2}\right]r\sin\theta}{\dfrac{r^2\cos^2\theta - a^2 + r^2\sin^2\theta}{r^2\cos^2\theta - a^2}} = \frac{-2ar\sin\theta}{r^2(\sin^2\theta + \cos^2\theta) - a^2}$$

$$= \tan\left(\frac{2\pi\psi}{q}\right) = \frac{-2ar\sin\theta}{r^2 - a^2} \qquad\qquad \because \ \sin^2\theta + \cos^2\theta = 1$$

or $$\frac{2\pi\psi}{q} = -\tan^{-1}\left(\frac{2ar\sin\theta}{r^2 - a^2}\right)$$

or $$\psi = -\frac{q}{2\pi}\tan^{-1}\frac{2ar\sin\theta}{r^2 - a^2}$$

For small values of the distance a, the tangent of an angle approaches the value of the angle for small angles.

$$\therefore \qquad\qquad \psi = -\frac{q}{2\pi} \times \frac{2ar\sin\theta}{(r^2 - a^2)}$$

$$\psi = -\frac{qar\sin\theta}{\pi(r^2 - a^2)} \qquad (iii)$$

A doublet is formed by letting a source and a sink approach one another $(a\to 0)$ while increasing the strength q $(q\to\infty)$ so that the product qa remains constant. In this case, since $\dfrac{r}{r^2 - a^2}\to\dfrac{1}{r}$, Eq. (iii) reduces to

$$\psi = \frac{-qa\sin\theta}{\pi r}$$

$$\psi = \frac{-K\sin\theta}{r} \qquad (iv)$$

where K is a constant is equal to qa/π, is called doublet strength.

The corresponding velocity potential for the doublet is

$$\phi = \frac{K\cos\theta}{r} \qquad (v)$$

Several streamlines and equipotential lines for a doublet are plotted in Fig. 5.52. If turms out that the streamlines are circles tangent to the axis, and the equipotential lines are circles tangent to the y-axis. The circles intersect at 90° angles everywhere except at the origin, which is a singularity point.

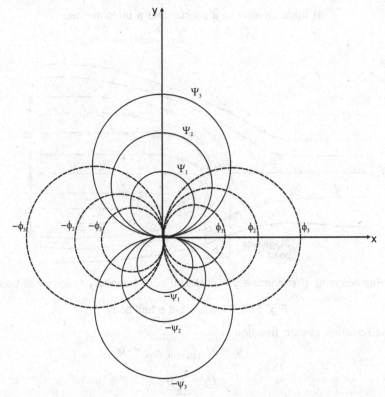

Fig. 5.52: Streamlines (firm) and equipotential lines (dashed) for a doublet of strength K located at the orgin in the xy-plane and aligned with the x-axis.

If K is -ve, the doublet is backwards, with the sink located at $x = 0^-$ (infinitesimally to the left of the origin) and the source located at $x = 0^+$ (infinitesimally to the right of the origin). In that case all the streamless in Fig. 5.52 would be identical in shape, but the flow would be in the opposite direction. It is left as an excercise to construct expressions for a doublet that is aligned at some angle α with x-axis.

5.24 HALF-BODY– SOURCE IN A UNIFORM STREAM

Flow around a half-body is obtained by the addition of a source to a uniform flow. Consider the superposition of a source and a uniform flow in Fig. 5.53 (a).

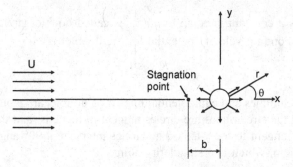

(a) Superposition of a source and a uniform flow.

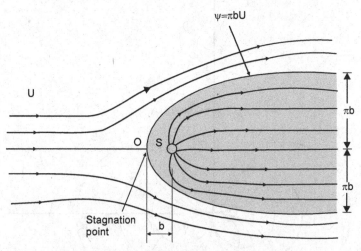

(b) Replacement of streamline y = pbU with solid boundary to form half-body.

Fig. 5.53: Flow around a half-body

The resulting stream function is

$$\psi = \psi_{\text{uniform flow}} + \psi_{\text{source}}$$

$$= Uy + \frac{q}{2\pi}\theta$$

$$= Ur \sin \theta + \frac{q\theta}{2\pi} \qquad (i)$$

and the corresponding velocity potential is

$$\phi = \phi_{\text{unifrom flow}} + \phi_{\text{source}}$$

$$= Ux + \frac{q}{2\pi} \log e\, r$$

$$= Ur \cos \theta + \frac{q}{2\pi} \log e\, r \qquad (ii)$$

It is clear that at some point along the negative x-axis the velocity due to the source will just cancel that due to the uniform flow and a stagnation point (point at which flow is stop) will be created. For the source alone

$$V_r = \frac{q}{2\pi r}$$

So that the stagnation point will source at x = –b where

$$U = \frac{q}{2\pi b}$$

$$\text{or} \quad b = \frac{q}{2\pi U} \qquad (iii)$$

The value of the stream function at the stagnation point can be obtained by evaluating ψ at r = b and $\theta = \pi$, which yields from Eq. (i)

$$\psi_{stagnation} = \frac{q}{2}$$

Since $\frac{q}{2} = \pi b U$ (from Eq (iii) it follows that the equation of the streamline passing through the stagnation point is

$$\pi b U = Ur \sin\theta + bU\theta \qquad \text{from Eq. } (i)$$

or $\qquad \pi b = r \sin\theta + b\theta$

or $\qquad r \sin\theta = b\,(\pi-\theta)$

or $\qquad r = \dfrac{b(\pi - \theta)}{\sin\theta} \qquad (iv)$

where θ can vary between 0 and 2π. A plot of this streamline is shown in Fig. 5.53 (b) If we replace this streamline with a solid boundary, as indicated in the figure, then it is clear that this combination of a uniform flow and a source can be used to describe the flow around a streamlined body placed in a uniform stream. The body is open at the downstream end, and thus is called a half-body. Other streamlines in the flow field can be obtained by setting ψ = constant in Eq (i) and plotting the resulting equation. A number of these streamlines are shown in Fig. 5.53 (b). Although the streamlines inside the body are shown, they are actually of no interest in this case, since we are concerned with the flow field outside the body. It should be noted that the singularity in the flow field (the source) occurs inside the body, and these are no singularities in the flow field of interest (outside the body).

The width of the half-body asymptotically approaches $2\pi b$. This follows from Eq. (iv), which can be written as

$$y = b\,(\pi-\theta)$$

so that as $\theta\to 0$ or $\theta\to 2\pi$ the half width approaches $\pm \pi b$. With the stream function (or velocity potential) known, the velocity components at any point can be obtained. For the half-body, using the stream function given by Eq (i),

$$V_r = \frac{1}{r}\frac{\partial\psi}{\partial\theta} = U\cos\theta + \frac{q}{2\pi r}$$

and

$$V_\theta = \frac{-\partial\psi}{\partial r} = -U\,Sin\,\theta$$

Thus, the square of the magnitude of the velocity V at any point is

$$V^2 = V_r^2 + V_\theta^2$$

$$= \left(U\cos\theta + \frac{q}{2\pi r}\right)^2 + U^2\sin^2\theta$$

$$= U^2\cos^2\theta + \left(\frac{q}{2\pi r}\right)^2 + 2U\cos\theta\times\frac{q}{2\pi r} + U^2\sin^2\theta$$

$$= U^2(\sin^2\theta + \cos^2\theta) + \left(\frac{q}{2\pi r}\right)^2 + 2U\cos\theta\times\frac{q}{2\pi r}$$

Since $\quad b = \dfrac{q}{2\pi U}$

$$\therefore \quad V^2 = U^2\left(1 + 2\frac{b}{r}\cos\theta + \frac{b^2}{r^2}\right)$$

or

$$V = U\sqrt{\left(1 + 2\frac{b}{r}\cos\theta + \frac{b^2}{r^2}\right)} \tag{v}$$

With the velocity known, the pressure at any point can be determined from the Bernoulli is equation, which can be written between any two points in the flow field since the flow is irrotational. Thus, applying the Bernoulli is equation between a point far from the body, the pressure is p_0 and the velocity is U, and some arbitrary point with pressure p and velocity V, if follows that

$$p_0 + \frac{1}{2}\rho U^2 = p + \frac{1}{2}\rho V^2 \tag{vi}$$

where change in datum heads is neglected. Eq. (v) can now be substituted into Eq. (vi) to obtain the pressure at any point in terms of the reference pressure p_0, and the free stream velocity U.

SUMMARY

1. Kinematics of fluid motion deals with the flow of particles without considering forces/pressure causing the motion.

2. Methods of describing fluid motion:

   (*a*) Langrangian method, and

   (*b*) Eulerian method.

       (**a**) **Langrangian Method:** By this method, we study the behaviour of single fluid particle at every instant of time as the particle moves to different positions.

       (**b**) **Eulerian Method:** It deals with flow pattern of all the particles simultaneously at one section. In mechanics of fluids, this method is commonly used, because of its mathematical simplicity. Moreover, in mechanics of fluids the movement of an individual fluid particle is not of much importance.

3. **Stream Line:** A stream line is an imaginary line drawn in a flow field such that a tangent drawn at any point on it indicates the direction of the velocity vector at that point and at that time.

4. **Path Line:** It is a line or path traced by a single fluid particle during its motion over some time period.

5. **Streak Line or Filament Line:** It is a line or curve which gives an instantaneous picture of the location of the fluid particles, which has passed through a prescribed point in the flow. For example, visible plume from a chimney represents such a locus because all the smoke particles have passed through the chimney's mouth.

   There is no distinction among streamline, path line and streak line for steady flow.

6. **Stream Tube:** It is a collection of a number of stream lines forming an imaginary tube. There is no flow through the walls of a stream tube. The properties of the flow are constant across the section of a stream tube. Therefore, the flow in a stream tube is one dimensional.

7. **Types of Fluid Flow:**

   (*a*) **Steady and Unsteady Flow:** A flow is considered to be steady if fluid flow parameters such as velocity: V, pressure: p, density: ρ, temperature: T, etc. at any point do not change with time. If any one of these parameters changes with time, the flow is said to be unsteady.

   Mathematically,

   $$\frac{\partial V}{\partial t} = 0, \ \frac{\partial p}{\partial t} = 0, \ \frac{\partial \rho}{\partial t} = 0 \ \text{ for steady flow.}$$

Contd...

$\dfrac{\partial V}{\partial t} \neq 0,\ \dfrac{\partial p}{\partial t} \neq 0,\ \dfrac{\partial \rho}{\partial t} \neq 0$ for unsteady flow.

(b) Uniform and Non-uniform Flow: A uniform flow is defined as that type of flow in which velocity: V, pressure : p, density: ρ, temperature: T, etc. at any given time do not change with respect to space (*i.e.,* length of direction of the flow).

Mathematically,

$\dfrac{\partial V}{\partial s} = 0,\ \dfrac{\partial p}{\partial s} = 0,\ \dfrac{\partial \rho}{\partial s} = 0$ for uniform flow.

In case of non-uniform flow, velocity, pressure, density etc. at given time change with respect to space.

Mathematically,

$\dfrac{\partial V}{\partial s} \neq 0,\ \dfrac{\partial p}{\partial s} \neq 0,\ \dfrac{\partial \rho}{\partial s} \neq 0$ for non-uniform flow.

(c) Laminar and Turbulent Flow:

Laminar flow: In this type of flow, the particles of fluid move in orderly and well defined path, laminar flow is also called viscous flow or streamline flow.

Turbulent flow: In this type of flow, the particles of fluid flow move in irregular and disorderly manner *i.e.,* fluid particles move in a zig-zag way.

Roynolds number: $\text{Re} = \dfrac{VL}{\nu}$ for flow over flat plate

$\text{Re} = \dfrac{VD}{\nu}$ for flow through pipe

For flow through pipe:

$\text{Re} \leq 2000$ for laminar flow

$\text{Re} \geq 4000$ for turbulent flow

$2000 < \text{Re} < 4000$ for transition flow

For flow over flat plate:

$\text{Re} \leq 3 \times 10^5$ for laminar flow

$\text{Re} \geq 5 \times 10^5$ for turbulent flow

$3 \times 10^5 < \text{Re} < 5 \times 10^5$ for transition flow.

(d) Compressible and Incompressible Flow:

Compressible Flow: If the density of the fluid changes from point to point in the fluid flow: it is referred to as compressible flow.

Mathematically,

Density: $\rho \neq c$

Contd...

For example: Gases are compressible.

Incompressible Flow: If the density of the fluid remains constant at every point in the fluid flow, it is referred to as an incompressible flow.

Mathematically,

Density : $\rho = c$

For example: Liquids are generally incompressible in nature.

Mach number : $M = \dfrac{V}{a}$

where V = velocity of fluid, a = speed of sound in the flowing fluid.

If M > 0.2, the fluid is considered to be compressible

M < 0.2, the fluid is assumed to be incompressible.

(e) **Rotational and Irrotational Flows:**

Rotational Flow: The rotational flow is that type of flow in which the fluid particles while flowing along stream lines, also rotate about their own axis.

Irrotational Flow: It is that type of flow in which the fluid particles while flowing along stream lines, do not rotate about their own axis.

(f) **One, two and three-dimensional Flow:**

One-dimensional Flow: For a steady one-dimensional flow, the velocity is a function of one space coordinate only. The variation of velocities in other mutually perpendicular direction is assumed negligible.

Mathematically;

$$u = f(x), \ v = 0 \text{ and } w = 0$$

Two-dimensional Flow: In this type of flow the velocity is a function of time and two rectangular space coordinates, say x and y.

Mathematically,

$$u = f_1(x, y), \ v = f_2(x, y) \text{ and } w = 0$$

Three-dimensional Flow: In this type of flow the velocity is a function of time and three mutually perpendicular direction.

Mathematically,

$$u = f_1(x, y, z), \ v = f_2(x, y, z) \text{ and } w = f_3(x, y, z)$$

8. **Rate of Flow:**

(i) Volume of the fluid flowing across the section per second is called discharge.

$$Q = AV = mv \ \ \text{m}^3/\text{s}$$

where A = cross-sectional area of pipe.

V = average velocity of fluid across the section

Contd...

$$v = \text{specific volume} = \frac{1}{\rho}$$

(ii) Mass of the fluid flowing across the section per second is called mass flow rate.

$$AV = mv$$

or

$$m = \frac{AV}{v} = \rho AV \quad \text{kg/s}$$

(iii) Weight of the fluid flowing across the section per second is weight flow rate.

$$\dot{W} = mg \quad \text{N/s} = \rho AVg \quad \text{N/s}$$

9. **Continuity Equation:** It is based on the law of conservation of mass. It means mass of fluid can neither be created nor be destroyed.

$$m = \rho AV = \text{constant}$$

$$\rho_1 A_1 V_1 = \rho_2 A_2 V_2$$

for compressible fluid

$$A_1 V_1 = A_2 V_2$$

for incompressible fluid

The continuity equation in three dimension in cartesian coordinates (x, y, z)

$$\left. \begin{array}{l} \dfrac{\partial \rho}{\partial t} + \dfrac{\partial (\rho u)}{\partial x} + \dfrac{\partial (\rho v)}{\partial y} + \dfrac{\partial (\rho w)}{\partial z} = 0 \\[4mm] \dfrac{\partial \rho}{\partial t} + \nabla.(\rho \vec{V}) = 0 \end{array} \right\}$$

...(i)

Equation. (i) is applicable to any type of flow for any fluid whether:

— Compressible or incompressible
— Steady or unsteady
— Uniform or non-uniform.

$$\frac{\partial u}{\partial x} + \frac{\partial v}{\partial y} + \frac{\partial w}{\partial z} = 0$$

...(ii)

Equation. (ii) is applicable to 3-D, steady and incompressible fluid flow.

$$\frac{\partial u}{\partial x} + \frac{\partial v}{\partial y} = 0$$

...(iii)

Equation. (iii) is applicable to 2-D, steady and incompressible fluid flow.

10. Continuity equation is cylindrical coordinate (r, θ, z)

$$\frac{\partial \rho}{\partial t} + \frac{1}{r}(\rho r V_r) + \frac{1}{r}\frac{\partial}{\partial \theta}(\rho V_\theta) + \frac{\partial}{\partial z}(\rho V_z) = 0$$

Contd...

The above 3-D continuity applicable for:
— steady or unsteady flow,
— compressible or incompressible

$$\frac{\partial}{\partial r}(rV_r) + \frac{\partial V_\theta}{\partial \theta} = 0 \qquad \text{for 2-D}$$

The above equation is applicable for steady and incompressible flow.

11. The components of acceleration in x, y, z directions are:

$$a_x = \frac{\partial u}{\partial t} + u\frac{\partial u}{\partial x} + v\frac{\partial u}{\partial y} + w\frac{\partial u}{\partial z}$$

$$a_y = \frac{\partial v}{\partial t} + u\frac{\partial v}{\partial x} + v\frac{\partial v}{\partial y} + w\frac{\partial v}{\partial z}$$

and $$a_z = \frac{\partial w}{\partial t} + u\frac{\partial w}{\partial x} + v\frac{\partial w}{\partial y} + w\frac{\partial w}{\partial z}$$

Acceleration vector: $\vec{a} = a_x i + a_y j + a_z k$

12. **Stream Function:** ψ : If is used for two-dimensional flow. The velocity components in terms of stream function is given by:

$$v = \frac{\partial \psi}{\partial x}, \quad -u = \frac{\partial \psi}{\partial y}$$

13. **Velocity Potential:** ϕ

The velocity components in terms of velocity potential is given by:

$$\left.\begin{aligned} -u &= \frac{\partial \phi}{\partial x} \\[2mm] -v &= \frac{\partial \phi}{\partial y} \\[2mm] -w &= \frac{\partial \phi}{\partial z} \end{aligned}\right\} \text{ in cartesian coordinates } (x,\ y,\ z)$$

$$\left.\begin{aligned} -V_r &= \frac{\partial \phi}{\partial r} \\[2mm] -V_\theta &= \frac{1}{r}\frac{\partial \phi}{\partial \theta} \\[2mm] -V_z &= \frac{\partial \phi}{\partial z} \end{aligned}\right\} \text{ in cylindrical coordinates } (r,\ \theta,\ z)$$

Contd...

14. Angular Deformation:

$$\varepsilon_{xy} = \frac{1}{2}\left(\frac{\partial u}{\partial y} + \frac{\partial v}{\partial x}\right) \text{ in the } (x - y) \text{ plane}$$

$$\varepsilon_{yz} = \frac{1}{2}\left(\frac{\partial v}{\partial z} + \frac{\partial w}{\partial y}\right) \text{ in the } (y - z) \text{ plane}$$

$$\varepsilon_{xz} = \frac{1}{2}\left(\frac{\partial u}{\partial z} + \frac{\partial w}{\partial x}\right) \text{ in the } (x - z) \text{ plane}$$

15. Rotation:

$$\omega_z = \frac{1}{2}\left(\frac{\partial v}{\partial x} - \frac{\partial u}{\partial y}\right) \text{ for } (x - y) \text{ plane}$$

$$\omega_x = \frac{1}{2}\left(\frac{\partial w}{\partial y} - \frac{\partial v}{\partial z}\right) \text{ for } (y - z) \text{ plane}$$

$$\omega_y = \frac{1}{2}\left(\frac{\partial u}{\partial z} - \frac{\partial w}{\partial x}\right) \text{ for } (z - x) \text{ plane}$$

16. Vorticity: ζ : Vorticity is equal to twice the value of angular velocity. Mathematically,

Vorticity : $\vec{\xi} = 2\vec{\omega}$

$$\xi_z = 2\omega_z = \frac{\partial v}{\partial x} - \frac{\partial u}{\partial y}$$

$$\xi_x = 2\omega_x = \frac{\partial w}{\partial y} - \frac{\partial v}{\partial z}$$

$$\xi_y = 2\omega_y = \frac{\partial u}{\partial z} - \frac{\partial w}{\partial x}$$

17. Vortex Flow: If fluid continuously flows along a curved path about a fixed axis, the flow is called vortex or whirling flow. The vortex flow is classified into two categories as:

(*i*) Forced vortex flow, and

(*ii*) Free vortex flow.

The relation between tangential velocity (*v*) and radius (*r*):

For forced vortex flow:

$$v = \omega r$$

or
$$\frac{v}{r} = \text{constant}$$

For free vortex flow:

$$vr = \text{constant}.$$

Contd...

The variation of pressure in r and z directions

$$dp = \frac{\rho v^2}{r} dr - \rho g dz$$

For forced vortex flow:

Height of paraboloid formed:

$$z = \frac{v_2^2}{2g} = \frac{\omega^2 r_2^2}{2g}$$

18. For a forced vortex flow in a rotating vertical open cylinder:
    Rise of liquid level at the ends = Fall of liquid level at centre.

19. **Flow Net:** The flow net is graphical representation of two-dimensional irrotational flow and consists of stream lines interesting orthogonally of equipotential lines.

20. **Source and Sink Flows:** The flow coming from a point moving out radially in all directions of a plane at uniform rate, is called source flow.
    The sink flow is just opposite to the source flow. In a sink flow, the flow is radially inward to a common point where it disappears at constant rate.

21. **Circulation (Γ):** The circulation is defined as the line integral of the tangential component of velocity around a closed contour in the flow field.

    Circulation: $\Gamma = \oint V_f ds = \oint \vec{V} . \overrightarrow{ds}$

22. **Doublet:** A doublet is formed by combining a source and a sink of equal strength in a special way, which are allowed to approach each other in such a manner that the product of their strength and a distance between them remains a constant.

23. **Half- body – Source in a uniform stream:** Flow around a half-body is obtained by the addition of a source to a uniform flow.

ASSIGNMENT - 1

1. What are the methods of describing fluid flow?
2. Explain the terms:
   (*i*) Path line (*ii*) streak line (*iii*) streamline.
3. Differentiate between steady flow and uniform flow.

(GGSIP University, Delhi, Dec. 2005)

4. Distinguish between:
   (*i*) Steady flow and unsteady flow
   (*ii*) Uniform and non-uniform flow
   (*iii*) Rotational and irrotational flow and

(*iv*) Laminar and turbulent flow.   *(GGSIP University, Delhi, Dec. 2001)*

5. Distinguish between:

    (*i*) Laminar and turbulent flow

    (*ii*) Uniform and non-uniform flow

    (*iii*) Rotational and irrotational flow

    (*iv*) Free and forced vortex.   *(GGSIP University, Delhi, Dec. 2002)*

6. Explain briefly steady and unsteady flow.

                                             (GGSIP University, Delhi, Dec. 2008)

7. Define the equation of continuity.

8. Derive an expression for continuity equation in the differential form as given below:

$$\frac{\partial \rho}{\partial t} + \nabla.(\rho \vec{V}) = 0 \quad \text{or}$$

Derive an expression for continuity equation in three-dimensions for cartesian coordinate which is applicable to any type of flow and for any fluid whether compressible or incompressible.

9.   (*i*) Define velocity potential function and stream function.

    (*ii*) Show that the streamlines and equipotential lines form a net of mutually perpendicular lines.   *(GGSIP University, Delhi, Dec. 2005)*

10. Derive an expression for continuity equation in three dimension for cylindrical coordinate which is applicable for steady and incompressible flow.

11. Prove that in case of forced vortex, the rise of liquid level at the ends is equal to the fall of liquid level at the axis of rotation.

                                           (GGSIP University, Delhi, Dec. 2005)

12. What is free vortex? Give some examples of its occurrence. Show how the velocity and pressure vary with radius in a free vortex flow.

                                           (GGSIP University, Delhi, Dec. 2008)

13. Define the following:

    (*i*) Flow net   (*ii*) Rotation

    (*iii*) Circulation   (*iv*) Source flow

    (*v*) Sink flow

14. Explain uniform flow with source and sink. Derive expressions for stream and potential functions.

15. Explain doublet and define the strength of the doublet.

ASSIGNMENT - 2

1. The diameter of the pipe at sections 1-1 and 2-2 are 50 mm and 100 mm respectively. If the discharge through the pipe is 0.05 m³/s, find the average velocities at the two sections.   **Ans.** V_1 = 25.47 m/s, V_2 = 6.36 m/s

2. A 250 mm diameter pipe conveying water branches into two pipes of diameter 150 mm and 100 mm respectively. If the average velocities in the 250 mm diameter pipe and 150 mm are 3 m/s and 2 m/s, determine the velocity in the 100 mm diameter pipe.   **Ans.** 14.25 m/s

3. The velocity vector in a fluid flow is given as:
$$V = 4x^3 i - 10x^2 yj + 2tk$$
Find the velocity and acceleration of fluid particle at (2, 1, 3) at time t = 1.
      Ans. 51.26 units, 1568.98 units.

4. The velocity potential function is given by $\phi = 5(x^2 - y^2)$. Find the velocity components at the point (4, 5).   **Ans.** u = – 40 units, v = 50 units.

5. In a 2-D incompressible flow, the fluid velocity components are given by:
$$u = x - 4y \text{ and } v = -y - 4x$$
Show that velocity potential exists and find its form. Find also the stream function.   **Ans.** $\phi = -\dfrac{x^2}{2} + 4xy + \dfrac{y^2}{2}$, $\psi = -xy - 2x^2 + 2y^2$

6. The velocity potential function is given by $x^2 - y^2$. Find the velocity components in x and y directions.   **Ans.** u = 2x, v = –2y

7. The stream function is given by expression $2x - 5y$.
Find the velocity components and, also magnitude and direction of the resultant velocity at any point.
      Ans. u = 5 units/s, v = 2 units/s, V = 5.385 units/s, θ = 21.80°

8. The potential function is $4x(3y - 4)$. Find the velocity at the point (2, 3). Find also the value of stream function at the point (2, 3).
      Ans. V = 40 units/s, $\psi = 6x^2 - 4\left(\dfrac{3}{2}y^2 - 4y\right), -18$

9. The stream function for 2-D flow is given as $8xy$, find the velocity at the point $P(4, 5)$. Find also the velocity potential function.
      Ans. u = –32 units/s, v = 40 units/s, $\phi = 4y^2 - 4x^2$

10. The velocity vector of 2-D fluid flow is given as:
$V = (ay \sin xy)i + (ax \sin xy)j$.
Find the expression for velocity potential function.   **Ans.** $\phi = a \cos xy$

11. The velocity vector of 2-D fluid flow is given as:

$$\vec{V} = \left(8x^2y - \frac{8}{3}y^2\right)i + \left(-8xy^3 + \frac{8}{3}x^3\right)j$$

Show that the velocity vector represents a possible cases of an irrotational flow.   **Ans.** Possible flow to satisfy the continuity equation,

irrotational flow *i.e.*, $\omega_z = 0$

12. An open circular tank of 200 mm diameter and 1000 mm length contains water upto a height of 600 mm. The tank is rotated about its vertical axis at 300 rpm, find the depth of parabola formed at the free surface of liquid.

     Ans. 0.5025 m or 50.25 cm

13. An open cylinder of 150 mm diameter and 1000 mm length contains water upto a height of 800 mm. Find the speed at which the cylinder is to be rotated about its vertical axis so that no water spills.   **Ans.** N = 356.84 rpm

14. An open circular cylinder of 130 mm diameter and 800 mm length contains water upto a height of 500 mm. Find the speed at which the cylinder is to be rotated about its vertical axis, so that the axial depth becomes zero.

     Ans. N = 582.33 rpm

15. The velocity components for 3-D flow are given as:

     $u = x^2 + z^2 + 5$

     $v = y^2 + z^2$

     $w = 4xyz$

     Show whether the flow is possible or not.   **Ans.** Flow is not possible.

16. The velocity components for 2-D flow are given as:

     $u = 3x + 4y$

     $v = 2x - 3y.$

     Show whether the flow is rotational or irrotational.   **Ans.** Rotational flow.

17. For the steady incompressible flow, are the following values of u and v possible?

     (*i*) $u = 4xy + y^2$, $v = 6xy + 3x$, and

     (*ii*) $u = 2x^2 + y^2$, $v = -4xy.$

     Ans. (*i*) Flow is not possible, (*ii*) Flow is possible.

Dynamics of Fluid Flow

6.1 INTRODUCTION

In this chapter, we will study various forces acting on a fluid and their effect causing fluid flow. The dynamic behaviour of the fluid flow is analysed by the application of Newton's 2nd law of motion. According to this law,

Force : $\qquad$ $F = $ mass (m) × acceleration (a)

or $\qquad$ $F = ma$

In this equation, both force (F) and acceleration (a) are vector quantities. The motion of a fluid element in a particular direction depends upon the forces in that direction. Thus, the Newton's 2nd law of motion in the three axes is written as:

$$\Sigma F_x = ma_x$$
$$\Sigma F_y = ma_y$$
and $\qquad$ $$\Sigma F_z = ma_z$$

where F_x, F_y and F_z are the forces on the fluid element along x, y and z axes respectively and a_x, a_y and a_z accelerations along x, y and z axes respectively.

6.2 TYPES OF FORCES INFLUENCING MOTION

The following forces may be acting on a fluid, influencing its motion:

1. Gravity force: F_g
2. Pressure force: F_p
3. Viscous force: F_v
4. Turbulent force: F_t
5. Elastic force: F_e
1. *Gravity Force* (F_g): It is equal to the product of mass and acceleration due to gravity of the flowing fluid.

   Mathematically,

   Gravity force: $\qquad$ $F_g = $ mass × acceleration due to gravity

   $\qquad\qquad\qquad\qquad = Mg = \rho Vg \qquad (\because M = \rho V)$

© The Author(s) 2023
S. Kumar, *Fluid Mechanics (Vol. 1)*,
https://doi.org/10.1007/978-3-030-99762-5_6

2. *Pressure Force* (F_p): It is equal to the product of pressure intensity and cross-sectional area of the flowing fluid.

Mathematically,

Pressure force : F_p = pressure intensity × area

$$= pA$$

3. *Viscous Force* (F_v): It is equal to the product of shear stress due to viscosity and surface area of the flow.

Mathematically,

Viscous force: (F_v) = shear stress × area

$$= \tau A$$

4. *Turbulent Force* (F_t): Turbulent forces are caused in highly turbulent flows in which the velocity fluctuates with time at all points and varies in an irregular manner at different points. This results in momentum transfer between adjacent layers causing normal and shear stress.

5. *Elastic Force* (F_e): It is equal to the product of elastic stress and area of the flowing fluid. It becomes significant only in compressible fluids and also called force due to compressibility.

Mathematically,

Elastic force: F_e = elastic stress × area

$$= KA$$

6.3 EQUATIONS OF MOTION

Let all the forces are present in fluid flow along x-axis. According to Newton's 2nd law of motion, the net force acting on a fluid element along x-axis is equal to the product of mass of the fluid element and acceleration along x-axis.

$$\Sigma F_x = ma_x$$

$$\boxed{F_{gx} + F_{px} + F_{vx} + F_{tx} + F_{ex} = ma_x} \qquad ...(6.3.1)$$

Case I: If the fluid is ideal *i.e.*, viscous force : $F_{vx} = 0$, incompressible *i.e.*, elastic force (due to compressibility) : $F_e = 0$ and streamline flow *i.e.*, turbulent force $F_t = 0$, then Eq. (6.3.1) becomes

$$\boxed{F_{gx} + F_{px} = ma_x} \qquad ...(6.3.2)$$

Thus, equation (6.3.2) called **Euler's equation of motion.**

Case II: If the real fluid flow *i.e.*, viscous force is predominant $(F_{vx} \neq 0)$, incompressible *i.e.*, elastic force : $F_e = 0$ and streamline flow *i.e.*, turbulent force: $F_t = 0$.

Then, Eq. (6.3.1) is become

$$\boxed{F_{gx} + F_{px} + F_{vx} = ma_x} \qquad ...(6.3.3)$$

Thus, equation (6.3.3) is called **Navier-Stokes equation of motion.**

Case III: If the real fluid flow *i.e.*, viscous force is predominant, incompressible *i.e.*, elastic force: $F_{ex} = 0$ and turbulent flow *i.e.*, turbulent force (F_{tx}) is predominant, then Eq. (6.3.3) becomes

$$F_{gx} + F_{px} + F_{vx} + F_{tx} = ma_x$$

...(6.3.4)

Thus, equation (6.3.4) is called **Reynolds equation of motion.**

Note: This chapter is based upon **Euler's equation of motion** only because we set up assumptions as:

(*i*) Ideal fluid flow.

(*ii*) Streamline flow.

(*iii*) Incompressible fluid flow.

(*iv*) Steady and irrotational fluid flow.

6.4 SYSTEM

A system is defined as the quantity of matter (or region in space) upon which we have to make study. The mass or region outside the system is called the surroundings (*i.e.*, everything external to the system is called the surroundings). The real or imaginary surface that separates the system from its surroundings is called the boundary as shown in Fig. 6.1. The boundary of a system can be fixed or moveable. Note that the boundary is the contact surface shared by both the system and the surroundings. Mathematically speaking, the boundary has no thickness and thus it can neither contain any mass nor occupy any volume in space.

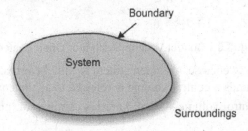

Fig. 6.1: System, surroundings and boundary.

6.5 TYPES OF SYSTEM

Systems are of following three types:

1. Control mass system (closed system).

2. Control volume system (open system).

3. Isolated system (dead system).

6.5.1 Control Mass System

A control mass system also known as closed system, it consists of a fixed amount of mass and no mass can cross its boundary. But energy in the form of heat or work, can cross the boundary and the volume of closed system is either fixed or variable.

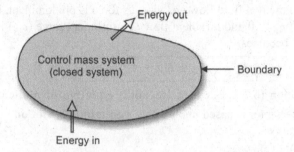

Fig. 6.2: Control Mass System or Closed System

6.5.2 Control Volume System

In this type of system, mass of fluid flows cross the boundary of system which remains fixed without any change in the volume of the system. This type of system is usually referred as open system.

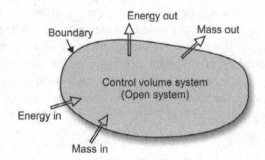

Fig. 6.3 : Control Volume System or Open System

The region usually encloses the open system is called **control volume (CV)**. The surface which surrounds a control volume is referred to as the **control surface (CS)**.

Examples of control volume system: Turbine, compressor, pump and nozzle etc.

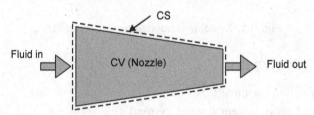

Fig. 6.4: Control Volume *(CV)* and Control Surface *(CS)*

6.5.3 Isolated System

The isolated system is one in which both mass and energy cannot cross the boundary of the system. Even there is no chemical reaction taking place within the system. For example, Thermos flask.

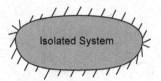

Fig. 6.5: Isolated System

6.6 REYNOLDS TRANSPORT THEOREM (RTT)

The principles of fluid mechanics are adopted from solid mechanics, where the physical laws dealing with the time, rates of change of extensive properties (mass, momentum, energy) are expressed for systems. In fluid mechanics, it is generally more convenient to work with control volumes and thus there is a need to relate the change in a control volume to the change in system. The relationship between the time rates of change of an extensive property (mass, momentum, energy) for a system and for a control volume is expressed by the Reynolds Transport Theorem (RTT), which provides the link between the system and control volume approaches.

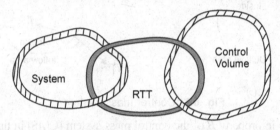

Fig. 6.6: Reynolds Transport Theorem (RTT) provides a link between the system and the control volume

Derivation of Reynolds Transport Theorem: The general form of the Reynolds transport theorem can be derived by considering a system with an arbitrary shape and arbitrary interactions but the derivation is rather involved. Consider any arbitrary control mass system as shown in fig. 6.7(a). At some initial time t, the control mass system coincides with the control volume and thus the control mass system and control volume are identical. At time $t + dt$, the control mass system has moved in the direction of flow, since each particle constituting the control mass system moves with the velocity associated with its location.

Let the volume of the control mass system and that of the control volume be $\forall_I$ at time t with both of them coinciding with each other.

At time $t + dt$, the volume of the control mass system changes and comprises volume $\forall_{III}$ and $\forall_{IV}$ and as shown in fig. 6.7 (b). Volumes $\forall_{II}$ and $\forall_{IV}$ are the intercepted regions between the control mass system and control volume at time $t + dt$.

Let B be the total amount of some extensive property (such as mass, momentum, energy) within control mass system at time t, and b be the amount of intensive property ($\frac{B}{m}$, extensive property per unit mass).

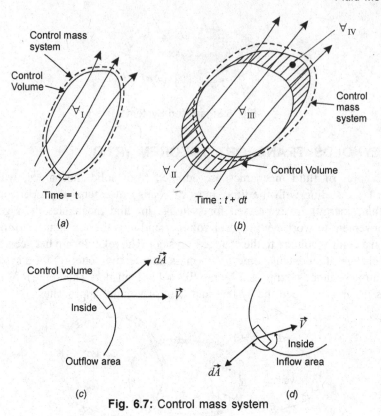

Fig. 6.7: Control mass system

The increase in property B of the control mass system (CMS) in time dt is given by

$$(B_{t+dt} - B_t)_{CMS} = \left[\iiint_{\forall_{III}} b\rho d\forall + \iiint_{\forall_{IV}} b\rho d\forall \right]_{t+dt} - \left[\iiint_{\forall_I} b\rho d\forall \right]_t$$

where subscript CMS refers to Control Mass System.

$$d\forall = \text{volume of small element.}$$

Adding and subtracting $\left[\iiint_{\forall_{II}} b\rho d\forall \right]_{t+dt}$ to the right hand side of the equation then dividing by dt on both sides, we get

$$\frac{(B_{t+dt} - B_t)_{CMS}}{dt} = \frac{\left[\iiint_{\forall_{III}} b\rho d\forall + \iiint_{\forall_{IV}} b\rho d\forall \right]_{t+dt} - \left[\iiint_{\forall_I} b\rho d\forall \right]_t}{dt}$$

$$+ \frac{\left[\iiint_{\forall_{IV}} b\rho d\forall \right]_{t+dt} - \left[\iiint_{\forall_{IV}} b\rho d\forall \right]_t}{dt} \qquad ...(6.6.1)$$

(*i*) The left side of Eq. (6.6.1) is the average time rate of increase in B within Control Mass System (CMS) during the time dt. In the limit as dt approaches zero, it becomes $\dfrac{dB}{dt}$ (the rate of change of B within the control mass system at a time t)

i.e.,
$$\lim_{dt \to 0} \frac{(B_{t+dt} + B_t)_{CMS}}{dt} = \left[\frac{dB}{dt}\right]_{CMS}$$

(*ii*) In the first term of the right side of Eq. (6.6.1), the first two integrals are the amount of B in the control volume at time $t+dt$, while the third integral is the amount of B in the control volume at time t. In the limit as dt approaches zero, this term represents the time rate of increase of the property B within the control volume and can be written as $\dfrac{\partial}{\partial t}\iiint_{CV} b\rho d\forall$

i.e.,
$$\lim_{dt \to 0} \frac{\left[\iiint_{\forall_{III}} b\rho d\forall + \iiint_{\forall_{IV}} b\rho d\forall\right]_{t+dt} - \left[\iiint_{\forall_I} b\rho d\forall\right]_t}{dt} = \frac{\partial}{\partial t}\iiint_{CV} b\rho d\forall$$

(*iii*) The next term, which is the time rate of flow of B out of the control volume may be written in the limit $dt \to 0$, as

$$\lim_{dt \to 0} \frac{\left[\iiint_{\forall_{IV}} b\rho d\forall\right]_{t+dt}}{dt} = \iiint_{\text{out flow area}} n\rho \vec{V}.d\vec{A}$$

where $\vec{V}$ = velocity vector

$d\vec{A}$ = elemental area vector on the control surface.

The sign of vector $d\vec{A}$ is positive if its direction is outward normal as shown in Fig. 6.7 (*c*).

(*iv*) Similarly, the last term of the Eq. (6.6.1) which is the rate of flow of B into the control volume is, in the limit $dt \to 0$.

$$\lim_{dt \to 0} \frac{\left[\iiint_{\forall_{II}} b\rho d\forall\right]_{t+dt}}{dt} = -\iint_{\text{in flow area}} b\rho \vec{V}.d\vec{A}$$

The minus sign is needed as $\vec{V}.d\vec{A}$ is negative for inflow.

Hence, Eq. (6.6.1) can be written as

$$\left]\frac{dB}{dt}\right[_{CMS} = \frac{d}{dt}\iiint_{CV} b\rho d\forall + \iint_{\substack{\text{out flow}\\\text{arrow}}} b\rho \vec{V}.d\vec{A} - \iint_{\substack{\text{in flow}\\\text{arrow}}} b\rho \vec{V}.d\vec{A}$$

The last two terms of above equation may be combined into a single one which is an integral over the entire surface of the control volume and is written as

$$\iint_{CS} b\eta \vec{V}.d\vec{A}.$$

This term indicates the net rate of outflow B from the control volume. Hence, above equation can be written as

$$\left[\frac{dB}{dt}\right]_{CMS} = \frac{\partial}{\partial t}\iiint_{CV} b\eta.d\forall + \iint_{CS} b\eta\vec{V}.d\vec{A} \qquad ...(6.6.2)$$

The Equation (6.2.2) is known as **Reynolds Transport Theorem (RTT)** is stated as the time rate of increase in extensive property B within a control mass system (CMS) is equal to the time rate of increase of property B within the control volume plus the net rate of efflux of the property B across the control surface.

6.7 EULER'S EQUATION OF MOTION

The following assumptions are set up in fluid flow:

(*i*) Fluid is ideal (or inviscid) *i.e.*, viscosity of the fluid is zero.

(*ii*) Stream line flow.

(*iii*) Flow is irrotational.

(*iv*) Flow is steady.

Let us consider cylindrical small fluid element AB along the direction of stream line flow (S) as shown in Fig 6.8. Let dA cross-sectional area of cylindrical fluid element and ds is length. The forces acting on the fluid element are

1. *Pressure force*: pdA at side A in the direction of flow.

2. *Pressure force*: $(p + dp)dA$ at side B in the opposite direction of fluid flow.

3. *Weight of fluid element*: $rgdA\ ds$

Let θ is the angle between the direction of fluid flow and the line of action of the weight of fluid element.

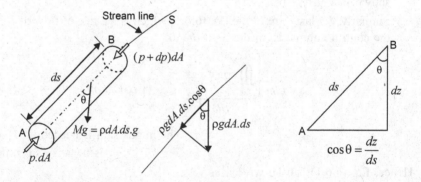

Fig. 6.8: Pressure and gravity forces on a cylindrical element along a stream line.

According to Newton's 2nd law of motion, the net force along the direction of flow (S) is equal to ma_s.

Mathematically,

$$\Sigma F_s = ma_s$$

$$pdA - (p+dp)dA - \rho gdA.ds.cos\ \theta = ma_s \qquad\qquad m = \text{density} \times \text{volume}$$

$$pdA - pdA - dp.dA - \rho gdAds \times \frac{dz}{ds} = \rho dA.ds.a_s \qquad\qquad m = \rho dA.ds$$

$$- dp.dA - \rho gdAdz = \rho dAds \times V\frac{dV}{ds} \qquad \cos\ \theta = \frac{dz}{ds}$$

$$- dp - \rho gdz = \rho VdV \qquad\qquad a_s = \frac{dV}{dt} = \frac{dV}{dt} \times \frac{ds}{ds}$$

or $$\qquad - \frac{dp}{\rho} - gdz = VdV \qquad\qquad a_s = \frac{ds}{dt} \times \frac{dV}{ds} = V\frac{dV}{ds}$$

$$\boxed{\frac{dp}{\rho} + VdV + gdz = 0} \qquad\qquad\qquad ...(6.7.1)$$

Equation (6.7.1) is known as **Euler's equation of motion.**

6.8 BERNOULLI'S EQUATION

Assumptions of Bernoulli's equation:

 (*i*) Fluid is ideal (or inviscid) *i.e.*, viscosity of the fluid is zero.

 (*ii*) Flow is steady.

 (*iii*) Streamline flow.

 (*iv*) Fluid is incompressible.

 (*v*) Flow is irrotational, and

 (*vi*) No shaft work and heat interaction.

We know that the Euler's equation of motion,

$$\frac{dp}{\rho} + VdV + gdz = 0$$

Bernoulli's equation is obtained by integration of Euler's equation of motion,

$$\int \frac{dp}{\rho} + \int VdV + \int gdz = \text{constant}$$

$$\frac{p}{\rho} + \frac{V^2}{2} + gz = \text{constant}$$

or $$\qquad \frac{p}{\rho g} + \frac{V^2}{2g} + z = \mathbf{C} \qquad\qquad ...(6.8.1)$$

Equation (6.8.1) is called **Bernoulli's equation**

where $$\qquad \frac{p}{\rho g} = \frac{p}{w} = \text{pressure energy per unit weight of fluid or}$$

$$\text{pressure head.}$$

also $\dfrac{p}{\rho g}$ = flow work (or flow energy) per unit weight.

                                $\dfrac{V^2}{2g}$ = kinetic energy per unit weight or kinetic head.

and   z = potential (or datum) energy per unit weight or
                                potential (or datum) head. It is also called elevation
                                head.

Statement of Bernoulli's Theorem:

It states that, *the total energy (i.e., sum of pressure, kinetic and potential energies) at every point in steady, irrotational, incompressible and non-viscous flow is constant.*

OR

The total mechanical energy* at every point in steady, irrotational, incompressible and non-viscous flow is constant.

Form of Bernoulli's equation:

Bernoulli's equation has three forms as:

Ist form $\dfrac{p}{\rho g} + \dfrac{V^2}{2g} + z = C$ ...(6.8.2)

IInd form $\dfrac{p}{\rho} + \dfrac{V^2}{2} + gz = C$ ...(6.8.3)

IIIrd form $p + \dfrac{\rho V^2}{2} + \rho gz = C$ ...(6.8.4)

The **Ist form of Bernoulli's equation** (6.8.2) state that *the total energy per unit weight of fluid at every point in the flow is constant.*

OR

The total head (i.e., sum of pressure, kinetic and datum heads) of fluid at every point in the flow is constant.

Units : $\dfrac{Nm}{N}$ or m

$\dfrac{p}{\rho g} = \dfrac{N}{m^2 \dfrac{kg}{m^3} \cdot \dfrac{m}{s^2}} = \dfrac{N}{\dfrac{kg}{s^2}} = \dfrac{Nm}{\dfrac{kg\,m}{s^2}} = \dfrac{Nm}{N}$ or   m   (1 N = 1 kgm/s^2)

$\dfrac{V^2}{2g} = \dfrac{m^2}{s^2 \dfrac{m}{s^2}} = m = \dfrac{Nm}{N}$

and $z = m = \dfrac{N.m}{N}$

---

* Mechanical energy: The hydraulic energy is also called high grade energy *i.e.,* mechanical energy

The **2nd form of Bernoulli's equation** (6.8.3) state that *the total energy per unit mass of fluid at every point in the flow is constant.*

Units : $\dfrac{Nm}{kg}$

$$\frac{p}{\rho} = \frac{N}{\dfrac{m^2 - kg}{m^3}} = \frac{Nm}{kg}$$

$$\frac{V^2}{2} = \frac{m^2}{s^2} = \frac{m^2}{s^2} \times \frac{kg}{kg} = \frac{kg\,m\,m}{s^2 kg} = \frac{Nm}{kg}$$

and

$$gz = \frac{m.m}{s^2} = \frac{m^2}{s^2} \times \frac{kg}{kg} = \frac{kg\,m\,m}{s^2 kg} = \frac{Nm}{kg}$$

The **3rd form of Bernaulli's equation** (6.8.4) states *that the total energy per unit volume of fluid at every point in the flow is constant.*

Unit : $\dfrac{Nm}{m^3}$ or $\dfrac{N}{m^2}$

$$p = \frac{N}{m^2} = \frac{N.m}{m^2.m} = \frac{Nm}{m^3} = \frac{N}{m^2}$$

$$\frac{\rho V^2}{2} = \frac{kg}{m^3} \cdot \frac{m^2}{s^2} = \frac{kg.m}{s^2} \cdot \frac{m}{m^3} = \frac{Nm}{m^3} = \frac{N}{m^2}$$

and

$$\rho gz = \frac{kg}{m^3} \cdot \frac{m.m}{s^2} = \frac{kg.m}{s^2} \cdot \frac{m}{m^3} = \frac{Nm}{m^3} = \frac{N}{m^2}$$

The Ist form of Bernoulli's equation is normally used in flow field.

$$\frac{p}{\rho g} + \frac{V^2}{2g} + z = C \qquad (\because \text{ Specific weight: } w = \rho g)$$

or

$$\frac{p}{w} + \frac{V^2}{2g} + z = C$$

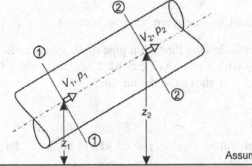

Fig. 6.9: Flow through inclined pipe

Application of the Bernoulli's equation at section (1) – (1) and (2) – (2), we get

$$\frac{p_1}{w} + \frac{V_1^2}{2g} + z_1 = \frac{p_2}{w} + \frac{V_2^2}{2g} + z_2$$

6.9 LIMITATIONS OF BERNOULLI'S THEOREM

The limitations of Bernoulli's theorem are implied in its statement and proof. We take assumptions:

(*i*) Fluid is ideal.

(*ii*) Flow is steady.

(*iii*) Streamline flow.

(*iv*) Fluid is incompressible.

(*v*) Flow is irrotational.

(*vi*) No shaft work and heat interaction.

But real fluid is not ideal, steady, streamline and irrotational flow, and also is not maintained adiabatic flow (*i.e.*, heat interaction zero). These are serious limitations. A real fluid does possess viscosity and consequently it offers resistance to its flow. In order to overcome this viscous resistance and other resistance due to surface roughness and turbulence, a part of the total energy of flow is lost. This energy loss, in this strict sense of the word, is not a lost energy, as from the first law of thermodynamics, energy can neither be created nor destroyed, but is converted into heat energy, thereby increasing the temperature of the fluid slightly. The increase of temperature of the fluid causes an increase in the internal energy. The increase in internal energy and the heat transfer from the fluid represent a loss of useful energy. The total loss of energy in real fluid due to frictional effect and heat transfer per unit weight is called the loss of head (h_L).

Mathematically,

$$\text{Loss of head: } h_L = \frac{u_2 - u_1 - q}{g} \quad \frac{\text{Nm}}{\text{N}}$$

The modified Bernoulli's equation may be used in real fluid flow system

$$\frac{p_1}{w} + \frac{V_1^2}{2g} + z_1 = \frac{p_2}{w} + \frac{V_2^2}{2g} + z_2 + h_L$$

where h_L is the head loss between the two sections.

Problem 6.1: Water is flowing through a pipe of 80 mm diameter under a gauge pressure of 60 kPa, and with a mean velocity of 2 m/s. Neglecting friction, find the total head, if the pipe is 7 m above the datum line.

Solution: Given data:

Diameter of pipe:	$d = 80$ mm $= 0.08$ m
Gauge pressure of water:	$p = 60$ kPa $= 60 \times 10^3$ Pa or N/m^2
Mean velocity of water:	$V = 2$ m/s
Datum head:	$z = 7$ m

According to Bernoulli's equation,

Total head of water: $H = \dfrac{p}{\rho g} + \dfrac{V^2}{2g} + z = \dfrac{60 \times 10^3}{1000 \times 9.81} + \dfrac{(2)^2}{2 \times 9.81} + 7$

$$= 6.11 + 0.20 + 7 = \mathbf{13.31 \ m \ of \ water.}$$

Problem 6.2: A pipe of 250 mm diameter carries an oil of specific gravity 0.8 at the rate of 130 litre/s and under a pressure of 3 kPa. Find the total energy per unit weight at a point which is 3 m above the datum line. Find also the total energies per unit mass and per unit volume.

Solution: Given data:

Diameter of pipe: $d = 250 \ mm = 0.250 \ m$

Specific gravity of oil: $S = 0.8$

∴ Density of oil: $\rho = 0.8 \times 1000 = 800 \ kg/m^3$

Discharge through pipe: $Q = 130 \ \text{litre/s} = \dfrac{130}{1000} \ m^3/s \ (\because \ 1 \ \text{litre} = \dfrac{1}{1000} \ m^3)$

$$= 0.13 \ m^3/s$$

Pressure of oil in pipe: $p = 3 \ kPa = 3 \times 10^3 \ Pa \ \text{or} \ N/m^2$

Datum head: $z = 3 \ m$

Cross-sectional area of pipe: $A = \dfrac{\pi}{4} \ d^2 = \dfrac{3.14 \times (0.25)^2}{4} = 0.0490 \ m^2$

We know,   Discharge: $Q = AV$

or $V = \dfrac{Q}{A} = \dfrac{0.13}{0.0490} = 2.65 \ m/s$

We know,

Total energy per unit weight $= \dfrac{p}{\rho g} + \dfrac{V^2}{2g} + z$

$$= \dfrac{3 \times 10^3}{800 \times 9.81} + \dfrac{(2.65)^2}{2 \times 9.81} + 3$$

$$= 0.3822 + 0.3579 + 3$$

$$= \mathbf{3.74} \ \dfrac{\mathbf{Nm}}{\mathbf{N}} \ \text{or} \ \mathbf{m}$$

and

Total energy per unit mass $= \dfrac{p}{\rho} + \dfrac{V^2}{2} + gz$

$$= g \times \text{total energy per unit weight}$$

$$= 9.81 \times 3.74 = \mathbf{36.68} \ \dfrac{\mathbf{Nm}}{\mathbf{kg}} \ \text{or} \ \mathbf{J/kg}$$

$$\text{Total energy per unit volume} = p + \frac{\rho V^2}{2} + \rho g z$$

$$= \rho \times \text{total energy per unit mass}$$
$$= 800 \times 36.68$$

$$= 29344 \ \frac{\text{N.m}}{\text{m}^3} \ \text{or N/m}^2$$

$$= 29344 \ \text{J/m}^3 = \mathbf{29.34 \ kJ/m^3}$$

Problem 6.3: A pipeline carrying oil of specific gravity 0.8, changes in diameter from 300 mm at a position A to 500 mm diameter of a position B, which is 5 m at a higher level. If the pressures at A and B are 200 kPa and 152 kPa respectively and discharge is 150 litre/s, determine the loss of head and direction of flow.

(GGSIP University, Delhi, Dec. 2008)

Solution : Given data:

Specific gravity of oil: $S = 0.8$

$\therefore$ Density of oil: $\rho = S \times 1000 \ \text{kg/m}^3 = 0.8 \times 1000 \ \text{kg/m}^3$
$$= 800 \ \text{kg/m}^3$$

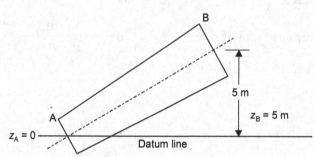

Fig. 6.10: Schematic for Problem 6.3

Diameter at position A: $d_A = 300 \ \text{mm} = 0.3 \ \text{m}$

$\therefore$ Cross-sectional area: $A_A = \frac{\pi}{4} d_A^2 = \frac{3.14}{4}(0.3)^2 = 0.07065 \ \text{m}^2$

Diameter at position B: $d_B = 500 \ \text{mm} = 0.5 \ \text{m}$

$\therefore$ Cross-sectional area: $A_B = \frac{\pi}{4} d_B^2 = \frac{3.14}{4}(0.5)^2 = 0.19625 \ \text{m}^2$

Datum head at point A: $z_A = 0$

Datum head at point B: $z_B = 5 \ \text{m}$

Pressure at point A: $p_A = 200 \ \text{kPa} = 200 \times 10^3 \ \text{N/m}^2$

Pressure at point B: $p_B = 152 \ \text{kPa} = 152 \times 10^3 \ \text{N/m}^2$

Discharge: $Q = 150 \ \text{litre/s} = \frac{150}{1000} \ \text{m}^3/\text{s} = 0.150 \ \text{m}^3/\text{s}$

also $$Q = A_A V_A = A_B V_B \text{ (By continuity equation)}$$

$$V_A = \frac{Q}{A_A} = \frac{0.150}{0.07065} = 2.12 \text{ m/s}$$

and $$V_B = \frac{Q}{A_B} = \frac{0.150}{0.19625} = 0.7643 \text{ m/s}$$

Total energy available at position A: E_A

$$E_A = \frac{p_A}{\rho g} + \frac{V_A^2}{2g} + z_A = \frac{200 \times 10^3}{800 \times 9.81} + \frac{(2.12)^2}{2 \times 9.81} + 0$$

$$= 25.48 + 0.229 = 25.709 \frac{\text{Nm}}{\text{m}} \text{ or m}$$

Total energy available at position B: E_B

$$E_B = \frac{p_B}{\rho g} + \frac{V_B^2}{2g} + z_B$$

$$= \frac{152 \times 10^3}{800 \times 9.81} + \frac{(0.7643)^2}{2 \times 9.81} + 5$$

$$= 19.36 + 0.029 + 5 = 24.389 \frac{\text{Nm}}{\text{N}} \text{ or m}$$

(i) Loss of head: $h_f = E_A - E_B = 25.709 - 24.389$
$$= \mathbf{1.32 \text{ m of oil.}}$$

(ii) Direction of flow: As $E_A > E_B$ and hence **flow is taking place from A to B.**

Problem 6.4: The oil is flowing through a pipe of 80 mm diameter under a gauge pressure of 100 kPa, and with a mean velocity of 2.5 m/s. The specific gravity of oil is 0.85. Neglecting friction, find the total head, if the pipe is 7 m above the datum line.

Solution: Given data :

Diameter of pipe: $d = 80$ mm $= 0.08$ m
Pressure of oil: $p = 100$ kPa $= 100 \times 10^3$ Pa or N/m²
Velocity of oil: $V = 2.5$ m/s
Specific gravity of oil: $S = 0.85$
∴ Density of oil: $\rho = 0.85 \times 1000 = 850$ kg/m³
Datum head: $z = 7$ m

We know that the total head (H) of oil in pipe, according to Bernoulli's equation (energy per unit weight):

$$\text{Total head: } H = \frac{p}{\rho g} + \frac{V^2}{2g} + z$$

$$= \frac{100 \times 10^3}{850 \times 9.81} + \frac{(2.5)^2}{2 \times 9.81} + 7 = \mathbf{19.31 \text{ m of oil.}}$$

Problem 6.5: A pipeline carrying oil of specific gravity 0.8, changes in diameter from 300 mm at a position A to 500 mm diameter to position B which is 5 m at a higher level. If the pressure at A and B are 19.62 N/cm^2 and 14.91 N/cm^2 respectively, and discharge is 150 litres/s, determine the loss of head and direction of flow.

Solution: Given data :

Specific gravity of oil : $\quad S = 0.8$

$\therefore$ Density of oil : $\qquad \rho = S \times 1000$ kg/m^3

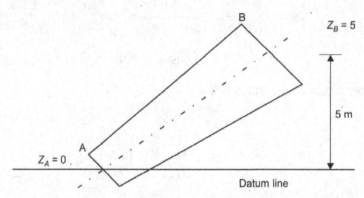

Fig. 6.11: Schematic for Problem 6.5

$$= 0.8 \times 1000 \text{ kg/m}^3 = 800 \text{ kg/m}^3$$

Diameter at position A; $d_A = 300$ mm $= 0.3$ m

$\therefore$ Cross-sectional area : $A_A = \dfrac{\pi}{4} d_A^2 = \dfrac{3.14}{4} \times (0.3)^2 = 0.07065$ m^2

Diameter at position B : $d_B = 500$ mm $= 0.5$ m

$\therefore$ Cross-sectional area : $A_B = \dfrac{\pi}{4} d_B^2 = \dfrac{3.14}{4} \times (0.5)^2 = 0.19625$ m^2

Datum head at point A : $z_A = 0$

Datum head at point B : $z_B = 5$ m

Pressure at point A : $\quad p_A = 19.62$ N/cm$^2 = 19.62 \times 10^4$ N/m^2

Pressure at point B : $\quad p_B = 14.91$ N/cm$^2 = 14.91 \times 10^4$ N/m^2

Discharge : $\qquad Q = 150$ litre/s

$$= \dfrac{150}{1000} \text{ m}^3/\text{s} = 0.150 \text{ m}^3/\text{s}$$

also $\qquad Q = A_A V_A = A_B V_B \qquad$ (by continuity equation)

$$V_A = \dfrac{Q}{A_A} = \dfrac{0.150}{0.07065} = 2.12 \text{ m/s}$$

and $\qquad V_B = \dfrac{Q}{A_B} = \dfrac{0.150}{0.19625} = 0.7643 \text{ m/s}$

Total energy per unit weight at position $A : E_A$

$$E_A = \frac{p_A}{\rho g} + \frac{V_A^2}{2g} + z_A = \frac{19.62 \times 10^4}{800 \times 9.81} + \frac{(2.12)^2}{2 \times 9.81} + 0$$

$$= 25 + 0.229 = 25.229 \ \frac{Nm}{N} \ \text{or m}$$

Total energy per unit weight at position $B : E_B$

$$E_B = \frac{p_B}{\rho g} + \frac{V_B^2}{2g} + z_B = \frac{14.91 \times 10^4}{800 \times 9.81} + \frac{(0.7643)^2}{2 \times 9.81} + 5$$

$$= 18.998 + 0.029 + 5 = 24.027 \ \frac{Nm}{N} \ \text{or m}$$

(i) Loss of head : $h_f = E_A - E_B = 25.229 - 24.027 = \textbf{1.202 m of oil.}$

Problem 6.6: Water flows through the combination of pipes and nozzle as shown in Fig. 6.12. Find the diameter of pipe at section (2). The suitable data at sections (1) and (2) are given. Neglecting fiction losses.

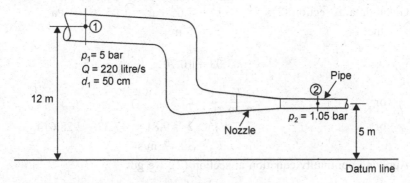

Fig. 6.12: Schematic for Problem 6.6

Solution: Given data:

Diameter of pipe at section (1): $d_1 = 50$ cm $= 0.50$ m

$\therefore$ Cross-sectional area at section (1): $A_1 = \frac{\pi}{4} d_1^2 = \frac{3.14}{4}(0.5)^2 = 0.19625$ m^2

Pressure at section (1): $p_1 = 5$ bar $= 5 \times 10^5$ N/m^2

Discharge: $Q = 220$ litre/s $= \dfrac{220}{1000}$ m^3/s

$$= 0.22 \ \text{m}^3/\text{s}$$

Applying continuity equation at section (1), we get

$$Q = A_1 V_1$$

or $\qquad$ Velocity: $V_1 = \dfrac{Q}{A} = \dfrac{0.22}{0.19625} = 1.121$ m/s

Datum head at section (1): $\qquad z_1 = 12$ m

We know that the total head at section (1), according to Bernoulli's equation:

$$\text{Total head: } H_1 = \frac{p_1}{\rho g} + \frac{V_1^2}{2g} + z_1$$

$$= \frac{5 \times 10^5}{1000 \times 9.81} + \frac{(1.125)^2}{2 \times 9.81} + 12$$

$$= 50.968 + 0.0645 + 12$$

$$= 63.03 \text{ m of water.}$$

Also we know that the total head at every section in the flow is constant [neglecting friction losses].

i.e., Total head of section (1): $\quad H_1 = H_2$: Total head at section (2)

$$63.03 = \frac{p_2}{\rho g} + \frac{V_2^2}{2g} + z_2$$

$$= \frac{1.05 \times 10^5}{1000 \times 9.81} + \frac{V_2^2}{2 \times 9.81} + 5$$

Given data at section (2): $\qquad p_2 = 1.05$ bar $= 1.05 \times 10^5$ Pa

and $\qquad\qquad\qquad z_2 = 5$ m

$\therefore \qquad\qquad\qquad$ $63.03 = 10.70 + \dfrac{V_2^2}{2 \times 9.81} + 5$

or $\qquad\qquad\qquad \dfrac{V_2^2}{2 \times 9.81} = 47.33$

or $\qquad\qquad\qquad V_2^2 = 2 \times 9.81 \times 47.33 = 928.614$

$\qquad\qquad\qquad\qquad V_2 = 30.47$ m/s

Applying continuity equation at section (2), we get

$$Q = A_2 V_2$$

$$0.22 = A_2 \times 30.47$$

or $\qquad\qquad\qquad A_2 = 7.22 \times 10^{-3}$ m^2

also $\qquad\qquad\qquad A_2 = \dfrac{\pi}{4} d_2^2$

$\therefore \qquad\qquad\qquad 7.22 \times 10^{-3} = \dfrac{3.14}{4} \times d_2^2$

or $\qquad\qquad\qquad d_2^2 = 9.1977 \times 10^{-3}$ m^2

or $\qquad\qquad\qquad d_2 = 0.0959$ m $= \mathbf{9.59}$ **cm**

Problem 6.7: The diameter of a pipe changes from 300 mm at a section 6 m above datum to 100 mm at a section 3 m above datum. The pressure of water at section (1) is 400 kPa. If the velocity flow at section (1) is 2 m/s. Find the pressure at section (2).

Solution: Given data:

Diameter at section (1): $\qquad d_1 = 300$ mm $= 0.30$ m

$\therefore$ Cross-sectional area at section (1): $A_1 = \dfrac{\pi}{4} \times d_1^2 = \dfrac{3.14}{4} \times (0.3)^2$

$\qquad\qquad\qquad\qquad\qquad\qquad = 0.07065$ m^2

Datum head at section (1): $\qquad z_1 = 6$ m

Diameter at section (2): $\qquad d_2 = 100$ mm $= 0.1$ m

$\therefore$ Cross-sectional area at section (2): $\qquad A_2 = \dfrac{\pi}{4} d_2^2 = \dfrac{3.14}{4} \times (0.1)^2 = 7.85 \times 10^{-3}$ m^2

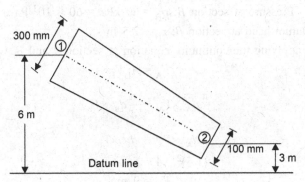

300 mm

6 m

Datum line

100 mm

3 m

Fig. 6.13: Schematic for Problem 6.7

Datum head at section (2): $z_2 = 3$ m

Pressure of water at section (1): $p_1 = 400$ kPa $= 400 \times 10^5$ Pa or N/m^2

Velocity of water at section (1): $V_1 = 2$ m/s

Applying the continuity equation at section (1) and (2), we get

$$A_1 V_1 = A_2 V_2$$
$$0.07065 \times 2 = 7.85 \times 10^{-3} \times V_2$$

or $\qquad\qquad\qquad\qquad V_2 = \textbf{18 m/s}$

Now applying the Bernoulli's equation at sections (1) and (2), we get

$$\frac{p_1}{\rho g} + \frac{V_1^2}{2g} + z_1 = \frac{p_2}{\rho g} + \frac{V_2^2}{2g} + z_2$$

$$\frac{400 \times 10^3}{1000 \times 9.81} + \frac{(2)^2}{2 \times 9.81} + 6 = \frac{p_2}{1000 \times 9.81} + \frac{(18)^2}{2 \times 9.81} + 3$$

$$40.77 + 0.2038 + 6 = \frac{p_2}{9810} + 16.5137 + 3$$

or $\qquad\qquad\qquad\qquad \dfrac{p_2}{9810} = 27.460$

or $\qquad\qquad\qquad\qquad p_2 = 9810 \times 27.46$ N/m^2 or Pa

$$= \frac{9810 \times 27.46}{1000} \text{ kPa} = \textbf{269.38 kPa}$$

Problem 6.8: At a certain section A of a pipeline carrying water, the diameter is 1.5 m, the pressure is 100 kPa and the velocity is 4 m/s. At another section B which is 2.5 m higher than A, the diameter is 1 m and the pressure is 60 kPa. What is the direction of flow?

Solution: Given data:

Diameter at section A: $d_A = 1.5$ m

Pressure at section A: $p_A = 100$ kPa $= 100 \times 10^3$ Pa or N/m²

Velocity at section A: $V_A = 4$ m/s

Diameter at section B: $d_B = 1$ m

Pressure at section B: $p_B = 60$ kPa $= 60 \times 10^3$ Pa or N/m²

Datum head at section B: $z_B = 2.5$ m

Now applying the continuity equation at section A and B, we get

$$A_A V_A = A_B V_B$$

$$\frac{\pi}{4} d_A^2 \times V_A = \frac{\pi}{4} d_B^2 \times V_B$$

$$d_A^2 \times V_A = d_B^2 \times V_B$$

$$(1.5)^2 \times 4 = 1^2 \times V_B$$

or $$V_B = 9 \text{ m/s}$$

Total energy per unit weight at section A,

$$= \frac{p_A}{\rho g} + \frac{V_A^2}{2g} + z_A$$

where $z_A = 0$, because section A lies on datum line

$$= \frac{p_A}{\rho g} + \frac{V_A^2}{2g} = \frac{100 \times 10^3}{1000 \times 9.81} + \frac{(4)^2}{2 \times 9.81}$$

$$= 10.19 + 0.81$$

$$= 11 \frac{\text{Nm}}{\text{N}} \text{ or m}$$

Total energy per unit weight at section B,

$$= \frac{p_B}{\rho g} + \frac{V_B^2}{2g} + z_B$$

$$= \frac{60 \times 10^3}{1000 \times 9.81} + \frac{(9)^2}{2 \times 9.81} + 2.5$$

$$= 6.116 + 4.128 + 2.5 = 12.744 \frac{\text{Nm}}{\text{N}} \text{ or m}$$

Since the total energy at section B is greater than at section A, so **the flow direction is from section B to section A.**

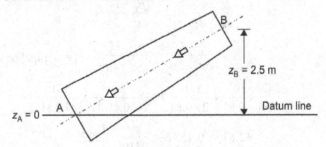

Fig. 6.14: Schematic for Problem 6.8

Problem 6.9: At a point in the pipeline the diameter is 300 mm, the velocity of water is 3 m/s and the pressure is 420 kN/m². At a point 16 m downstream the diameter gradually reduces to 150 mm. Find the pressure at this point, if the pipe is (*i*) horizontal, (*ii*) vertical with flow downward, and (*iii*) vertical with flow upward.

Solution: Given data:

At section (1),

$$d_1 = 300 \text{ mm} = 0.3 \text{ m}$$
$$V_1 = 3 \text{ m/s}$$
$$p_1 = 420 \text{ kN/m}^2 = 420 \times 10^3 \text{ N/m}^2$$

At section (2),

Distance between two sections $= 16$ m

$$d_2 = 150 \text{ mm} = 0.15 \text{ m}$$
$$p_2 = ?$$

Case I: When pipeline is horizontal:

Applying continuity equation at section (1) and (2), we get

$$A_1 V_1 = A_2 V_2$$

$$\frac{\pi}{4} d_1^2 \times V_1 = \frac{\pi}{4} d_2^2 \times V_2$$

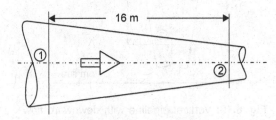

Fig. 6.15: Horizontal pipeline.

$$d_1^2 \times V_1 = d_2^2 \times V_2$$

$$(0.3)^2 \times 3 = (0.15)^2 \times V_2$$

or

$$V_2 = 12 \text{ m/s}$$

Now applying the Bernoulli's equation at section (1) and (2), we get

$$\frac{p_1}{\rho g}+\frac{V_1^2}{2g}+z_1 = \frac{p_2}{\rho g}+\frac{V_2^2}{2g}+z_2$$

$$\frac{p_1}{\rho g}+\frac{V_1^2}{2g} = \frac{p_2}{\rho g}+\frac{V_2^2}{2g} \; (\because z_1 = z_2, \text{pipeline is horizontal})$$

$$\frac{420\times10^3}{1000\times9.81}+\frac{(3)^2}{2\times9.81} = \frac{p_2}{1000\times9.81}+\frac{(12)^2}{2\times9.81}$$

$$42.81 + 0.4587 = \frac{p_2}{9810}+7.337$$

or
$$\frac{p_2}{9810} = 35.931$$

or
$$p_2 = 35.931 \times 9810 \text{ N/m}^2$$

$$= \frac{35.931\times9810}{1000} \text{ kN/m}^2 = \mathbf{352.48 \text{ kN/m}^2}$$

Case-II : Pipeline vertical with downward flow:

Applying the Bernoulli's equation at sections (1) and (2), we get

$$\frac{p_1}{\rho g}+\frac{V_1^2}{2g}+z_1 = \frac{p_2}{\rho g}+\frac{V_2^2}{2g}+z_2$$

$$\frac{p_1}{\rho g}+\frac{V_1^2}{2g}+(z_1-z_2) = \frac{p_2}{\rho g}+\frac{V_2^2}{2g}$$

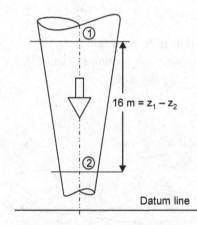

Fig. 6.16: Vertical pipeline with downward flow

$$\frac{420\times10^3}{1000\times9.81}+\frac{(3)^2}{2\times9.81}+16 = \frac{p_2}{1000\times9.81}+\frac{(12)^2}{2\times9.81}$$

$$42.81 + 0.4587 + 16 = \frac{p_2}{9810} + 7.337$$

or $$\frac{p_2}{9810} = 51.931$$

or $$p_2 = 51.931 \times 9810 \text{ N/m}^2$$

$$= \frac{51.931 \times 9810}{1000} \text{ kN/m}^2 = \textbf{509.44 kN/m}^2$$

Case-III: Pipeline vertical with upward flow:

Applying the Bernoulli's equation at sections (1) and (2), we get

$$\frac{p_1}{\rho g} + \frac{V_1^2}{2g} + z_1 = \frac{p_2}{\rho g} + \frac{V_2^2}{2g} + z_2$$

$$\frac{p_1}{\rho g} + \frac{V_1^2}{2g} = \frac{p_2}{\rho g} + \frac{V_2^2}{2g} + (z_2 - z_1)$$

$$\frac{420 \times 10^3}{1000 \times 9.81} + \frac{(3)^2}{2 \times 9.81} = \frac{p_2}{1000 \times 9.81} + \frac{(12)^2}{2 \times 9.81} + 16$$

$$42.81 + 0.4587 = \frac{p_2}{9810} + 7.337 + 16$$

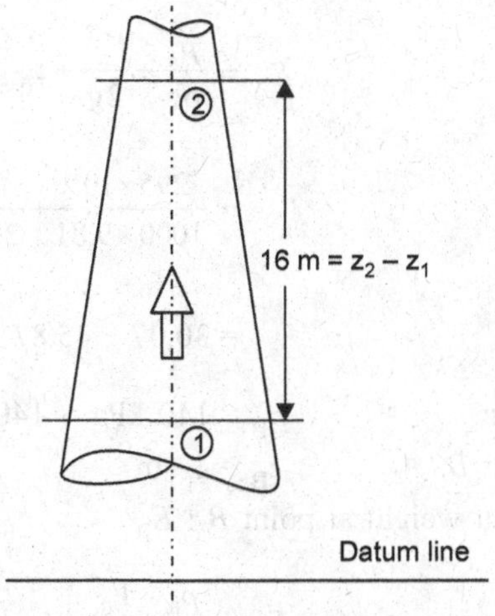

16 m = $z_2 - z_1$

Datum line

Fig. 6.17: Vertical pipeline with upward flow

or $$\frac{p_2}{9810} = 19.931$$

or $$p_2 = 19.931 \times 9810 \text{ N/m}^2$$

$$= \frac{19.931 \times 9810}{1000} \text{ kN/m}^2 = \textbf{195.52 kN/m}^2$$

Problem 6.10: A pipe of diameter 300 mm carries water at a velocity of 30 m/s. The pressure at the points A to B are given as 295 kPa and 140 kPa respectively while the datum head at A and B are 30 m and 42 m. Find the loss of head between A and B, the direction of flow from point A to B.

Solution: Given data:

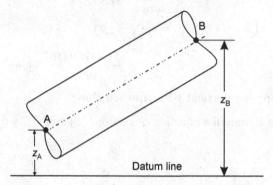

Fig. 6.18: Vertical pipeline with downward flow

Diameter of pipe: $\qquad d = 300$ mm $= 0.3$ m

Velocity of water: $\qquad V = 30$ m/s

Pressure at point A: $\qquad p_A = 295$ kPa $= 295 \times 10^3$ Pa

Datum heat at point A : $\qquad z_A = 30$ m

Total energy per unit weight at point A: E_A

$$E_A = \frac{p_A}{\rho g} + \frac{V_A^2}{2g} + z_A \qquad (\because V_A = V = 30 \text{ m/s})$$

$$= \frac{295 \times 10^3}{1000 \times 9.81} + \frac{(30)^2}{2 \times 9.81} + 30$$

$$= 30.07 + 45.87 + 30 = 105.94 \; \frac{\text{Nm}}{\text{N}} \text{ or m}$$

Pressure at point B: $\qquad p_B = 140$ kPa $= 140 \times 10^3$ Pa or N/m^2

Datum head at point B: $\qquad z_B = 42$ m

Total energy per unit weight at point B : E_B

$$E_B = \frac{p_B}{\rho g} + \frac{V_B^2}{2g} + z_B$$

$$= \frac{140 \times 10^3}{1000 \times 9.81} + \frac{(30)^2}{2 \times 9.81} + 42$$

$$(\because V_B = V = 30 \text{ m/s})$$

$$= 14.2 + 45.87 + 42 = 102.07 \; \frac{\text{N.m}}{\text{N}} \text{ or m}$$

Applying the modified Bernoulli's equation for real fluid at point A and B, we get total energy per unit weight at point A : E_A = total energy per unit weight at point B: E_B + loss of head: h_L

$$E_A = E_B + h_L$$
$$105.94 = 102.07 + h_L$$

or
$$h_L = 105.94 - 102.07 = \textbf{3.87 m of water}$$

Problem 6.11:   A conical tube of length 2.5 m is fixed vertically with its smaller and upwards. The velocity of flow at the smaller end is 6 m/s while at the lower end is 3.5 m/s. The pressure head at the smaller end is 3 m of oil (specific gravity of oil is 0.85). The loss of head in the tube is $\dfrac{0.45(V_1 - V_2)^2}{2g}$, where V_1 is the velocity at the smaller end and V_2 is velocity at the lower end respectively. Find the pressure at the lower end. The flow of oil takes place in the downward direction.

Solution: Given data:

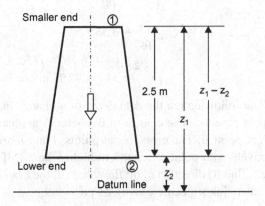

Fig. 6.19: Schematic for Problem 6.11

Length of tube: $\quad\quad\quad\quad\quad\quad\quad\quad$ $L = 2.5$ m
Also datum head different: $\quad$ $(z_1 - z_2) = L = 2.5$ m
Velocity of flow at smaller end: $\quad$ $V_1 = 6$ m/s
Velocity of flow at lower end: $\quad$ $V_2 = 3.5$ m/s

Pressure head at smaller end: $\quad$ $\dfrac{p_1}{\rho g} = 3$ m of oil

Specific gravity of oil: $\quad\quad\quad$ $S = 0.85$
∵ Density of oil: $\quad\quad\quad\quad\quad$ $\rho = S \times 1000 = 0.85 \times 1000 = 850$ kg/m^3

Loss of head: $\quad\quad\quad\quad\quad\quad$ $h_L = \dfrac{0.45 \times (V_1 - V_2)^2}{2g}$

$$= \dfrac{0.45 \times (6 - 3.5)^2}{2 \times 9.81} = 0.143 \text{ m of oil}$$

Applying the modified Bernoulli's equation for real fluid at sections (1) and (2), we get

$$\frac{p_1}{\rho g} + \frac{V_1^2}{2g} + z_1 = \frac{p_2}{\rho g} + \frac{V_2^2}{2g} + z_2 + h_L$$

$$\frac{p_1}{\rho g} + \frac{V_1^2}{2g} + (z_1 - z_2) = \frac{p_2}{\rho g} + \frac{V_2^2}{2g} + h_L$$

$$3 + \frac{(6)^2}{2 \times 9.81} + 2.5 = \frac{p_2}{\rho g} + \frac{(3.5)^2}{2 \times 9.81} + 0.143$$

$$3 + 1.834 + 2.5 = \frac{p_2}{\rho g} + 0.624 + 0.143$$

$$7.334 = \frac{p_2}{\rho g} + 0.767$$

or

$$\frac{p_2}{\rho g} = 6.567 \text{ m}$$

or

$$p_2 = \rho g \times 6.567 = 850 \times 9.81 \times 6.567$$
$$= 54758.92 \text{ N/m}^2 \text{ or Pa} = \textbf{54.758 kPa}$$

Problem 6.12: The following are the data given of a change in diameter effected in laying an oil supply pipeline. The change in diameter is gradual from 200 mm at point A to 500 mm at point B. The pressures at points A and B are 78.5 kN/m² and 58.9 kN/m² respectively with point B being 3 m higher than A. If the flow of oil in pipeline is 200 litre/s, find (*i*) direction of oil flow, and (*ii*) the head lost due to friction between point A and B. Take specific gravity of oil is 0.9.

Solution: Given data:

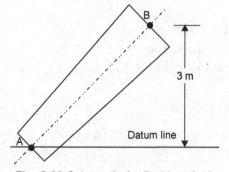

Fig. 6.20 Schematic for Problem 6.12

Diameter at point A: $d_A = 200$ mm $= 0.2$ m

Diameter at point B: $d_B = 500$ mm $= 0.5$ m

The pressure at point A: $p_A = 78.5$ kN/m² $= 78.5 \times 10^3$ N/m²

The pressure at point B: $p_B = 58.9$ kN/m² $= 58.9 \times 10^3$ N/m²

Datum head at point A: $z_A = 0$

Datum head at point B: $z_B = 3$ m

$$\text{Discharge: } Q = 200 \text{ litre/s} = \frac{200}{1000} = 0.2 \text{ m}^3/\text{s}$$

Specific gravity of the oil: $S = 0.9$

$\therefore$ Density of oil: $\rho = S \times 1000 = 0.9 \times 1000 = 900$ kg/m^3

According to continuity equation: $Q = A_A \, V_A = A_B V_B$

$Q = A_A V_A$ $Q = A_B V_B$

$0.2 = \dfrac{\pi}{4} d_A^2 \times V_A$ $0.2 = \dfrac{\pi}{4} d_B^2 \times V_B$

$0.2 = \dfrac{3.14}{4} \times (0.2)^2 V_A$ $0.2 = \dfrac{3.14}{4} \times (0.5)^2 V_B$

or $\quad V_A = 6.369$ m/s   or $\quad V_B = 1.02$ m/s

Total energy per unit weight at point A : E_A

$$E_A = \frac{p_A}{\rho g} + \frac{V_A^2}{2g} + z_A = \frac{78.5 \times 10^3}{900 \times 9.81} + \frac{(6.369)^2}{2 \times 9.81} + 0$$

$$= 8.89 + 2.06 = 10.95 \, \frac{\text{Nm}}{\text{N}} \text{ or m of oil.}$$

Total energy per unit weight at point B: E_B

$$E_B = \frac{p_B}{\rho g} + \frac{V_B^2}{2g} + z_B = \frac{58.9 \times 10^3}{900 \times 9.81} + \frac{(1.02)^2}{2 \times 9.81} + 3$$

$$= 6.67 + 0.053 + 3$$

$$= 9.723 \, \frac{\text{Nm}}{\text{N}} \text{ or m of oil.}$$

(*i*) We know the flow always takes place from higher energy level to lower energy level, since the energy at point A is greater than point B, so **the direction of oil flow is from A to B.**

(*ii*) Applying the modified Bernoulli's equation for real fluid at point A and B, we get

Total energy per unit weight at point A :

E_A = total energy per unit weight at point B :

E_B + head lost due to friction between A and B : h_L

$E_A = E_B + h_L$

or $\quad h_L = E_A - E_B = 105.94 - 102.07$

$\quad\quad = $ **3.87 m of oil.**

6.10 GRAPHICAL REPRESENTATION OF BERNOULLI'S EQUATION

The first form of Bernoulli's equation:

$$\frac{p}{w} + \frac{V^2}{2g} + z = C$$

or

$$\frac{p}{w} + z + \frac{V^2}{2g} = C \qquad\qquad ...(6.10.1)$$

The sum of pressure head and datum head is called **Piezometric head.** Mathematically,

$$\text{Piezometric head} = \frac{p}{w} + z$$

Equation (6.10.1) can be written as

$$\text{Piezometric head} + \frac{V^2}{2g} = C$$

The line joining the points represented by Piezometric head, along the direction of flow is called Piezometric head line or **Hydraulic Grade Line (HGL).**

The sum of pressure, datum and velocity head is called total head. *i.e.,* constant C is called total head. The total head is also defined as the sum of Piezometric head and velocity head.

The line joining the points represented by total head, along the direction of flow is called **Total Head Line** or **Total Energy Line (TEL).** In case of ideal fluid flow, the total energy line is horizontal as the total head remains constant along the direction of flow. Whereas in real fluid flow, the TEL is continually fall as the total head decreases along the direction of flow.

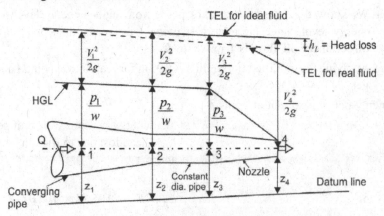

Fig. 6.21: Graphical representation of Bernoulli's equation.

(*i*) Graphical representation of Bernoulli's equation for venturimeter:

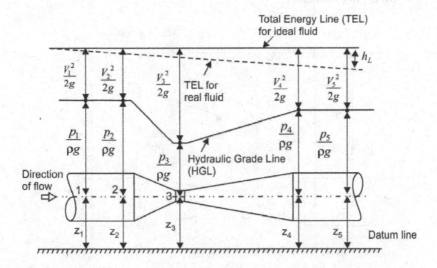

Fig. 6.22: Graphical representation of Bernoulli's equation for venturimeter.

(*ii*) Graphical Representation of Bernoulli's equation for converging-diverging nozzle:

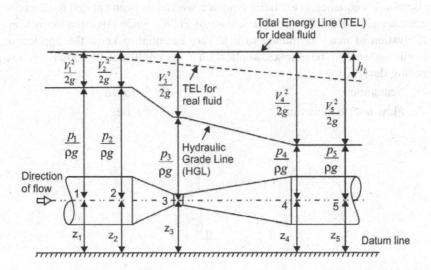

Fig. 6.23: Graphical representation of Bernoulli's equation for converging-diverging nozzle.

(*iii*) Graphical Representation of Bernoulli's equation for converging-diverging diffuser:

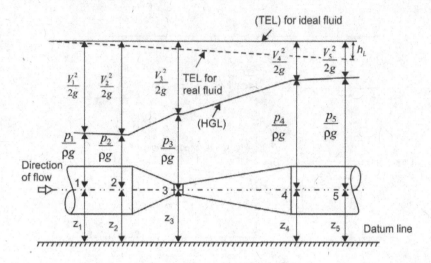

Fig. 6.24: Graphical representation of Bernoulli's equation for converging-diverging diffuser.

6.11 PRACTICAL APPLICATION OF BERNOULLI'S EQUATION

The Bernoulli's equation is the basic equation which has been applied in all problems of incompressible fluid flow in 'Mechanics of Fluid'. Since this equation is used for the derivation of many formulae, so it is very essential to know the application of Bernoulli's equation. The practical application of Bernoulli's equation to the following measuring devices :

1. Venturimeter
2. Orifice meter
3. Flow nozzle meter, and
4. Pitot tube.

6.12 IMPULSE MOMENTUM EQUATION

Momentum equation is based on Newton's 2nd law of motion. According to Newton's 2nd law of motion, the force applied on the body is equal to the rate of change of momentum in the direction of force.

As we know, momentum is the product of mass and velocity of the fluid.

Mathematically,

$$\text{Momentum} = MV$$

where M = mass of the fluid and

V = velocity of the fluid

The rate of change of momentum

$$= \frac{d}{dt}(MV)$$

According to Newton's 2nd law of motion, the force acting on fluid,

$$F = \text{Rate of change of momentum} = \frac{d}{dt}(MV)$$

Mass of the fluid: M = constant

$$F = M\frac{dV}{dt} \qquad \qquad ...(6.12.1)$$

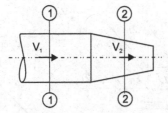

Fig. 6.25: Flow through horizontal pipe of varying cross-section

Eq. (6.12.1) for section 1-1 and section 2-2, we get

$$F = M\;\frac{(\text{Final velocity} - \text{Initial velocity})}{\text{Time taken}}$$

$$= \frac{M}{t}(V_2 - V_1) = m\Delta V \qquad \qquad ...(6.12.2)$$

where m = mass flow rate = $\dfrac{M}{t}$ kg/s

$$= \rho Q \quad \text{(by continuity equation)}$$

and ΔV = change in velocity of the fluid

$\therefore \qquad\qquad F = \rho Q.\Delta V \qquad\qquad ...(6.12.3)$

Equation (6.12.3) is known as the momentum principle and it expresses that the rate of change in linear momentum of flow in any direction is equal to the net force acting on the fluid mass in that direction. It is usually known as the momentum equation.

Equation (6.12.2) may be written as

$$F.dt = M.dV \qquad ...(6.12.4)$$

Left-hand side of the Eq. (6.12.4) represents the impulse of force [Force (F) acting over a short period of time (dt)] and right-hand sides gives the change in momentum in the direction of the force and hence it is known as **impulse-momentum equation.**

6.12.1 Application of Momentum Equation or Momentum Principle

The momentum equation is used in the following types of problems:

    (*i*) To determine the resultant force acting on a pipe-bend caused by the fluid flowing through it.

    (*ii*) The force exerted by jet of fluid striking on moving or fixed plate surface.

   (*iii*) Thrust on a propeller.

   (*iv*) Reaction of a jet *etc.*

6.13 FORCE EXERTED BY A FLOWING FLUID ON A PIPE-BEND

Let section (1)-(1) before and section (2)-(2) after the pipe bend as shown in Fig. 6.26.

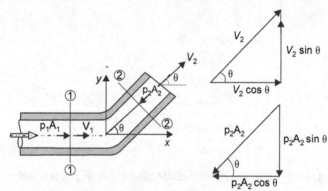

Fig. 6.26: Top view of a pipe bend.

Let V_1, A_1, p_1 = velocity of flow, cross-sectional area and pressure at section (1)-(1)

                       V_2, A_2, p_2 = velocity of flow, cross-sectional area and pressure at section (2)-(2)

and θ = angle of pipe bend

Let F_x and F_y be the components of the force exerted by the flowing fluid on the bend in x and y-directions respectively.

Then the force exerted by the bend (*i.e.*, resistance force) on the fluid is x and y directions be F_x and F_y respectively but in the opposite direction according to Newton's IIIrd law of motion. Hence component of force acting by bend on the fluid in x-direction = $-F_x$ and the component of the force acting by bend on the fluid in y-direction = $-F_y$.

p_1A_1, p_2A_2 = pressure forces acting on the sections (1)-(1) and sections (2)-(2) respectively.

Applying momentum equation is x-direction:

Net force acting in x-direction = rate of change of momentum in x-direction.

$\therefore p_1A_1 - p_2A_2 \cos \theta - F_x$ = mass flow rate × (change in velocity in x-direction)

$p_1A_1 - p_2A_2 \cos \theta - F_x = m$ (final velocity of the fluid in x-direction – initial velocity of the fluid in x-direction)

$p_1A_1 - p_2A_2 \cos \theta - F_x = \rho Q(V_2\cos \theta - V_1)$ $\because$ By continuity equation

$$Q = mv, \ sp. \ \text{volume}: v = \frac{1}{\rho}$$

$$\therefore \ m = \rho Q$$

$F_x = \rho Q \ (V_1 - V_2 \cos \theta) + (p_1A_1 - p_2A_2\cos \theta)$ ...(6.12.1)

or   = change in dynamics force in x-direction
                 + change in static force in x-direction

Similarly the momentum equation in y-direction gives.

$0 - p_2A_2 \sin \theta - F_y$ = mass flow rate × (change in velocity in y-direction)

$0 - p_2A_2 \sin \theta - F_y = \rho Q \ (V_2 \sin \theta - 0)$

or $F_y = \rho Q(0 - V_2 \sin \theta) + (0 - p_2A_2 \sin \theta)$ ...(6.12.2)

F_y = change in dynameter force in y-direction
    + change in static force in y-direction.

Equations (6.12.1) and (6.12.2), both are the standard equations

Now the resultant force (F) acting on the bend:

$$F = \sqrt{F_x^{\,2} + F_y^{\,2}}$$

The angle made by the resultant force with horizontal direction is given by

$$\tan \phi = \frac{F_y}{F_x}$$

$$\phi = \tan^{-1}\left(\frac{F_y}{F_x}\right)$$

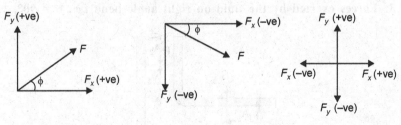

6.14 SUMMARY

Forces along x-direction and y-direction exerted on the pipe bend at different value of angle of bend (θ).

According to the standard linear momentum equation of the forces in x-direction and y-direction.

$$F_x = \rho Q(V_1 - V_2 \cos\theta) + (p_1 A_1 - p_2 A_2 \cos\theta)$$

and

$$F_y = \rho Q(0 - V_2 \sin\theta) + (0 - p_2 A_2 \sin\theta)$$

1. **Forces exerted by the fluid on the pipe:** At angle of bend, $\theta = 0$

$$F_x = \rho Q(V_1 - V_2) + (p_1 A_1 - p_2 A_2) \text{ and}$$
$$F_y = 0$$

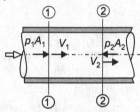

Fig. 6.27 (a): $\theta = 0$

2. **Forces exerted by the fluid on the pipe bend when the angle of bend greater than zero and less than 90°.**

$$F_x = \rho Q(V_1 - V_2 \cos\theta) + p_1 A_1 - p_2 A_2 \cos\theta$$

and

$$F_y = \rho Q(0 - V_2 \sin\theta) + (0 - p_2 A_2 \sin\theta)$$
$$= -\rho Q V_2 \sin\theta - p_2 A_2 \sin\theta$$

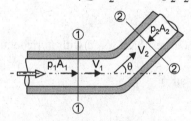

Fig. 6.27 (b): $0° < \theta < 90°$.

3. **Forces exerted by the fluid on right angle bend** *i.e.,* $\theta = 90°$.

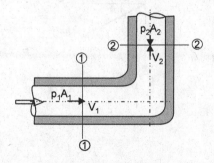

Fig. 6.27 (c): $\theta = 90°$.

$$F_x = \rho Q(V_1 - 0) + p_1 A_1 - 0$$
$$= \rho Q V_1 + p_1 A_1$$

and
$$F_y = \rho Q(0 - V_2) + (0 - p_2 A_2)$$
$$= -\rho Q V_2 - p_2 A_2$$

4. **Forces exerted by the fluid on the pipe bend when the angle of bend greater than 90° and less than 180°.**

$$F_x = \rho Q(V_1 - V_2 \cos\theta) + p_1 A_1 - p_2 A_2 \cos\theta$$

and $$F_y = \rho Q(0 - V_2 \sin\theta) + (0 - p_2 A_2 \sin\theta)$$

OR

$$F_x = \rho Q[(V_1 - V_2(-\cos(180° - \theta))] + p_1 A_1 - p_2 A_2(-\cos(180° - \theta))$$
$$= \rho Q[V_1 + V_2 \cos(180° - \theta)] + p_1 A_1 - p_2 A_2 \cos(180° - \theta)$$

and $$F_y = \rho Q[0 - V_2 \sin(180° - 0)] + [0 - p_2 A_2(180° - 0)]$$
$$= -\rho Q\ V_2 \sin(180° - \theta) - p_2\ A_2 \sin(180° - \theta)$$

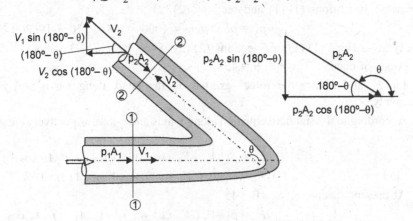

Fig. 6.27 (d): 90° < θ < 180°.

5. **Forces exerted by the fluid on the semi circular bend.** *i.e.*, **angle of bend, θ = 180°.**

$$F_x = \rho Q(V_1 - V_2 \cos 180°) + (p_1 A_1 - p_2 A_2 \cos 180°)$$
$$= \rho Q(V_1 + V_2) + (p_1 A_1 + p_2 A_2)$$
$$F_y = 0$$

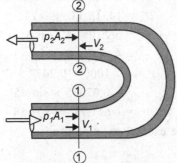

Fig. 6.27 (e): θ = 180°.

Problem 6.13: A 300 mm diameter pipe carries water under a head of 20 metre with a velocity of 3.5 m/s . If the axis of the pipe turns through 45° and the pipe is laid horizontally, find the magnitude and direction of the resultant force on the bend.

(GGSIP, University, Delhi, Dec. 2005)

Solution: Given data:

Diameter of pipe: $d_1 = d_2 = d = 300$ mm $= 0.3$ m

$\therefore$ Cross-sectional area of pipe:

$$A = \frac{\pi}{4} d^2$$

$$= \frac{3.14}{4} \times (0.3)^2$$

$$= 0.07065 \text{ m}^2$$

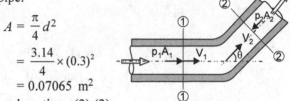

Fig. 6.28: Schematic for Problem 6.13

Water head at sections (1)-(1) and sections (2)-(2):

$$h_1 = h_2 = h = 20 \text{ m}$$

$\therefore$ Pressure at sections (1)-(1) and sections (2)-(2):

$$p_1 = p_2 = \rho g h = 1000 \times 9.81 \times 20 = 196200 \text{ N/m}^2$$

Velocity at sections (1)-(1) and sections (2)-(2): $V_1 = V_2 = V = 3.5$ m/s

Angle of bend: $\theta = 45°$

Let F_x and F_y are the forces exerted on the bend along x-axis and y-axis respectively.

According to the standard equation along x-axis and y-axis respectively are given below.

$$F_x = \rho Q(V_1 - V_2 \cos \theta) + (p_1 A_1 - p_2 A_2 \cos \theta)$$

and $\quad\quad F_y = \rho Q(0 - V_2 \sin \theta) + (0 - p_2 A_2 \sin \theta)$

At present case: $\theta = 45°$

$\therefore$ $\quad\quad F_x = \rho Q(V_1 - V_2 \cos 45°) + (p_1 A_1 - p_2 A_2 \cos 45°)$

where $\quad\quad Q = AV = \frac{\pi}{4} d^2 \times V = \frac{3.14}{4} \times (0.3)^2 \times 3.5 = 0.2472 \text{ m}^3/\text{s}$

$\therefore$ $\quad\quad F_x = 1000 \times 0.2472(3.5 - 3.5 \cos 45°)$

$$+ (196200 \times 0.07065 - 196200$$

$$\times 0.07065 \cos 45°)$$

$$= 247.2(1.025) + (13861.53 - 9801.58)$$

$$= 14313.33 \text{ N}(\rightarrow)$$

and $\quad\quad F_y = \rho Q (0 - V_2 \sin 45°) + (0 - p_2 A_2 \sin 45°)$

$$= - \rho Q V_2 \sin 45° - p_2 A_2 \sin 45°$$

$$= - 1000 \times 0.2472 \times 3.5 \times \sin 45° - 196200$$

$$\times 0.07065 \times \sin 45°$$

$$= - 611.78 - 9801.58 = -10413.36 \text{ N} (\downarrow)$$

$\therefore$ The resultant force exerted on pipe bend:

$$F = \sqrt{F_x^2 + F_y^2} = \sqrt{(4313.33)^2 + (-10413.36)^2}$$

$$= \sqrt{127042882.2} = 11271.33 \text{ N}$$

Angle made by resultant force with x-axis on the bend: ϕ

$$\tan \phi = \frac{F_y}{F_x} = \left(\frac{F_y}{F_x}\right) = \tan^{-1}\left(\frac{-20602.07}{49752.14}\right)$$

$$= -65.50°$$

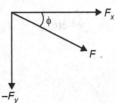

The $-ve$ sign shows that resultant force (F) lies in fourth quadrant and angle $65.50°$ made by resultant force with x-axis.

Problem 6.14: A 60° bend in a horizontal plane has its cross-sectional area at inlet and outlet equal to 1 m² and 0.5 m² respectively. Calculate the magnitude and direction of the force required to hold the bend in position if the velocity of water at inlet is 10 m/s and the pressure at inlet and outlet is 40 kN/m² and 30 kN/m² respectively.

Solution. Given data:

Angle of bend: $\theta = 60°$

$$A_1 = 1 \text{ m}^2$$
$$A_2 = 0.5 \text{ m}^2$$
$$V_1 = 10 \text{ m/s}$$
$$p_1 = 40 \text{ kN/m}^2 = 40 \times 10^3 \text{ N/m}^2$$
$$p_2 = 30 \text{ kN/m}^2 = 30 \times 10^3 \text{ N/m}^2$$

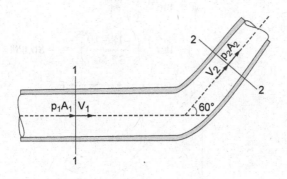

Fig. 6.29: Schematic for Problem 6.14

Applying the continuity equation between sections (1) and (2), we get

$$A_1 V_1 = A_2 V_2$$
$$1 \times 10 = 0.5 \times V_2$$

or $\qquad V_2 = 20 \text{ m/s}$

Let F_x and F_y are the forces exerted on the bends along x–axis and y–axis respectively.

According to the standard equation along x–axis and y–axis are given below:

$$F_x = \rho Q \left(V_1 - V_2 \cos \theta\right) + \left(p_1 A_1 - p_2 A_2 \cos \theta\right)$$

and $$F_y = \rho Q \left(0 - V_2 \sin \theta\right) + \left(0 - p_2 A_2 \sin \theta\right)$$

At present case: $\theta = 60°$

$\therefore$ $$F_x = \rho Q \left(V_1 - V_2 \cos 60°\right) + \left(p_1 A_1 - p_2 A_2 \cos 60°\right)$$

where $Q = A_1 V_1 = 1 \times 10 = 10 \ m^2/s$

$\therefore$ $$F_x = 1000 \times 10 \ (10 - 20 \times 0.5) + 40 \times 10^3 \times 0.5 \times 0.5$$

$$= 0 + \ 40000 - 7500$$

$$= 32500 \ N = 32.50 \ kN \ (\rightarrow)$$

and $$F_y = - \rho Q \ V_2 \sin 60° - p_2 A_2 \sin 60°$$

$$= - 1000 \times 10 \times 20 \times 0.866 - 30 \times 10^3 \times 0.5 \times 0.866$$

$$= - \ 186190 \ N = - \ 186.19 \ kN \ (\downarrow)$$

The resultant force exerted on bend,

$$F = \sqrt{F_x^2 - F_y^2} = \sqrt{\left(32.50\right)^2 + \left(-186.19\right)^2} = \textbf{189 kN}$$

Angle made by resultant force with x–axis on the bend: ϕ

$$\tan \phi = \frac{F_y}{F_x}$$

$$\phi = \tan^{-1} \left(\frac{F_y}{F_x}\right)$$

$$= \tan^{-1} \left(\frac{-186.19}{32.50}\right) = \textbf{- 80.09°}$$

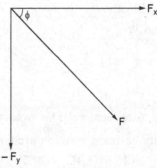

The $-ve$ sign shows that resultant force (F), lies in fourth quadrant and angles 80.09° made by resultant force with x–axis.

Problem 6.15: A pipe of 300 mm diameter converging 0.30 m³/s of water has a right angles bend in a horizontal plane. Find the resultant force exerted on the bend if the pressure at inlet and outlet of the bend are 24.525 N/cm² and 23.544 N/cm². Also find the angle made by resultant force with x-axis.

Solution : Given data:

Diameter of pipe: $d_1 = d_2 = d = 300$ mm $= 0.3$ m

∴ Cross-sectional area at section (1)-(1) = cross-sectional area at section (2)-(2)

$$\therefore A_1 = A_2 = \frac{\pi}{4}d^2 = \frac{3.14}{4} \times (0.3)^2 = 0.07065 \text{ m}^2$$

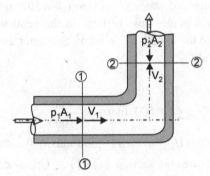

Fig. 6.30: Right angle bend.

Pressure at inlet: $p_1 = 24.525$ N/cm² $= 24.525 \times 10^4$ N/m²
Pressure at outlet: $p_2 = 23.544$ N/cm² $= 23.544 \times 10^4$ N/m²

Discharge: $Q = 0.30$ m³/s

Velocity at section (1)-(1): $V_1 = \dfrac{Q}{A_1} = \dfrac{0.30}{0.07065} = 4.246$ m/s

At same cross-sectional area:

$$V_1 = V_2 = 4.246 \text{ m/s}$$

Let F_x & F_y be forces exerted on pipe bend along x-axis and y-axis respectively

F_x = change in dynamic force + change in static
force along x-axis
$= \rho Q(V_1 - 0) + (p_1 A_1 - 0) = \rho Q V_1 + p_1 A_1$
$= 1000 \times 0.3 \times 4.246 + 24.525 \times 10^4 \times 0.07065$
$= 18600.71$ N $= 18.60$ kN $(\rightarrow)$

F_y = change in dynamic force along y-axis +
change in static force along y-axis
$= \rho Q(0 - V_2) + (0 - p_2 A_2) = -\rho Q V_2 - p_2 A_2$
$= -1000 \times 0.3 \times 4.246 - 23.544 \times 10^4 \times 0.07065$
$= -17907.636$ N $(\downarrow) = -17.907$ kN$(\downarrow)$

The resultant force exerted on the bend:

$$F = \sqrt{F_x^2 + F_y^2} = \sqrt{(18.60)^2 + (-17.907)^2} = 25.81 \text{ kN}.$$

Angle made by resultant force with x-axis: ϕ

$$\tan \phi = \frac{F_y}{F_x} = \tan^{-1}\left(\frac{-17.907}{18.60}\right) = -\mathbf{43.91°}$$

The $-ve$ sign shows that resultant force (F) lies in fourth quadrant and 43.91° angle made by resultant force with x-axis.

Problem 6.16: 500 liter/s of water flows through a 300 mm diameter pipe having a 90° bend in a horizontal plane. Loss of head in the bend is 2 m of water and the pressure at the entrance to the bend is 0.2 N/mm². Diameter the resultant force exerted by the water on the bend.

Solution: Given data:

$$Q = 500 \text{ liter/s} = \frac{500}{1000} \text{ m}^3/\text{s} = 0.5 \text{ m}^3/\text{s}$$

Diameter of pipe: $d_1 = d_2 = d = 300 \text{ mm} = 0.3 \text{ m}$

∴ Cross–sectional area at section (1)–(1) = Cross-sectional area at section (2)–(2)

$$A_1 = A_2$$

$$= \frac{\pi}{4} d^2 = \frac{3.14}{4} \times (0.3)^2 = 0.07065 \text{ m}^2$$

Angle of bend : $\theta = 90°$

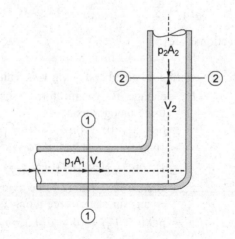

Fig. 6.31: Schematic for Problem 6.16

Loss of head in the bend:

$$h = 2 \text{ m of water}$$

∴ Loss of pressure in the bend:

$$p = \rho g h$$
$$= 1000 \times 9.81 \times 2 \text{ N/m}^2$$
$$= 19620 \text{ N/m}^2 = 0.1962 \times 10^5 \text{ N/m}^2$$

Pressure at entrance of the bend :

$$p_1 = 0.2 \text{ N/mm}^2$$
$$= 0.2 \times 10^6 \text{ N/m}^2$$
$$= 2 \times 10^5 \text{ N/m}^2$$

Pressure at outlet of the bend:

$$p_2 = p_1 - p$$
$$= 2 \times 10^5 - 0.1962 \times 10^5$$
$$= 1.8038 \times 10^5 \text{ N/m}^2$$

Velocity at section (1)–(1) :

$$V_1 = \frac{Q}{A_1} = \frac{0.5}{0.07065}$$
$$= 7.077 \text{ m/s} = V_2 \qquad\qquad \because A_2 = A_1$$

Let F_x and F_y be force exerted on pipe bend along x-axis and y-axis respectively.

$$F_x = \text{Change in dynamic force}$$
$$+ \text{ change in static force along x-axis.}$$
$$= \rho Q (V_1 - 0) + (p_1 A_1 - 0)$$
$$= \rho Q V_1 + p_1 A_1$$
$$= 1000 \times 0.5 \times 7.077 + 2 \times 10^5 \times 0.07065$$
$$= 17668.5 \text{ N } (\rightarrow) = 17.668 \text{ kN } (\rightarrow)$$

And $\qquad\qquad F_y = \text{Change in dynamic force along y-axis}$
$$+ \text{ change in static force along y-axis.}$$
$$= \rho Q (0 - V_2) + (0 - p_2 A_2)$$
$$= - \rho Q V_2 - p_2 A_2$$
$$= - 1000 \times 0.5 \times 7.077 - 1.8038$$
$$\times 10^5 \times 0.07065$$
$$= - 3538.5 - 12743.84$$
$$= - 16282.34 \text{ N } (\downarrow) = - 16.282 \text{ kN } (\downarrow)$$

The resultant force exerted on the bend :

$$F = \sqrt{F_x^2 + F_y^2}$$

$$= \sqrt{(17.668)^2 + (16.282)^2} = \textbf{24.02 kN}$$

Angle made by resultant force with x-axis : ϕ

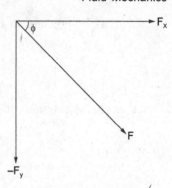

$$\tan \phi = \frac{F_y}{F_x}$$

$$\phi = \tan^{-1}\left(\frac{F_y}{F_x}\right)$$

$$= \tan^{-1}\left(\frac{-16.282}{17.668}\right) = -\,\mathbf{42.66°}$$

The –*ve* sign shows that resultant force (*F*) lies in fourth quadrant and 42.66° angle made by resultant force with *x*-axis.

Problem 6.17: Water is flowing through a pipe of diameter 300 mm with a rate of flow as 250 litre per second. If the pipe is bent by 135° find the magnitude and direction of the resultant force on the bend. The pressure of water flowing in the pipe is 400 kPa.   *(GGSIP University, Delhi, Dec. 2004)*

Solution: Given data:

Diameter pipe: $d = 300$ mm $= 0.30$ m

Rate of flow: $Q = 250$ litre/s $= \dfrac{250}{1000}$ m³/s $= 0.25$ m³/s

Angle of pipe bend: $\theta = 135°$

Pressure at sections (1)-(1) and at sections (2)-(2) is same:

i.e., $p_1 = p_2 = 400$ kPa $= 400 \times 10^3$ Pa
                            $= 400 \times 10^3$ N/m $\because$ 1 Pa = 1 N/m²

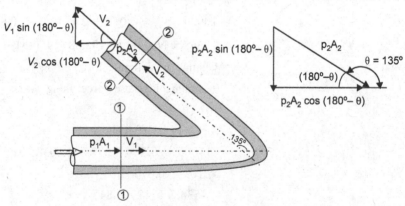

Fig. 6.32: Schematic for Problem 6.17

Cross-sectional area at sections (1)-(1), and at sections (2)-(2) is same.

i.e., $A_1 = A_2 = A = \dfrac{\pi}{4}d^2 = \dfrac{3.14}{4} \times (0.3)^2 = 0.07065$ m²

$\therefore$ Velocity at sections (1)-(1) = velocity at section (2)-(2)

$$V_1 = V_2 = \frac{Q}{A} = \frac{0.25}{0.07065} = 3.538 \text{ m/s}$$

Let F_x and F_y are the forces exerted on the bend along x-axis and y-axis respectively.

According to the standard equation along x-axis and y-axis respectively are given below:

$$F_x = \rho Q(V_1 - V_2 \cos \theta) + (p_1 A_1 - p_2 A_2 \cos \theta)$$

and
$$F_y = \rho Q(0 - V_2 \sin \theta) + (0 - p_2 A_2 \sin \theta)$$

At present case, $\theta = 135°$

$$\cos \theta = - \cos (180 - \theta)$$
$$\cos 135° = - \cos (180 - 135°)$$
$$= - \cos 45°$$
$$- 0.7071 = - 0.7071$$

Similarly
$$\sin \theta = \sin (180° - \theta)$$
$$\sin 135° = \sin (180° - 135°)$$
$$= \sin 45°$$
$$0.7071 = 0.7071$$

$\therefore$
$$F_x = \rho Q(V_1 - V_2 \cos 135°) + (p_1 A_1 - p_2 A_2 \cos 135°)$$
$$= 1000 \times 0.25[3.538 - 3.538 \times (-0.7071)] +$$
$$[400 \times 10^3 \times 0.07065 - 400 \times 103 \times 0.07065$$
$$\times (- 0.7071)]$$
$$= 250(3.538 + 2.501) + 28260 + 19982.64)$$
$$= 49752.14 \text{ N} (\rightarrow)$$

and
$$F_y = \rho Q (- V_2 \sin 135°) - p_2 A_2 \sin 135°$$
$$= - 1000 \times 0.25 \times 3.538 \times 0.7071$$
$$- 400 \times 10^3 \times 0.07065 \times 0.7071 = - 20602.07 \text{ N} (\downarrow)$$

The resultant force (F) exerted on pipe bend:

$$F = \sqrt{F_x^2 + F_y^2}$$
$$= \sqrt{(49752.14)^2 + (-20602.07)^2} = \mathbf{53849.05 \text{ N}}$$

Angle made by resultant force with x-axis on the bend: ϕ

$$\tan \phi = \frac{F_y}{F_x}$$

$$\phi = \tan^{-1}\left(\frac{F_y}{F_x}\right)$$

$$= \tan^{-1}\left(\frac{-20602.07}{49752.14}\right) = \mathbf{- 22.49°}$$

The −ve sign shows that resultant force (F) lies in fourth quadrant and angle 22.49° made resultant force with x-axis.

Problem 6.18 : A reducer bend having an outlet diameter of 15 cm discharges freely. The bend connected to a pipe of 20 cm diameter, has a deflection of 60° and lies in a horizontal plane. Determine the magnitude and direction of force on the anchor block supporting the pipe when a discharge of 0.3 m³/s passes through the pipe.

Solution : Given data:

Outside diameter: $d_2 = 15$ cm $= 0.15$ m

∴ Cross-sectional area: $A_2 = \dfrac{\pi}{4}d_2^2 = \dfrac{3.14}{4} \times (0.15)^2 = 0.01766$ m^2

Pipe diameter: $d_1 = 20$ cm $= 0.2$ m

∴ Cross-sectional area: $A_1 = \dfrac{\pi}{4}d_1^2 = \dfrac{3.14}{4} \times (0.2)^2 = 0.0314$ m^2

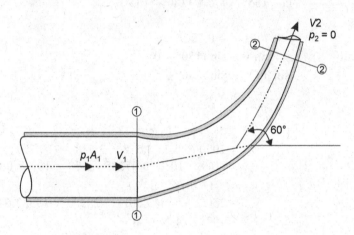

Fig. 6.33: Schematic for Problem 6.18

Angle of bend: $\theta = 60°$

Discharge: $Q = 0.3$ m^3/s

also $Q = A_1 V_1 = A_2 V_2$

or $V_1 = \dfrac{Q}{A_1}$

$$V_1 = \dfrac{0.3}{0.0314} = 9.55 \text{ m/s}$$

and $V_2 = \dfrac{Q}{A_2} = \dfrac{0.3}{0.01766} = 16.98$ m/s

Applying Bernoulli's equation at sections (1) and (2), we have

$$\dfrac{p_1}{\rho g} + \dfrac{V_1^2}{2g} + z_1 = \dfrac{p_2}{\rho g} + \dfrac{V_2^2}{2g} + z_1$$

As $z_1 = z_2$, $p_2 = 0$, atmospheric pressure

∴ $\dfrac{p_1}{\rho g} + \dfrac{V_1^2}{2g} = \dfrac{V_2^2}{2g}$

$$\frac{p_1}{\rho g} = \frac{V_2^2}{2g} - \frac{V_1^2}{2g}$$

or
$$p_1 = \frac{\rho g}{2g}\left[V_2^2 - V_1^2\right] = \frac{\rho}{2}\left[V_2^2 - V_1^2\right]$$

$$= \frac{1000}{2}\left[(16.98)^2 - (9.55)^2\right]$$

$$= 98558.95 \text{ Nm}^2$$

Let F_x and F_y are the forces exerted on the bend (or anchor block supporting the pipe) alone x-axis and y-axis respectively are give below :

$$F_x = \rho Q \,(V_1 - V_2 \cos\theta) + (p_1 A_1 - p_2 A_2 \cos\theta)$$

and $\qquad F_y = -\rho Q\, V_2 \sin\theta - p_2 A_2 \sin\theta$

Now $\qquad F_x = 1000 \times 0.3\,(9.55 - 16.98 \times \cos 60°) + 98558.95 \times 0.0314 - 0$

$$= 318 + 3094.75 = 3412.75 \text{ N } (\rightarrow)$$

and $\qquad F_y = -1000 \times 0.3 \times 16.98 \sin 60°$

$$= -4411.53 \ (\downarrow)$$

$\therefore$ Resultant force exerted on the bend : F

$$F = \sqrt{F_x^2 + F_y^2}$$

$$= \sqrt{(3412.75)^2 + (-4411.53)^2}$$

$$= \mathbf{5577.49 \text{ N}}$$

Angle made by resultant force with x-axis on the bend : ϕ

$$\tan\phi = \frac{F_y}{F_x}$$

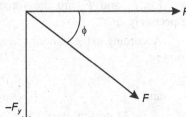

or

$$\phi = \tan^{-1}\left(\frac{F_y}{F_x}\right) = \tan^{-1}\left(\frac{-4411.53}{3412.75}\right)$$

$$= \tan^{-1}(-1.2926) = \mathbf{-52.29°}$$

The −ve sign shows that resultant force (F) lies in forth quadrant and angle 52.29 made by resultant force with x-axis.

Problem 6.19: Water flowing through the reducing elbow shown at the rate of 1 m³/s. The gauge pressure at section 1 is 0.1 MPa and that at section 2 is 0.09 MPa. What is the resultant force on the elbow ? Neglect the weight of water.

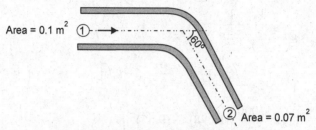

Fig. 6.34: Schematic for Problem 6.19

Solution: Given data:

Discharge: $Q = 1$ m³/s

Gauge pressure at section 1:

$$p_1 = 0.1 \text{ MPa} = 0.1 \times 10^6 \text{ N/m}^2$$

Cross-sectional area: $A_1 = 0.1$ m²

Gauge pressure at section 2: $p_2 = 0.09$ MPa $= 0.09 \times 10^6$ N/m²

Cross-sectional area: $A_2 = 0.07$ m²

We know, $Q = A_1 V_1 = A_2 V_2$

Velocity at section 1: $V_1 = \dfrac{Q}{A_1} = \dfrac{1}{0.1} = 10$ m/s

Velocity at section 2: $V_2 = \dfrac{Q}{A_2} = \dfrac{1}{0.07} = 14.28$ m/s

Let F_x and F_y are the forces exerted on the bend along x-axis and y-axis respectively.

According to the standard equation along x-axis and y-axis respectively are given below:

$$F_x = \rho Q(V_1 - V_2 \cos\theta) + (p_1 A_1 - p_2 A_2 \cos\theta)$$

and
$$F_y = \rho Q(0 - V_2 \sin\theta) + (0 - p_2 A_2 \sin\theta)$$

At present case : $\theta = 360° - 60° = 300°$ (anti clockwise)

OR

$\theta = -60°$ (clockwise)

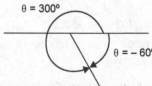

∴ $F_x = 1000[10 - 14.28 \cos(-60°)] + [0.1 \times 10^6 \times 0.1 - 0.09 \times 10^6$
$\times 0.07 \cos(-60°)]$

$= 1000\,(2.86) + (1000 - 3150) = 2860 + 6850$ N $= 9710$ N($\rightarrow$)

and $F_y = 1000 \times 1[-14.28 \sin(-\sin 60°)] - 0.09 \times 10^6 \times 0.07 \sin(-60°)$

$= 1000\,(12.36) + 5455.96 = 17815.96$ N ($\uparrow$)

∴ Resultant force on the elbow: $F = \sqrt{F_x^2 + F_y^2} = \sqrt{(9710)^2 + (17815.96)^2}$
$$= 20290.20 \text{ N} = \mathbf{20.29 \text{ kN}}$$

Problem 6.20: A converging pipe bend, with its centerline in a horizontal plane, charge the direction of pipeline by 60° in the clockwise direction and reduce the pipeline diameter from 30 cm to 20 cm in the direction of flow. If the pressure indicated by a Bourden gauge, at the centerline of the 30 cm diameter entrance to bend is 140 kN/m² and the flow of water through the pipeline is 0.10 m³/s. Determine the magnitude and direction of force on the bend due to the moving water.

(GGSIP University Delhi, Dec. 2008)

Solution: Given data:

Angle of bend = 60° (clockwise)

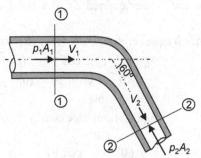

Fig. 6.35: Schematic for Problem 6.20

Diameter at section (1)-(1):
$$d_1 = 30 \text{ cm} = 0.3 \text{ m}$$
∴ Cross-sectional area at section (1)-(1):
$$A_1 = \frac{\pi}{4}d^2 = \frac{3.14}{4} \times (0.3)^2 = 0.07065 \text{ m}^2$$
Diameter at section (1)-(1):
$$A_2 = \frac{\pi}{4}d^2 = \frac{3.14}{4} \times (0.2)^2 = 0.0314 \text{ m}^2$$
Pressure at section (1)-(1):
$$p_1 = 140 \text{ kN/m}^2 = 140 \times 10^3 \text{ N/m}^2$$
Discharge through pipe : $Q = 0.10 \text{ m}^3/\text{s}$
also
$$Q = A_1 V_1 = A_2 V_2$$
∴
$$V_1 = \frac{Q}{A_1} = \frac{0.10}{0.07065} = 1.415 \text{ m/s}$$
$$V_2 = \frac{Q}{A_2} = \frac{0.10}{0.0314} = 3.184 \text{ m/s}$$
Applying Bernoulli's equation at sections (1)-(1) and (2)-(2), we get
$$\frac{p_1}{\rho g} + \frac{V_1^2}{2g} + z_1 = \frac{p_2}{\rho g} + \frac{V_2^2}{2g} + z_2$$

$$\frac{p_2}{\rho} + \frac{V_1^2}{2} = \frac{p_2}{\rho} + \frac{V_2^2}{2}$$

$\because z_1 = z_2$, pipeline in a horizontal plane.

$$\frac{140 \times 10^3}{1000} + \frac{(1.415)^2}{2} = \frac{p_2}{1000} + \frac{(3.184)^2}{2}$$

$$140 + 1 = \frac{p_2}{1000} + 5.068$$

$$141 = \frac{p_2}{1000} + 5.068$$

or $\qquad \dfrac{p_2}{1000} = 141 - 5.068 = 135.932$

or $\qquad p_2 = 135.932 \times 10^3 \text{ N/m}^2$

Let F_x and F_y are the forces exerted on the bend along x-axis and y-axis respectively.

According to the standard equation along x-axis and y-axis respectively are given below:

$$F_x = \rho Q(V_1 - V_2 \cos \theta) + (p_1 A_1 - p_2 A_2 \cos \theta)$$

and $\qquad F_y = \rho Q(0 - V_2 \sin \theta) + (0 - p_2 A_2 \sin \theta)$

At present case: $\qquad \theta = 360° - 60°$

$\qquad\qquad = 300°$ (anti clockwise)

OR

$\qquad\qquad = -60°$ (clockwise)

$\therefore \qquad F_x = \rho Q[(V_1 - V_2 \cos (-60)] + [(p_1 A_1 - p_2 A_2 \cos (-60°)]$

$\qquad = 1000 \times 0.10 \, (1.415 - 3.184 \times 0.5)$

$\qquad + (140 \times 10^3 \times 0.07065$

$\qquad - 135.932 \times 10^3 \times 0.0314 \times 0.5)$

$\qquad = 100(-0.177) + 9891 - 2134.13$

$\qquad = -17.7 + 9891 - 2134.13$

$\qquad = 7756.87 \text{ N } (\rightarrow)$

and $\qquad F_y = \rho Q[0 - V_2 \sin (-60°)] + [0 - p_2 A_2 \sin (-60°)]$

$\qquad = 1000 \times 0.1 \, [-3.184 \times (-0.866)] - 135.932$

$\qquad \times 10^3 \times 0.0314 \times (-0.866)$

$\qquad = 100 \, (2.757) + 3696.31 = 3972.01 \text{ N } (\uparrow)$

$\therefore$ The magnitude of resultant force on the pipe bend is

$$F = \sqrt{F_x^2 + F_y^2} = \sqrt{(7756.87)^2 + (3972.01)^2} = \mathbf{87146.69\,N}$$

Direction of resultant force with x-axis: ϕ

$$\tan \phi = \frac{F_y}{F_x}$$

$$\tan \phi = \frac{3972.01}{7756.87} = 0.5120$$

$$\phi = \tan^{-1} (0.5120) = \mathbf{27.11°}$$

Problem 6.21: A pipe of 0.3 m diameter carrying a flow of 0.3 m³/s of water has a right angle bend lying in a horizontal plane. The pressure at the inlet and outlet of the bend are 24.54 N/cm² and 23.53 N/cm². Find the resultant force exerted on the bend and the angle of the resultant force.

Solution: Given data:

Diameter of pipe: $d_1 = d_2 = d = 0.3$ m

∴ Cross-sectional area at section (1) – (1) = cross-sectional area at section (2) – (2).

$$A_1 = A_2 = \frac{\pi}{4}d^2 = \frac{3.14}{4} \times (0.3)^2 = 0.07065 \text{ m}^2$$

Discharge: $Q = 0.3$ m³/s

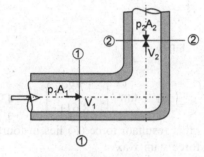

Fig. 6.36 Right angle bend

Pressure at inlet: $p_1 = 24.54$ N/cm² $= 24.54 \times 10^4$ N/m²

Pressure at outlet: $p_2 = 23.53$ N/cm² $= 23.53 \times 10^4$ N/m²

Velocity at section (1) – (1):

$$V_1 = \frac{Q}{A_1} = \frac{0.3}{0.07065} = 4.246 \text{ m/s}$$

$$= V_2 \qquad\qquad \because A_1 = A_2$$

Let F_x and F_y be forces exerted on pipe bend along x-axis and y-axis respectively

F_x = change in dynamic force + change in static force along x-axis

$= \rho Q(V_1 - 0) + (p_1 A_1 - 0)$

$= \rho Q V_1 + p_1 A_1$

$= 1000 \times 0.3 \times 4.246 + 24.54 \times 10^4 \times 0.07065$

$= 18611.31$ N $= 18.611$ kN $(\rightarrow)$

and F_y = change in dynamic force along y-axis + change in static force along y-axis

$= \rho Q(0 - V_2) + (0 - p_2 A_2)$

$= -\rho Q V_2 - p_2 A_2$

$= -1000 \times 0.3 \times 4.246 - 23.53 \times 10^4 \times 0.07065$

$= -1273.8 - 16623.94 = -17.89774$ N$(\downarrow)$

$= -17.897$ kN$(\downarrow)$

The resultant force exerted on the bend:

$$F = \sqrt{F_x^2 + F_y^2} = \sqrt{(18.611)^2 + (-17.897)^2} = \mathbf{25.819\,kN}$$

Angle made by resultant force with x-axis: ϕ

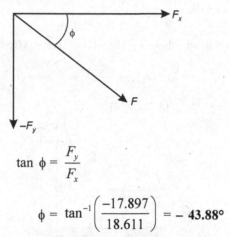

$$\tan\phi = \frac{F_y}{F_x}$$

$$\phi = \tan^{-1}\left(\frac{-17.897}{18.611}\right) = -\,\mathbf{43.88°}$$

The $-ve$ sign shows that resultant force (F) lies in fourth quadrant and 43.88° angle made by resultant force with x-axis.

Problem 6.22: Water flows at the rate of 0.71 m³/s through the pipe transition as shown in Fig. 6.55. If the pressure intensity at the centre line of the 900 mm section is 9.81 kN/m², what will be the centre line pressure in the 600 mm section ? What force will be required to produce the change in momentum of water as it passes through the transition?

Solution: Given data:

Diameter at section 1: D_1 = 900 mm = 0.9 m

Cross-sectional area: $A_1 = \dfrac{\pi}{4}D^2 = \dfrac{3.14}{4} \times (0.9)^2 = 0.6358$ m²

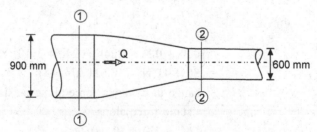

Fig. 6.37: Schematic for Problem 6.22

Pressure at section: p_1 = 9.81 kN/m² = 9.81 × 10³ N/m²

Diameter at section 2: D_2 = 600 mm = 0.6 m

∴   Cross-sectional area at section 2: $A_2 = \dfrac{\pi}{4}D^2 = \dfrac{3.14}{4} \times (0.6)^2 = 0.2826$ m²

Applying continuity equation at sections (1) and (2), we get

$$Q = A_1 V_1 = A_2 V_2$$
$$Q = A_1 V_1 \qquad\qquad\qquad Q = A_2 V_2$$

or $\qquad V_1 = \dfrac{Q}{A_1} = \dfrac{0.71}{0.6358} \qquad\qquad V_2 = \dfrac{Q}{A_2} = \dfrac{0.71}{0.2826}$

$$= 1.116 \text{ m/s} \qquad\qquad\qquad = 2.51 \text{ m/s}$$

Now applying the Bernoulli's equations at section (1) and (2), we get

$$\frac{p_1}{\rho g} + \frac{V_1^2}{2g} + z_1 = \frac{p_2}{\rho g} + \frac{V_2^2}{2g} + z_2$$

$$\frac{p_1}{\rho g} + \frac{V_1^2}{2g} = \frac{p_2}{\rho g} + \frac{V_2^2}{2g} \qquad\qquad (\because z_1 = z_2, \text{ pipe is horizontal})$$

or $\qquad \dfrac{p_1}{\rho} + \dfrac{V_1^2}{2} = \dfrac{p_2}{\rho} + \dfrac{V_2^2}{2}$

$$\frac{9.81 \times 10^3}{1000} + \frac{(1.116)^2}{2} = \frac{p_2}{1000} + \frac{(2.15)^2}{2}$$

$$9.81 + 0.622 = \frac{p_2}{1000} + 3.15$$

or $\qquad\qquad \dfrac{p_2}{1000} = 7.282$

$$p_2 = 7.282 \times 1000 = 7.282 \times 10^3 \text{ N/m}^2 = \mathbf{7.282 \ kN/m^2}$$

Let F_x is force exerted by the fluid on the pipe along x-axis.

According to the standard equation along x-axis is given below:

$$F_x = \rho Q(V_1 - V_2 \cos \theta) + (p_1 A_1 - p_2 A_2 \cos \theta)$$

where $\qquad\qquad \theta = 0$, became pipe is horizontal

$$F_x = \rho Q(V_1 - V_2) + (p_1 A_1 - p_2 A_2)$$
$$= 1000 \times 0.71(1.116 - 2.51)$$
$$+ (9.81 \times 10^3 \times 0.6358 - .282 \times 10^3 \times 0.2826)$$
$$= 710(-1.394) + (6237.19 - 2057.89)$$
$$= -989.74 + 4179.3 = 3189.56 \text{ N}$$

The force: $F_x = 3189.56$ N is exerted by water to the pipe, the same force but opposite (*i.e.*, $F = -3189.56$ N) is required to produce the desired change is momentum and acts from right to left.

Problem 6.23: Water at the rate of 40 liters in flowing though a 150 mm diameter fire hose at the end of which a 50 mm diameter nozzle is fixed. Calculate the force exerted by the nozzle if the pressure at the inlet of the nozzle is 200 kPa.

Solution: Given data:

$$\text{Discharge} : Q = 40 \text{ liter/s} = \frac{40}{1000} \ \frac{\text{m}^3}{\text{s}} = 0.04 \text{ m}^3\text{/s}$$

Diameter of fire house: $d_1 = 150$ mm $= 0.15$ m

$\therefore$ Cross–sectional area: $A_1 = \dfrac{\pi}{4} d_1^2 = \dfrac{3.14}{4} \times (0.15)^2 = 0.01766$ m^2

Diameter at the nozzle end:

$$d_2 = 50 \text{ mm} = 0.05 \text{ m}$$

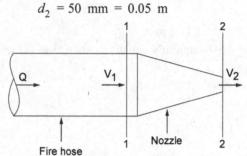

Fire hose
Fig. 6.38: Schematic for Problem 6.23

$\therefore$ Cross-sectional area,

$$A_2 = \frac{\pi}{4} d_2^2$$

$$= \frac{3.14}{4} \times (0.05)^2$$

$$= 0.00196 \text{ m}^2$$

Pressure at the inlet of the nozzle,
$$p_1 = 200 \text{ kPa} = 200 \times 10^3 \text{ Pa}$$
Applying continuity equation at section (1) and (2), we get,

$$Q = \qquad A_1 V_1 = A_2 V_2$$

$$Q = A_1 V_1 \qquad\qquad\qquad\qquad\qquad Q = A_2 V_2$$

or $V_1 = \dfrac{Q}{A_1} = \dfrac{0.04}{0.01766} = 2.265$ m/s $\qquad$ or $\quad V_2 = \dfrac{Q}{A^2} = \dfrac{0.04}{0.00196}$

$$= 20.408 \text{ m/s}$$

Since, the pressure at nozzle exit is atmospheric, so $p_2 = 0$

Let F_x is force exerted by the fluid on the nozzle along x-axis,

$$F_x = \rho Q (V_1 - V_2 \cos \theta) + (p_1 A_1 - p_2 A_2 \cos \theta)$$

where $\theta = 0$, nozzle is horizontal

$$p_2 = 0, \text{ nozzle exit at atmospheric pressure}$$

$\therefore \qquad\qquad F_x = \rho Q (V_1 - V_2) + p_1 A_1$

$$= 1000 \times 0.04 \,(2.265 - 20.408) + 200$$
$$\times 10^3 \times 0.01766$$
$$= -725.72 + 3532 = 2806.28 \text{ N}$$

The force : F_x = 2806.28 N is exerted by water to the nozzle, the same force but opposite *i.e.,* F_x = –286.28 N exerted by the nozzle to the fluid.

6.15 FLUID JET OR JET

A fluid jet is a stream of fluid issuing from nozzle with high velocity or kinetic energy $\left[i.e., \frac{1}{2}mV^2 \right]$

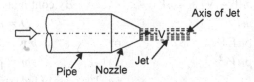

Fig. 6.39: Jet

6.16 IMPACT OF JET

When a jet impinges on a plate or vane, it exerts a force on the plate (due to change in momentum) called *dynamic force* or *hydrodynamic force* or *Impact of Jet.*

or

Jet exerts a dynamic force on the plate, which is known as the 'Impact of Jet. Because the force is induced by the motion of the liquid, therefore, it is also known as *dynamic force* or *hydrodynamic force.*

6.17 FORCE EXERTED BY THE JET ON A STATIONARY FLAT PLATE

(*a*) Plate is vertical to the jet (*b*) Plate is inclined to the jet (*c*) Plate is curved.

6.17.1 Plate is Vertical to the Jet

Assumption: Smooth plate, there is no friction loss, impact loss, the magnitude of leaving and initial velocity of the jet is same.

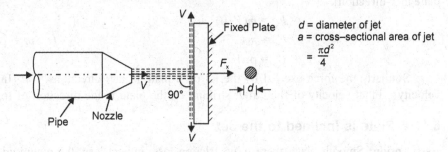

Fig. 6.40: Plate is vertical to the jet

Let F_x be the force exerted by flowing fluid on fixed plate in x-direction. The force exerted by fixed plate on the fluid in the x-direction and will be equal to F_x but in the opposite direction (according Newton's IIIrd law of motion).

Hence the force acting by plate on the fluid in x-direction $= -F_x$

Applying momentum equation in x-direction:

Force acting on the fluid in x-direction = Rate of change of momentum in x-direction.

$- F_x =$ Mass flow rate [Final velocity of fluid after striking on the plate in x-direction – Initial

velocity of fluid before striking on the plate in x-direction].

$$-F_x = m[0 - V] = -mV \qquad \text{By continuity equation,}$$

or $\qquad F_x = mV \qquad\qquad\qquad\qquad\qquad Q = mv = aV$

$$F_x = \rho aV . V \qquad\qquad \text{or} \qquad\qquad mv = aV$$

$$\boldsymbol{F_x = \rho aV^2} \qquad\qquad\qquad\qquad\qquad m = \rho aV$$

$\qquad\qquad\qquad\qquad\qquad\qquad\qquad$ *i.e.* The mass of fluid striking on the plate per

$\qquad\qquad\qquad\qquad\qquad\qquad\qquad\qquad\qquad$ second $= \rho aV$

where $\qquad \rho =$ density of fluid, kg/m^3

$\qquad\qquad a =$ cross-sectional area of jet, m^2

$\qquad\qquad V =$ velocity of jet, m/s

$\therefore \qquad\qquad$ Unit of $F_x = \dfrac{\text{kg}}{\text{m}^3} . \text{m}^2 . \dfrac{\text{m}^2}{\text{s}^2} = \text{kg} . \dfrac{\text{m}^2}{\text{s}^2} = \text{N}$

Thus the force exerted by the jet on fixed plate in x-direction is ρaV^2

Notes:

1. The work done on the plate is zero, because the plate is fixed *i.e.,* no displacement of the plate, no work.

2. In this chapter, we are required to determine the force produced on a solid body by the action of flowing fluid. The magnitude of such a force is determined by using momentum equation, which may be expressed in a somewhat different and more useful form.

The force exerted by the jet on a plate in x-direction,

$F_x =$ Mass flow rate [Initial velocity - Final velocity of fluid after striking on the plate in x-direction].

$$F_x = m(V-0)$$

$$F_x = mV \qquad\qquad\qquad\qquad\qquad (\because \ m = \rho aV)$$

$$F_x = \rho aV^2$$

Similarly the force exerted by the jet on a plate in any directions $= m$ [Initial velocity – Final velocity of fluid after striking on the plate in the direction of force]

6.17.2 Plate is Inclined to the Jet:

Assumption: Smooth plate, there is no friction loss, impact loss, the magnitude of leaving and initial velocity of the jet is same.

Let the stationary flat plate be inclined at angle θ to the axis of horizontal jet as shown in Fig. 6.41. If a and V are the cross-sectional area and velocity of the jet respectively, then the mass of liquid per second striking the plate, $m = \rho aV$.

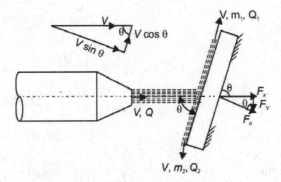

Fig. 6.41: Plate is inclined to the jet

Applying impulse-momentum equation in the direction normal to the plate.

Normal force on the plate, $F_n = m\,\Delta V_n$

where $\qquad\qquad \Delta V_n$ = change in velocity normal to the plate,

$\qquad\qquad\qquad F_n = m$ [Initial velocity – Final velocity normal to the plate]

$\qquad\qquad\qquad\quad = m\,[V\sin\theta - 0]$

$\qquad\qquad\qquad\quad = m\,V\sin\theta = \rho a V^2 \sin\theta$

This normal force can be resolved into two components. One along the direction of the jet (x-direction) and another perpendicular to it (y-direction).

$$F_x = F_n\sin\theta = \rho a V^2 \sin^2\theta$$

and $\qquad\qquad F_y = F_n\cos\theta = \rho a V^2 \sin\theta\,\cos\theta$

$$= \frac{\rho a V^2 \sin 2\theta}{2}$$

Applying impulse-momentum equation parallel to the plate.

Final momentum – Initial momentum = 0

$$(m_1 V - m_2 V) - mV\cos\theta = 0$$

$$m_1 - m_2 - m\cos\theta = 0$$

$$\rho\,Q_1 - \rho\,Q_2 - \rho\,Q\cos\theta = 0$$

$$Q_1 - Q_2 - Q\cos\theta = 0 \qquad\qquad ...(6.17.1)$$

According to continuity equation, we get

$$Q = Q_1 + Q_2$$

$$Q_1 = Q - Q_2$$

Substituting the value of Q_1 in Eq. (6.17.1), we get

$$Q - Q_2 - Q_2 - Q\cos\theta = 0$$

$$-2\,Q_2 + Q\,(1 - \cos\theta) = 0$$

$$Q_2 = \frac{Q(1-\cos\theta)}{2} \qquad\qquad ...(6.17.2)$$

and $\qquad\qquad\qquad\qquad Q_1 = Q - \dfrac{Q(1-\cos\theta)}{2}$

$$Q_1 = \frac{2Q - Q(1-\cos\theta)}{2}$$

$$= \frac{2Q - Q + Q\cos\theta}{2}$$

$$= \frac{Q + Q\cos\theta}{2}$$

$$Q_1 = \frac{Q(1+\cos\theta)}{2} \qquad \dots(6.17.3)$$

Discharge ratio, $\qquad \dfrac{Q_1}{Q_2} = \dfrac{\dfrac{Q(1+\cos\theta)}{2}}{\dfrac{Q(1-\cos\theta)}{2}} = \dfrac{1+\cos\theta}{1-\cos\theta}$

6.17.3 Plate is Curved

(i) Jet strikes at the centre.   (ii) Jet strikes at one end.

(i) Jet Strikes at the Centre

Assumption: Smooth plate, there is no friction loss, impact loss, the magnitude of leaving and initial velocity of the jet is same.

Let a fluid jet be striking at the centre of fixed plate as shown in Fig. 6.42

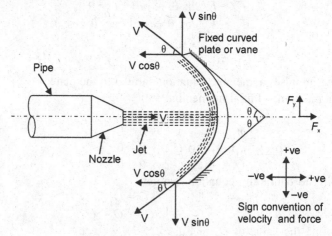

Fig. 6.42: Jet strikes at the centre

θ = angle between the direction of the incoming and the leaving jets or vane angle at the end or tip on each side.

The velocity of the jet at the leaving end of the vane can be resolved into two components.

Component of velocity in the direction of the jet $= - V \cos \theta$

(− *ve* sign indicates that the direction is opposite to the direction of the jet)

Component of the velocity perpendicular to the jet $= V \sin \theta$

Let m = Mass flow rate of water striking on the curved plate; is divided into m_1 and m_2 as shown in Fig. 6.43

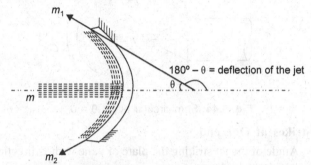

Fig. 6.43: Mass m is split into two parts m_1 and m_2

According to the principle of impulse-momentum equation the force exerted by the jet along *x*-direction or in the direction of jet.

$$F_x = \text{Rate of change of momentum}$$
$$\text{(in the direction of force)}$$
$$= \text{Initial momentum} - \text{Final momentum}$$
$$= mV - (- m_1 \, V \cos \theta - m_2 \, V \cos \theta)$$
$$= mV + (m_1 + m_2) \, V \cos \theta \qquad (\because \; m = m_1 + m_2)$$
$$= mV + mV \cos \theta$$
$$= mV \, (1 + \cos \theta) \qquad\qquad ...(6.17.4)$$

And due to symmetry of the plate, the force exerted by the jet along *y*-direction or perpendicular to the direction of the jet will be zero.

i.e.,
$$\bullet \; F_y = \text{Initial momentum} - \text{Final momentum}$$
$$= 0 - [m_1 \, V \sin \theta - m_2 \, V \sin \theta \,]$$
$$= - m_1 \, V \sin \theta + m_2 \, V \sin \theta$$
$$= (- m_1 + m_2) \, V \sin \theta$$
$$= 0 \qquad\qquad (\because \; m_1 = m_2, \text{ due to symmetry})$$

Case 1: When $\theta = 90°$, it becomes the case of a flat plate held perpendicular to the direction of the jet or along *x*-direction Eq. (*iv*) becomes,
$$F_x = mV$$
$$F_x = \rho a V^2 \qquad\qquad (\because \; m = \rho a V)$$

Case 2: When the plate is semicircular, the jet is deflected back in the incoming direction.

i.e., $\qquad\qquad\qquad\qquad \theta = 0$

Equation (*6.17.4*) becomes,

$$F_x = mV(1+\cos 0) = 2\ mV$$

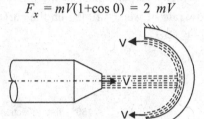

Fig. 6.44: Semicircular plate, $\theta = 0$.

(*ii*) Jet Strikes at One end

Let $\quad \theta_i$ = Angle of the jet striking the plate or vane with x-direction or angle of vane at the inlet

$\qquad \theta_0$ = Angle of the jet at the outlet with x-direction or angle of vane at the outlet.

The velocity of the jet at the inlet and at outlet can be resolved into two components;

Component of velocity at the inlet in x-direction = $V \cos \theta_i$

Component of velocity at the inlet in y-direction = $- V \sin \theta_i$

Component of velocity at the outlet in x-direction = $- V \cos \theta_0$

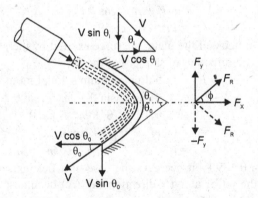

Fig. 6.45: Jet strikes at one end of unsymmetrical curved plate.

Component of velocity at the outlet in y-direction = $- V \sin \theta_0$

According to the principle of impulse-momentum equation;

The force on the vane along x-direction:

$$F_x = m\Delta V_x$$
$$F_x = m\ [V\cos\theta_i - (-V\cos\theta_0)]$$
$$= mV[\cos\theta_i + \cos\theta_0] \qquad \because m = \rho a V$$
$$= \rho a V^2[\cos\theta_i + \cos\theta_0] \qquad ...\ (6.17.5)$$

And the force on the vane along y-direction:

$$F_y = m\ \Delta V_y$$
$$= m\ [-V\sin\theta_i - (-V\sin\theta_0)]$$
$$= m\ V\ [\sin\theta_0 - \sin\theta_i)]$$
$$= \rho a V^2[\sin\theta_0 - \sin\theta_i] \qquad ...(6.17.6)$$

Case-I: If the vane is symmetrical,

$$\theta_i = \theta_0 = \theta$$

Then the Eqs. $(6.17.5)$ and $(6.17.6)$ become,

$$F_x = 2\rho a V^2\ \cos\theta$$

and
$$F_y = 0$$

Case-II: If jet strikes at inlet along x-axis, *i.e.,* $\theta_i = 0$

Then the Eqs. $(6.17.5)$ and $(6.17.6)$ become,

$$F_x = \rho a V^2\ [1 + \cos\theta_0]$$

and
$$F_y = \rho a V^2\ \sin\theta_0$$

Case-III: If the vane is semicircular or both the angle θ_i and θ_0 are zero.

Then the Eqs. $(6.17.5)$ and $(6.17.6)$ become,

$$F_x = 2\ \rho a V^2$$

and
$$F_y = 0$$

Resultant force on the vane:

$$F_R = \sqrt{F_x^2 + F_y^2}$$

And the direction will be

$$\phi = \tan^{-1}\frac{F_y}{F_x}\ \text{with } x\text{-direction}$$

Total angle of deflection of the jet $= 180° - (\theta_i + \theta_0)$. In case of fixed vanes, no work is done by the jet.

Note: If force exerted by jet:

$$F = m[V_1 - V_2]\ \text{N}$$

where
$$m = \text{mass flow rate} = \rho a V \quad \text{kg/s}$$
$$a = \text{cross-sectional area of the jet,}$$
$$V = \text{absolute velocity of the jet.}$$

Let
$$\dot{W} = \text{weight of water flow per second}$$
$$= \rho g a V \ \text{N/s}$$

Force exerted by jet:

$$F = \frac{\dot{W}}{g} \; [V_1 - V_2] \; N$$

Force exerted by jet per unit weight of water or Force exerted by jet per unit Newton of water

$$= \frac{1}{g} [V_1 - V_2] \; N/N \; \text{of water}$$

If plate is moving with velocity u, then work done per second per N or weight of water per second

$$= \frac{1}{g} [V_1 - V_2] u \; N\text{-}m/N$$

Problem 6.24. The water jet shown in figure strikes normal to a fixed plate. Neglecting gravity and friction and fraction, determine to hold the plate fixed.

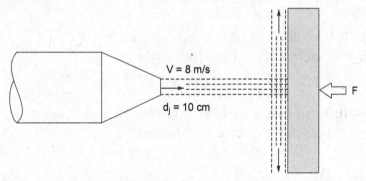

V = 8 m/s

d$_j$ = 10 cm

F

Fig. 6.46: Schematic for Problem 6.24

Solution: Given data:

Velocity of jet: $V = 8$ m/s

Diameter of jet: $d_j = 10$ cm $= 0.10$ m

$\therefore$ Cross–sectional area of jet:

$$a = \frac{\pi}{4} \; d_j^2 = \frac{3.14}{4} \times (0.10)^2 = 7.85 \times 10^{-3} \; m^2$$

The force required to hold the plane fixed is given by

$$F = \rho a V^2$$
$$= 1000 \times 7.85 \times 10^{-3} \times (8)^2 = \textbf{502.4 N}$$

Problem 6.25: A jet water issues from a nozzle with a velocity of 50 m/s. If the flow rate is 0.22 m³/s, what is thee power of the jet?

Solution: Given data:

Velocity of jet : $V = 50$ m/s

Flow rate : $Q = 0.22$ m³/s

Power of the jet = Kinetic energy jet per second.

$$= \frac{1}{2} \, mV^2$$

$$= \frac{1}{2} \, \rho Q \, V^2 \qquad\qquad\qquad \because \, m = \rho Q$$

$$= \frac{1}{2} \times 1000 \times 0.22 \times (50)^2$$

$$= 275000 \text{ W}$$

$$= \textbf{275 kW}$$

Problem 6.26: Find the force exerted by a jet against water of flat plate held normal to the axis of the jet, if the diameter and the velocity of the jet are 5 cm and 32 m/s respectively.

Solution: Given data:

The velocity of the jet: $V = 32$ m/s

Diameter of the jet: $d = 5$ cm $= 0.05$ m

Now, cross-sectional area of jet:

$$a = \frac{\pi}{4} d^2 \;=\; \frac{3.14}{4} \times (0.05)^2 \;=\; 1.962 \times 10^{-3} \text{ m}^2$$

Mass flow rate strikes on the flat plate:

$$m = \rho a V \qquad\qquad [\because \text{ for water, } \rho = 1000 \text{ kg/m}^3]$$

$$= 1000 \times 1.962 \times 10^{-3} \times 32 \; \frac{\text{kg}}{\text{m}^3} \times \text{m}^2 \times \frac{\text{m}}{\text{s}}$$

$$= 62.8 \text{ kg/s}$$

Force exerted by a jet on the flat plate:

$$F = mV$$

$$= 62.8 \times 32 \; \frac{\text{kg}}{\text{m}^3} \times \frac{\text{m}}{\text{s}} \;=\; 2009 \text{ N} = \textbf{2.009 kN}$$

Problem 6.27: A jet of water of diameter 10 mm impinges on a fixed plate with a velocity of 30 m/s. Find the normal force on the plate if it is include at 30° to the jet.

Solution: Given Data.

Diameter of the jet : $d = 10$ mm $= 0.01$ m

∴ Cross-sectional area of the jet:

$$a = \frac{\pi}{4} d^2 \;=\; \frac{3.14}{4} \times (0.01)^2$$

$$= 7.85 \times 10^{-5} \text{ m}^2$$

Velocity of jet : $V = 30$ m/s

Inclination of the plate to the jet :
$$\theta = 30°$$
Normal force on the plate :
$$F_n = \rho a V^2 \sin \theta$$
$$= 1000 \times 7.85 \times 10^{-5} \times (30)^2 \sin 30° = \mathbf{35.325 \ N}$$

Problem 6.28: A jet of water strikes with a velocity of 25 m/s on a flat plate inclined at 45° with the axis of the jet. If the cross-sectional area of the jet is 30 cm^2, determine

(a) maximum force of the jet on the plate.

(b) components of the force in the direction normal to the jet.

(c) ratio in which the discharge gets divided after striking the plate.

Solution: Given data:

Velocity of jet: $V = 25$ m/s

Inclination of the plate with the jet axis:
$$\theta = 45°$$
Cross-sectional area of the jet:
$$a = 30 \text{ cm}^2 = 30 \times 10^{-4} \text{ m}^2$$

(a) Force normal to the plate is the maximum force of jet on the plate:
$$F_n = mV \sin \theta = \rho a V^2 \sin \theta \qquad (\because \ m = \rho a V)$$
$$= 1000 \times 30 \times 10^{-4} \times 25^2 \sin 45° \quad \frac{kg}{m^3} \times m^2 \times \frac{m}{s}$$
$$= \mathbf{1325.82 \ N}$$

(b) Components of the force: F_n
$$F_x = F_a \sin \theta = 1325.82 \sin 45° = \mathbf{937.49 \ N}$$
$$F_y = F_n \cos \theta = 1325.82 \cos 45° = \mathbf{937.49 \ N}$$

(c) Ratio in which the discharge gets divided
$$\frac{Q_1}{Q_2} = \frac{1 + \cos \theta}{1 - \cos \theta} = \frac{1 + \cos 45}{1 - \cos 45} = \frac{1 + 0.707}{1 - 0.707} = \mathbf{5.826}$$

Problem 6.29: A jet of water of diameter 40 mm moving with a velocity of 60 m/s, strikes a curved fixed symmetrical plate at the centre. Find the force exerted by the jet of water in the direction of the jet, if the jet is deflected through an angle of 125° at the outlet of the curved plate.

Solution: Given data:

Diameter of the jet: $d = 40$ mm $= 0.04$ m

$\therefore$ Cross-sectional area of the jet:
$$a = \frac{\pi}{4} d^2 = \frac{3.14}{4} \times (0.04)^2 = 1.256 \times 10^{-3} \text{ m}^2$$
Velocity of jet: $V = 60$ m/s
Angle of deflection $= 125°$
The angle of deflection $= 180° - \theta$

$\therefore$ $180° - \theta = 125°$

                $\theta = 55°$

Force exerted by the jet on the curved plate in the direction of the jet,

$$F_x = \rho a V^2 [1 + \cos \theta]$$
$$= 1000 \times 1.256 \times 10^{-3} \times (60)^2 [1 + \cos 55°]$$
$$= \textbf{7115.08 N}$$

Problem 6.30: A jet of water of diameter 65 mm moving with a velocity of 45 m/s, strikes a curved plate tangentially at one end at an angle of 35° to the horizontal. The jet leaves the plate at an angle 25° to the horizontal. Find the force exerted by the jet on the plate in the horizontal and vertical directions.

Solution: Given data:

Diameter of the jet: $d = 65$ mm $= 0.065$ m

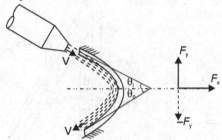

Fig. 6.47: Schematic 1 for Problem 6.30

$\therefore$ Cross-sectional area of the jet:

$$a = \frac{\pi}{4} d^2 = \frac{3.14}{4} \times (0.065)^2 = 3.316 \times 10^{-3} \text{ m}^2$$

Velocity jet: $V = 45$ m/s

           $\theta_i = 35°$

           $\theta_0 = 25°$

Now, (i) $F_x = \rho a V^2 [\cos \theta_i + \cos \theta_0]$ from Fig. 6.47

              $= 1000 \times 3.316 \times 10^{-3} \times (45)^2 [\cos 35° + \cos 25°]$

              $= \textbf{11586.29 N}$

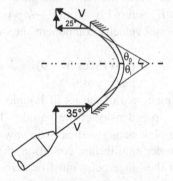

Fig. 6.48: Schematic 2 for Problem 6.23

(ii) $F_y = \rho\, a\, V^2\, [\sin\theta_0 - \sin\theta_i\,]$
              $= 1000 \times 3.316 \times 10^{-3} \times (45)^2\, [\sin 25° - \sin 35°]$
              $= -1013.669$ N [$- ve$ indicates, force F_y acts downward]
                              OR
              $F_y = \rho a V^2\, [\sin\theta_i - \sin\theta_0\,]$ from Fig. 6.48
              $= 1000 \times 3.316 \times 10^{-3} \times (45)^2\, [\sin 35° - \sin 25°]$
              $= +1013.669$ N   [Fig. 6.48, F_y acts upward]

6.18 FORCE EXERTED BY THE JET ON A HINGED PLATE

Let a jet of liquid strikes a vertical plate at point C, distance $x = l/2$ from A at which the plate is hinged. The plate can swing through an angle θ about the hinge point A as shown in Fig. 6.49

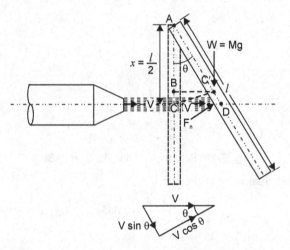

Fig. 6.49: Jet strikes on a vertical hinged plate

When jet strikes on the plate, point C on the plate shift to C'

So distance $AC = AC' = \dfrac{1}{2}$

(a) Force due to jet of water, normal to the plate:
$$F_n = m(\Delta V)_n = m(V_{n1} - V_{n2})$$
where V_{n1}, V_{n2} are normal velocity components before impact and after impact respectively.
$$F_n = \rho a V(V\cos\theta - 0)$$
$$F_n = \rho a V^2 \cos\theta \qquad\qquad ...(6.18.1)$$
where a = cross-sectional area of the jet of liquid,
          ρ = density of the liquid = 1000 kg/m^3 (for water).

(b) Weight 'W' of the plate acting vertically downward at the centre of gravity C' of the plate, under equilibrium condition, the moment of these forces, (W and F_n) about the hinge point must be zero.

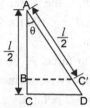

Fig. 6.50: Demonstrating of law of triangle

i.e., $F_n \times AD = W \times BC$ $\angle CAD : \cos \theta = \dfrac{AC}{AD}$

$$\rho a V^2 \cos \theta \times \frac{l/2}{\cos \theta} = W \times \frac{l}{2} \sin \theta \qquad \text{or} \qquad AD = \frac{AC}{\cos \theta}$$

or $\sin \theta = \dfrac{\rho a V^2}{W}$ ... (6.18.2) $\angle BAC : \sin \theta = \dfrac{BC'}{l/2}$

or $BC' = \dfrac{l}{2} \sin \theta$

From above Equation. (6.18.2), the angle of swing can be found out.

Problem 6.31: *A square plate of mass 13.5 kg, uniform thickness and 300 mm length of edge, is hung so that it can swing freely about its upper horizontal edge. A horizontal jet 16 mm in diameter strikes the plate with a velocity of 20 m/s. When the plate is vertical the jet strikes the plate normally at the centre.*

(*a*) What force must be applied at the lower edge of the plate to keep it vertical?

(*b*) What inclination of the vertical the plate will assume under the action of the jet if allowed to swing freely ?

Solution: Given data:

Mass of plate: $M = 13.5$ kg

∴   Weight of plate: $W = Mg = 13.5 \times 9.81 = 132.435$ N

Diameter of the jet: $d = 16$ mm $= 0.016$ m

∴ Cross-sectional area of the jet:

$$a = \frac{\pi}{4}d^2 = \frac{3.14}{4} \times (0.016)^2 = 2.0096 \times 10^{-4} \ m^2$$

∴   Velocity of the jet: $V = 20$ m/s

(*a*) Let p = horizontal force applied at lower edge of the plate to keep it vertical.

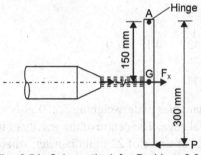

Fig. 6.51: Schematic 1 for Problem 6.31

Force exerted by a jet, normal to the plate:

$$F_x = \rho a V^2 = 1000 \times 2.0096 \times 10^{-4} \times (20)^2 = 80.384 \text{ N}$$

Taking moment at hinge point A

Anti-clockwise moment due to force, F_x about hinge point A

$$\qquad = \text{clockwise moment due to applied force, } P \text{ about}$$
$$\qquad\qquad \text{hinge point } A + \text{moment due to weight } (W) \text{ of plate.}$$

$$F_x \times 150 = P \times 300 + 0$$
$$80.384 \times 150 = P \times 300$$
$$P = \textbf{40.192 N}$$

or

(*b*) Let θ = angle of swing

Force exerted by a jet normal to the plate,

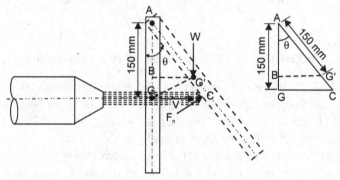

Fig. 6.52: Schematic 2 for Problem 6.31

i.e., $\quad F_n = \rho a V^2 \cos\theta$

$$\qquad = 1000 \times 2.0096 \times 10^{-4}$$
$$\qquad\qquad \times (20)^2 \cos\theta$$
$$\qquad = 80.384 \cos\theta$$

In $\angle BAG$: $\sin\theta = \dfrac{BG'}{150 \text{ mm}}$

$$BG' = 150 \sin\theta \text{ mm}$$
$$\qquad = 0.15 \sin\theta \text{ m}$$

Taking moment at hinge point A

In $\angle BAG$: $\cos\theta = \dfrac{150}{AC}$

$$F_n \times AC = W \times BG'$$

$$AC = \dfrac{150}{\cos\theta} \text{ mm}$$

$$80.384 \cos\theta \times \dfrac{0.150}{\cos\theta} \qquad\qquad = \dfrac{0.150}{\cos\theta} \text{ m}$$

$$\qquad = 132.435 \times 0.15 \sin\theta$$

or $\quad \sin\theta = 0.6069$

$$\theta = \textbf{37.37}°$$

Problem 6.32: A rectangular plate weighing 55.50 N is suspected vertically by a hinge on the top horizontal edge. The centre of the gravity of the plate is 125 mm from the hinge. A horizontal jet of water of 22 mm diameter, where axis is 150 mm below the hinge, impinges normally to the plate with a velocity of 8 m/s. Find:

(a) The horizontal force applied at the centre of gravity to maintain the plate in vertical position.

(b) The change in velocity of set if the plate is deflected through 35° and the same horizontal force continues to act at the centre of gravity of the plate.

Solution: Given data:

Weight of the plate: $W = 55.50$ N

Diameter of the jet: $d = 22$ mm $= 0.022$ m

∴ Cross-sectional area of the jet:

$$a = \frac{\pi}{4}d^2 = \frac{3.14}{4} \times (0.022)^2 = 3.7994 \times 10^{-4} \text{ m}^2$$

Velocity of the jet: $V = 8$ m/s

(a) Let $P =$ horizontal force applied at the centre of gravity to maintain the plate is vertical position.

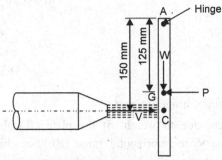

Fig. 6.53: Schematic 1 for Problem 6.32

Force exerted by a jet on the vertical plate at point C:

$$F_x = m \text{ [Initial velocity − Final velocity in direction of force, } F_x]$$
$$= m (V-0) = \rho aV.V = \rho aV^2 \qquad (\because m = \rho aV)$$
$$= 1000 \times 3.7994 \times 10^{-4} \times (8)^2 = \mathbf{24.316 \text{ N}}$$

Taking moment about hinge point A

Moment due to F_x in anti-clockwise = moment due to P in clockwise + moment due to W is zero

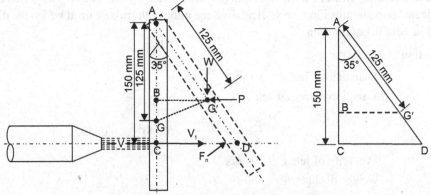

Fig. 6.54: Schematic 2 for Problem 6.32

$$F_x \times AC = P \times AG$$
$$24.316 \times 0.150 = P \times 0.128$$
$$P = \textbf{29.179 N}$$

$\because AC = 150$ mm $= 0.150$ m
$\because AG = 125$ mm $= 0.125$ m

Let $V_1 =$ absolute velocity of the jet.

Force exerted by the jet, normal to the plate:

In $\angle BAG'$ $\sin 35° = \dfrac{BG'}{AG'}$

$BG' = AG' \sin 35° = 125 \sin 35°$
$= 71.69$ mm $= 0.07169$ m

$$F_n = \rho a V_1^2 \cos \theta$$
$$= 1000 \times 3.7994 \times 10^{-4} \times V_1^2 \cos 35°$$
$$F_n = 0.3322\ V_1^2$$

and $\cos 35° = \dfrac{AB}{AG'}$

$AB = AG' \cos 35° = 125 \cos 35°$
$= 102.394$ mm $= 0.1024$ m

In $\angle CAD$ $\cos 35° = \dfrac{AC}{AD}$

$$AD = \dfrac{AC}{\cos 35°} = \dfrac{150}{\cos 35°}$$
$$= 183.116 \text{ mm} = 0.1831 \text{ m}$$

Taking moment about hinge point A.

Clockwise moment due to weight of the plate (W) about hinge point A + Clockwise moment due to the horizontal force (P) about hinge point A = Anticlockwise moment due to the normal force on the plate about hinge point A

$$W \times BG' + P \times AB = F_n \times AD$$
$$55.50 \times 0.07169 + 29.179 \times 0.1024 = 0.31122\ V_1^2 \times 0.1831$$
$$V_1^2 = 122.256$$
$$V_1 = 11.057 \text{ m/s}$$

Velocity increase of the jet $= V_1 - V = 1.057 - 8 = \textbf{3.057 m/s}$

Problem 6.33: A vertical jet of water 80 mm in diameter leaving the nozzle, with a 10 m/s velocity strikes a horizontal and movable disc weighing 180 N. The jet is then deflected horizontally. Find the vertical distance y above the nozzle tip at which the disc will be held in equilibrium.

Solution: Given data:

Diameter of jet: $d = 80$ mm $= 0.08$ m

$\therefore$ Cross sectional area of jet:

$$a = \frac{\pi}{4}d^2 = \frac{3.14}{4} \times (0.08)^2 = 0.005024 \text{ m}^2$$

Velocity of jet: $V = 10$ m/s

Weight of disc: $W = 180$ N

Discharge through nozzle: $Q = aV = 0.005024 \times 10 = 0.05024$ m³/s

Let u be the jet velocity at an elevation y above the nozzle.

Applying the Bernoulli's equation between the jet at the nozzle exit and the jet at an elevation y.

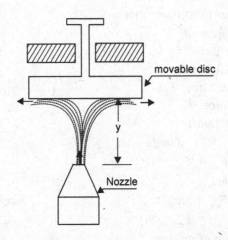

Fig. 6.55: Schematic for Problem 6.33

$$0 + \frac{V^2}{2g} + 0 = 0 + \frac{u^2}{2g} + y$$

where V = the velocity at the nozzle exit.

∴ $u^2 = V^2 - 2gy$

Applying the momentum equation for equilibrium:

Force exerted by the jet on the disc = weight of the disc.

$$\rho au = 180$$

$$u = \frac{180}{1000 \times 0.05024} = 3.58 \text{ m/s}$$

∴ $(3.58)^2 = 10^2 - 2 \times 9.81 \times y$

or $y = \textbf{4.44 m}$

SUMMARY

1. The dynamic behaviour of the fluid flow is analysed by the application of Newton's 2nd law of motion.

2. Types of forces influencing motion:
   - (*i*) Gravity force: F_g
   - (*ii*) Pressure force: F_p
   - (*iii*) Viscous force: F_v
   - (*iv*) Turbulent force: F_t
   - (*v*) Elastic force: F_e

3. Equation of motions:

   $\Sigma F_x = ma_x$ along *x*-axis

   $F_{gx} + F_{px} + F_{vx} + F_{tx} + F_{ex} = ma_x$

   $F_{gx} + F_{px} = ma_x$ ...Euler's equation

   $F_{gx} + F_{px} + F_{vx} = ma_x$ ...Navier-Stokes equation

   $F_{gx} + F_{px} + F_{vx} + F_{tx} = ma_x$ ...Reynolds equation

4. Euler's equation of motion:

   $$\frac{dp}{\rho} + VdV + gdz = 0$$

5. As assumptions of Bernoulli's equation:
   - (*i*) Fluid is ideal *i.e.*, viscosity of the fluid is zero,
   - (*ii*) Flow is steady,
   - (*iii*) Stream line flow,
   - (*iv*) Fluid is incompressible,
   - (*v*) Flow is irrotational, and
   - (*vi*) No shaft work and heat interaction.

6. The Bernoulli's equation is obtained by integration of Euler's equation of motion: $\int \frac{dp}{\rho} + \int VdV + \int gdz = $ constant

   $$\frac{p}{\rho} + \frac{V^2}{2} + gz = C$$

   or

   $$\frac{p}{\rho g} + \frac{V^2}{2g} + z = C$$

   where $\dfrac{p}{\rho g}$ = pressure energy per unit weight or pressure head.

   $\dfrac{V^2}{2g}$ = kinetic energy per unit weight or kinetic head.

   z = potential energy per unit weight or potential head.

Contd...

7. Statement of Bernoulli's theorem.

   It states that the total mechanical energy at every point in steady, irrotational, incompressible and non-viscous flow is constant.

8. Form of Bernoulli's equation:

   Ist form: Energy per unit weight $\qquad \dfrac{p}{\rho g} + \dfrac{V^2}{2g} + z = C$

   2nd form: Energy per unit mass $\qquad \dfrac{p}{\rho} + \dfrac{V^2}{2} + gz = C$

   3rd form: Energy per unit volume $p + \dfrac{\rho V^2}{2} + \rho gz = C$

9. The modified Bernoulli's equation may be used in real fluid flow system.

$$\frac{p_1}{\rho g} + \frac{V_1^2}{2g} + z_1 = \frac{p_2}{\rho g} + \frac{V_2^2}{2g} + z_2 + h_L$$

   where h_L = head loss between the two sections.

10. Practical application of Bernoulli's equation:

   (i) Venturimeter, $\qquad\qquad$ (ii) Orifice meter,

   (iii) Flow nozzle meter, and (iv) Pitot tube.

11. The momentum equation states that the net force acting on a fluid mass is equal to the change in momentum per second in that direction:

$$F = \frac{d}{dt}(mV)$$

   The impulse – momentum equation is given by

   $\qquad\qquad\qquad F.dt = M.dV$

12. Forces exerted by a flowing fluid on pipe bend:

   Force along x-axis: $F_x = \rho Q(V_1 - V_2 \cos \theta) + (p_1 A_1 - p_2 A_2 \cos \theta)$

   Force along y-axis: $F_y = \rho Q(0 - V_2 \sin \theta) + (0 - p_2 A_2 \sin \theta)$

   $\qquad\qquad\qquad = -\rho Q V_2 \sin \theta) - p_2 A_2 \sin \theta$

   Resultant force: $\qquad F = \sqrt{F_x^2 + F_y^2}$

   The direction of the resultant force with horizontal direction is given by

$$\tan \phi = \frac{F_y}{F_x}$$

13. Force exerted by the Jet on a stationary flat plate.

   (a) $\qquad\qquad\qquad F_x = \rho a V^2 \qquad\qquad$ } for a vertical flat plate.

   (b) $\qquad\qquad\qquad F_n = \rho a V^2 \sin \theta$
   $\qquad\qquad\qquad\quad F_x = \rho a V^2 \sin \theta \qquad$ } for an inclined flat plate.

   $\qquad\qquad\qquad F_y = \dfrac{1}{2}\rho a V^2 \sin^2 \theta$

Contd...

(c) $F_x = \rho a V^2 (1 + \cos\theta)$ for a curved plate and jet
                       $F_y = 0$ strikes at the centre.

14. Force exerted by a jet of water normal to one end hinged, other end free of plate

$$F_n = \rho a V^2 \cos\theta \qquad \text{where} \quad \theta = \text{angle of swing}$$

$$\sin\theta = \frac{\rho a V^2}{W} \qquad \text{where} \quad W = \text{weight of the plate}$$

a = cross-sectional area of jet
V = velocity of jet.

ASSIGNMENT - 1

1. Name the force involved in a fluid flow. Which forces are taken in the following equations:
   (a) Euler's Equation.
   (b) Navier-Stokes equation, and
   (c) Reynolds equation of motion.

2. What is Euler's equation of motion? How will you obtain Bernoulli's equation from it?   (*GGSIP University, Delhi, Dec.2007*)

3. State Bernoulli's theorem and write down all the assumptions made in its derivations.   (*GGSIP University, Delhi, Dec.2007, Dec.2008*)

4. Write the Bernoulli's equation and Euler's equation of motion.
                                              (*GGSIP University, Delhi, Dec.2005*)

5. Derive Euler's equation of motion along a streamline for an ideal fluid stating clearly the assumptions made.   (*GGSIP University, Delhi, Dec.2007*)

6. Define continuity equation and Bernoulli's equation. Explain briefly hydraulic grade and total energy lines.   (*GGSIP University, Delhi, Dec.2008*)

7. Graphical representation of the Bernoulli's equation in following devices:
   (i) Venturimeter,
   (ii) Converging-diverging nozzle, and
   (iii) Converging-diverging diffuser.

8. What is the limitation of the Bernoulli's equation?

9. State the momentum equation. How will you apply momentum equation for determining the force exerted by a flowing liquid in the pipe bend?

10. Define impulse momentum equation, where this equation is used?

11. Derive an expression for the force exerted by a Jet of water on fixed vertical flat plate in the direction of the jet.

12. Why work done by the jet of water on a fixed plate is zero?

13. Derive an expression for the force exerted by a jet of water on fixed inclined flat plate in normal to the plate.

ASSIGNMENT - 2

1. Water is flowing through a pipe of 70 mm diameter under pressure of 50 kPa, and with velocity of 3 m/s. Neglecting friction, find the total head, if the pipe is 10 m above the datum line. **[Ans. 15.55 m of water]**

2. A pipe of 300 mm diameter carries an oil of specific gravity 0.85 at the rate of 150 litre/s and under a pressure 4 kPa. Find the total energy per unit weight at a point which is 3.5 m above the datum line. **[Ans. 4.13 Nm/N or m of oil]**

3. The oil of specific gravity 0.8, is flowing through a pipe of 100 mm diameter under a pressure of 50 kN/m², and with a velocity of 4 m/s. Neglecting friction, find the total head, if the pipe is 5 m above the datum line.

   [Ans. 12.18 m of oil]

4. A pipe of 300 mm diameter carries an oil of specific gravity 0.85 at the rate of 140 litre/s and under a pressure of 4 kPa. Find the total energy per unit weight at a point which at 4 m above the datum line. Find also the total energies per unit mass per unit volume.

   [Ans. 4.60 Nm/N, 42.126 N.m/kg, 38.357 kJ/m³]

5. The diameter of a pipe change from 200 mm at a specific 5 m above datum to 50 mm at a section 3 m above datum. The pressure of water at section (1) is 500 kPa. If the velocity of flow at section (1) is 1 m/s. Find the pressure at section (2). **[Ans. V_2 = 16 m/s, p_2 = 392.40 kPa]**

6. At a certain section A of a pipeline carrying water, the diameter is 1 m, the pressure is 98.10 kN/m² and the velocity is 3 m/s. At another section B which is 2 m higher than section A, the diameter is 0.7 and the pressure is 59.2 kN/m². What is the direction of flow?

   [Ans. Total energy at section A is greater than at section B and hence the flow direction is from section A to section B**]**

7. At a point in the pipeline where the diameter is 200 mm, the velocity of water is 4 m/s and the pressure is 343 kN/m². At a point 15 m downstream the diameter gradually reduces to 100 mm. Find the pressure at this point, if the pipe is (*i*) horizontal, (*ii*) vertical with flow downward (*iii*) vertical with flow upward. **[Ans.** (*i*) 223.5 kN/m² (*ii*) 370.60 kN/m² (*iii*) 76.37 kN/m²**]**

8. A pipe of diameter 400 mm carries water at a velocity of 25 m/s. The pressure at the points A and B are given as 29.43 N/cm² and 22.563 N/cm² respectively while the datum head at A and B are 28 m and 30 m. Find the loss of head between A and B. **[Ans. 5 m]**

9. A conical tube of length 2 m is fixed vertically with its smaller end upwards. The velocity of flow at the smaller end is 5 m/s while at the lower end is 2 m/s. The pressure head at the smaller end is 2.5 m of liquid. The loss of head

   in the tube is $\dfrac{0.35(V_1 - V_2)^2}{2g}$, where V_1 is velocity at the smaller end and V_2

   is the velocity at the lower end respectively. Find the pressure head at the lower end. The direction of flow is downward. **[Ans. 5.407 m of liquid]**

10. A pipe line carrying oil of specific gravity 0.87, change in diameter from 200 mm diameter at a position A to 500 mm diameter at a position B which is 4 m at a higher level. If the pressure at A and B are 98.10 kPa and 58.86 kPa respectively and the discharge is 200 litre/s, find the loss of head and direction of oil flow.

[**Ans.** $h_L = 2.609$ m, total energy at position A is greater than at position B and hence oil flow is taking place from A to B]

7

Dimensional and Model Analysis

7.1 INTRODUCTION

Dimensional analysis is a mathematical technique that helps us to set up the empirical relation about a fluid flow or several engineering problems. Each physical process can be expressed by an equation composed of variables which may be dimensional and non-dimensional parameters. Dimensional analysis helps in determining a systematic arrangement of the variables in the physical relationship. It also helps in combining dimensional variables to form non-dimensional parameters.

For over a hundred years, engineers have used small scale models of the actual structure or machine or mechanism. **The actual structure or machine or mechanism is called prototype and the small scale replica of the actual structure is called model,** which help to provide information that will make their design more effective. For example; some prototypes *i.e.,* airplanes, motor vehicles, ships and dam spillways are tested in laboratories [as small-scale models] to determine the details of the fluid flow associated with their use. Physical modelling of these flows in a laboratory setting is often a necessary step in designing the prototype (full-scale device)].

Dimensional analysis provides an engineer, the tools to design, conduct and analyze the results of model testing and to predict the important flow properties that will be encountered by the full-scale structure (*i.e.,* prototype). The equations governing the flow of fluid are so complex even when derived under the simplified assumptions of constant fluid properties, that analytical solution can be obtained only for some very simple geometries. In actual designing, the calculation of flow and geometries *(i.e.,* a turbine, a rocket engine, a jet airliner *etc.)* are very difficult. To sort out such difficulties, the designer resorts to the testing of scaled models before the construction of a prototype. The models are frequently scaled down version and therefore, can be redesigned and tested at comparatively much less cost. In modern day engineering practice, any new type of machine or structure involving interaction with fluids is seldom constructed until a model test has been made *i.e.,* check is made to determine the feasibility of design by modelling.

© The Author(s) 2023
S. Kumar, *Fluid Mechanics (Vol. 1),*
https://doi.org/10.1007/978-3-030-99762-5_7

In this chapter, we describe how dimensional analysis can be applied to any fluid flow problem and used to simplify the dependence of important flow properties on the flow variables. Then we show how to derive useful information from model experiments by applying the principle of dimensional analysis. Both the dimensional analysis and modelling are essential tools for the engineer in designing of a complex system.

7.2 PRIMARY (OR BASIC OR FUNDAMENTAL) QUANTITIES AND SECONDARY (OR DERIVED) QUANTITIES

The mass, length, time, and temperature are independent of each other and their units of measurement are prescribed by International Standards (SI). These four quantities may be regarded as fundamental quantities. The fundamental quantities are described by symbols such as M for mass, L for length, T for time and θ for the temperature. The symbols L, M, T and θ are known as primary dimensions of the respective quantities.

The quantities that are derived from the fundamental quantities are called derived quantities or secondary quantities. The dimensions of derived quantities are called derived dimensions. For example, dimension of velocity, acceleration, density etc.

Velocity: $$V = \frac{\text{Displacement}}{\text{Time}} = \frac{L}{T} = LT^{-1}$$

Acceleration: $$a = \frac{\text{Velocity}}{\text{Time}} = \frac{LT^{-1}}{T} = LT^{-2}$$

Density: $$\rho = \frac{\text{Mass}}{\text{Volume}} = \frac{M}{L^3} = ML^{-3}$$

The dimensions that are commonly used in mechanics of fluid and heat transfer are given in Table 7.1.

Table 7.1: Quantities used in Fluid Mechanics and Heat Transfer and their Dimensions.

S. No.	Quantity	Nature of Quantity	Unit	Dimensions
1.	Length, breadth, height and depth	Geometric	m	L
2.	Area	Geometric	m^2	L^2
3.	Volume	Geometric	m^3	L^3
4.	Area moment of inertia	Geometric	m^4	L^4
5.	Velocity: V	Kinematic	m/s	LT^{-1}
6.	Velocity potential: ϕ	Kinematic	m^2/s	$L^2 T^{-1}$
7.	Acceleration: a	Kinematic	m/s^2	LT^{-2}
8.	Angular velocity: ω	Kinematic	rad/s	T^{-1}
9.	Stream function: ψ	Kinematic	m^2/s	$L^2 T^{-1}$

(Contd...)

S. No.	Quantity	Nature of Quantity	Unit	Dimensions
10.	Rotational speed: N	Kinematic	rev/s	T^{-1}
11.	Circulation: Γ	Kinematic	m²/s	$L^2 T^{-1}$
12.	Vorticity: ξ	Kinematic	s⁻¹	T^{-1}
13.	Rate of flow: Q	Kinematic	m³/s	$L^3 T^{-1}$
14.	Kinematic viscosity: v	Kinematic	m²/s	$L^2 T^{-1}$
15.	Density: ρ	Dynamic	kg/m³	ML^{-3}
16.	Force: F	Dynamic	N or kg m/s²	MLT^{-2}
17.	Mass flow rate: m	Dynamic	kg/s	MT^{-1}
18.	Pressure: p, Shear stress: τ	Dynamic	N/m²	$ML^{-1} T^{-2}$
19.	Specific weight: w	Dynamic	N/m³	$ML^{-2} T^{-2}$
20.	Dynamic viscosity: μ	Dynamic	N s/m²	$ML^{-1} T^{-1}$
21.	Surface tension: σ	Dynamic	N/m	MT^{-2}
22.	Bulk modulus of elasticity: K	Dynamic	N/m²	$ML^{-1} T^{-2}$
23.	Momentum, Impulse	Dynamic	N s	MLT^{-1}
24.	Energy, Work	Dynamic	Nm	$ML^2 T^{-2}$
25.	Torque	Dynamic	Nm	$ML^2 T^{-2}$
26.	Power	Dyanmic	N m/s or J/s or W	$ML^2 T^{-3}$
27.	Mass moment of inertia	Dynamic	kgm²	ML^2
28.	Temperature	Thermodynamic	°C	θ
29.	Absolute temperature	Thermodynamic	K	θ
30.	Thermal conductivity	Thermodynamic	W/m K	$MLT^{-3} \theta^{-1}$
31.	Entropy: S	Thermodynamic	J/K	$ML^2 T^{-2} \theta^{-1}$
32.	Specific enthalpy: h	Thermodynamic	J/kg	$L^2 T^{-2}$
33.	Specific internal energy: u	Thermodynamic	J/kg	$L^2 T^{-2}$
34.	Gas constant: R	Thermodynamic	J/kg K	$L^2 T^{-2} \theta^{-1}$
35.	Heat transfer	Thermodynamic	J/s	$ML^2 T^{-3}$

7.3 DIMENSIONAL HOMOGENEITY

According to the principle of dimensional homogeneity, a mathematical equation which describes a process of fluid flow or some other kind of process, the dimensions of each side of the equation must be same. Thus, the given equation is known as dimensionally homogeneous equation. The powers of fundamental dimensions [*i.e., L, M, T*] on both sides of the equation will be same for a dimensionally homogeneous equation.

For Example: Discharge flow through a passage is given by the continuity equation:

Discharge: $Q = AV$ ...(1)

where A = cross-sectional area of fluid flow.

                       V = average velocity of fluid flow.

Dimension of R.H.S.   = cross-sectional area × average velocity

$$= L^2 \times \frac{L}{T} = L^3 T^{-1}$$

Dimension of L.H.S. = Q

$$= \frac{\text{Volume of fluid}}{\text{Time taken}} = \frac{L^3}{T} = L^3 T^{-1}$$

Dimension of L.H.S. = dimension of R.H.S. = $L^3 T^{-1}$

Thus we find dimensional homogeneity in Eq. (1) because both sides of the equation have the same powers of the fundamental units.

Problem 7.1: Determine whether the following equations are dimensionally homogeneous:

(*i*) Discharge through a rectangular notch is given by an equation:

$$Q = \frac{2}{3}\sqrt{2g}\, LH^{3/2}$$

(*ii*) Head loss due to friction when fluid flow through pipes:

$$h_f = \frac{4f\,LV^2}{2Dg}$$

(*iii*) Chezy's formula for velocity of flow in pipes:

$$V = C\sqrt{mi}$$

Solution: (*i*) Discharge: $Q = \frac{2}{3}\sqrt{2g}\, LH^{3/2}$

Dimension of the variables involved in given problem.

Quantity	Dimension
Q	$L^3 T^{-1}$
g	LT^{-2}
L	L
H	L

Dimension of L.H.S. $= Q = L^3 T^{-1}$

Dimension of R.H.S. $= \sqrt{g}\, LH^{3/2} = (LT^{-2})^{1/2}.L\, L^{3/2}$

$$= L^{1/2}T^{-1}.L.L^{3/2} = L^3 T^{-1}$$

Thus, dimension of L.H.S. = dimension of R.H.S. = $L^3 T^{-1}$

Hence, given equation is homogeneous.

(ii) Head loss due to friction when fluid flows through pipes:

$$h_f = \frac{4f\,LV^2}{2Dg}$$

Dimension of the variables involved in given problem.

Quantity	Dimension
h_f	L
f	None
L	L
V	LT^{-1}
D	L
g	LT^{-2}

Dimension of L.H.S. $= h_f = L$

Dimension of R.H.S. $= \dfrac{4f\,LV^2}{2Dg} = \dfrac{L.L^2T^{-2}}{L.LT^{-2}} = L$

Thus, dimension of L.H.S. = dimension of R.H.S. = L

Hence, given equation is homogeneous.

(iii) Chezy's formula for velocity of flow in pipes:

$$V = C\sqrt{mi}$$

Dimension of the variables involved in given problem.

Quantity	Dimension
V	LT^{-1}
C	None
m	L
i	None

Dimension of L.H.S. $= V = LT^{-1}$

Dimension of R.H.S. $= C\sqrt{mi} = L^{\frac{1}{2}}$

Thus, dimension of L.H.S. $\neq$ dimension of R.H.S.

Hence, given equation is not homogenous.

7.4 METHOD USED FOR DIMENSIONAL ANALYSIS

The following two methods are used for dimensional analysis:

    1. Rayleigh's method, and   2. Buckingham's π (Pi)–theorem.

7.4.1 Rayleigh's Method

Suppose $A, B, C, D, E,$ ......etc. variables are present in a process. Variable A depends on B, C, D, E. Then according to Rayleigh's method, A is function of B, C, D and E and mathematically it is written as

$$A = f[B, C, D, E]$$

This can also be expressed in the form as

$$A = K B^a C^b D^c E^d$$

where K is dimensionless constant and a, b, c, d are arbitrarily powers. The value of a, b, c and d are obtained by comparing the power of the fundamental dimension on both sides. Thus, the expression is obtained for dependent variable.

This method is more efficiently used if a maximum three or four variables are involved in the process. If number of variables are more than four, then it is very difficult to find the expression by this method.

Problem 7.2: The velocity of propagation V of a pressure wave through a liquid depends on the mass density ρ and the bulk modulus K. Obtain an expression for the velocity of propagation of the wave.

Solution: The velocity of propagation V is function of mass density ρ and bulk modulus K.

i.e., $\qquad\qquad V = f(\rho, K)$

According to Rayleigh's method,

$$V = C\rho^a K^b \qquad\qquad\qquad\qquad ...(1)$$

where C is dimensionless constant.

Performing a dimensional analysis, substituting the dimension of given variables in Equation (1), we get

Dimension of variables involved in given problem.

Variables	Dimension
V	LT^{-1}
ρ	ML^{-3}
K	$ML^{-1}T^{-2}$
C	None

$$LT^{-1} = (ML^{-3})^a (ML^{-1}T^{-2})^b$$
$$LT^{-1} = M^a L^{-3a} M^b L^{-b} T^{-2b}$$
$$LT^{-1} = M^{a+b} L^{-3a-b} T^{-2b}$$

Now comparing the power of L: $1 = -3a - b$

Comparing the power of M: $\qquad 0 = a + b$

or $\qquad\qquad\qquad\qquad\qquad \boxed{a = -b}$

Comparing the power of T: $-1 = -2b$

or $\boxed{b = \dfrac{1}{2}}$

By solving, we get $\boxed{a = -\dfrac{1}{2}}$

Substituting the valve of 'a' and 'b' in Eq. (1), we get

$$V = C\rho^{-\frac{1}{2}}K^{\frac{1}{2}}$$

$$= C\sqrt{K/\rho}$$

Problem 7.3: A pump develops a hydraulic power P which is a function of discharge Q, head H and specific weight w of the liquid. Show that $P = K Q w H$ where K is a dimensionless constant.

Solution: The hydraulic power (P) developed by the pump is function of discharge (Q), head (H) and specific weight (w).

i.e., $P = f (Q, H, w)$

According to Rayleigh's method,

$$P = K Q^a H^b w^c \qquad \qquad ...(1)$$

Performing a dimensional analysis, substituting the dimension of given variable in Eq. (1), we get

Dimension of variables involved in given problem.

Variables	Dimension
P	ML^2T^{-3}
Q	$L^3 T^{-1}$
H	L
w	$ML^{-2}T^{-2}$

$$ML^2 T^{-3} = (L^3T^{-1})^a L^b (ML^{-2}T^{-2})^c$$
$$ML^2T^{-3} = L^{3a} T^{-a} L^b M^cL^{-2c}T^{-2c}$$
$$ML^2 T^{-3} = L^{3a+b-2c} M^cT^{-a-2c}$$

Now comparing the power of M:

$$1 = c$$

or $\boxed{c = 1}$

Comparing the power of L:

$$2 = 3a+b - 2c \qquad \qquad ...(2)$$

Comparing the power of T:

$$-3 = -a -2c = -a - 2 \times 1$$

or $\boxed{a = 1}$

Substituting the value of 'a' and 'c' in Eq. (2), we get

$$2 = 3 \times 1 + b - 2 \times 1$$

$$\boxed{b = 1}$$

Substituting the value of a, b and c in Eq. (1), we get

$$P = KQHW$$

Problem 7.4: Find an expression for the drag force on smooth sphere of diameter D, moving with a uniform velocity V in a fluid of density ρ and dynamic viscosity μ.

Solution: Drag force (F) is function of diameter (D), velocity (V), density (ρ) and dynamic viscosity (μ).

i.e., $F = f(D, V, \rho, \mu)$

According to Rayleigh's method,

$$F = K\, D^a\, V^b\, \rho^c\, \mu^d \qquad \qquad ...(1)$$

where K is dimensionless constant.

Performing a dimensional analysis, substituting the dimension of given variables in Equation (1), we get

Dimension of variables involved in given problem.

Variables	Dimension
F	MLT^{-2}
D	L
V	LT^{-1}
ρ	ML^{-3}
μ	$ML^{-1}T^{-1}$

$$MLT^{-2} = L^a\, (LT^{-1})^b\, (MT^{-3})^c\, (ML^{-1}T^{-1})^d$$
$$MLT^{-2} = L^a\, L^bT^{-b}\, M^cL^{-3c}\, M^dL^{-d}T^{-d}$$
$$MLT^{-2} = L^{a+b-3c-d}\, M^{c+d}\, T^{-d}$$

Now comparing the power of L: $1 = a + b - 3c - d$ ...(2)

Comparing the power of M: $1 = c + d$ ...(3)

Comparing the power of T: $-2 = -b - d$ ...(4)

There are four unknowns (*i.e.*, a, b, c and d) but equations are three [Eqs. (2). (3) and (4)]. Hence, it is impossible to find the value of a, b, c and d. But one way to solve this problem is to express all the three variables in terms of fourth variable, which is most important. Here viscosity plays more important role and hence a, b, c are expressed in terms of d which is the power to viscosity.

$\therefore$ From Eq. (3) $\boxed{c = 1 - d}$

From Eq. (4) $\boxed{b = 2 - d}$

From Eq. (2)
$$a = 1 - b + 3c + d$$
$$a = 1 - 2 + d + 3(1 - d) + d$$
$$a = 1 - 2 + d + 3 - 3d + d$$

$$\boxed{a = 2 - d}$$

Substituting the values of a, b and c in Eq. (1), we get

$$F = KD^{2-d}\,V^{2-d}\,\rho^{1-d}\,\mu^{d}$$

$$F = KD^{2}D^{-d}V^{2}V^{-d}\,\rho^{1}\,\rho^{-d}\,\mu^{d}$$

$$F = KD^{2}V^{2}\rho\left[V^{-d}D^{-d}\rho^{-d}\mu^{d}\right]$$

$$F = KD^{2}V^{2}\rho\left[\frac{\mu}{\rho VD}\right]^{d}$$

Instead of this form where K and d are unknown quantities, the above equation is written in the functional form

$$\frac{F}{\rho D^{2}V^{2}} = \phi\left[\frac{\mu}{\rho VD}\right] = \phi\,[Re] \qquad \because \text{ Reynolds number}: Re = \frac{\rho VD}{\mu}$$

where the right hand side represents a function of the Reynolds number $\dfrac{\rho VD}{\mu}$.

Problem 7.5:   A fluid of density ρ and viscosity μ flows through a pipe of diameter d. Show by dimensional analysis the resistance per unit area of surface is given by, $F = \rho V^{2}\phi\,(Re)$ where V is the mean velocity of flow and Re is the Reynolds number.

Solution: According to given problem,

The resistance force per unit area, (F) is function of ρ, V, D and μ.

i.e., $$F = f\left(\rho, V, D, \mu\right)$$

According to Rayleigh's method, $F = K\,\rho^{a}V^{b}D^{c}\mu^{d}$ ...(1)

Dimensions of variables involved in given problem.

Variables	Dimension
F	$ML^{-1}T^{-2}$
ρ	ML^{-3}
V	LT^{-1}
D	L
μ	$ML^{-1}T^{-1}$

Performing a dimensional analysis, substituting the dimension of given variables in Eq. (1), we get

$$ML^{-1}T^{-2} = (ML^{-3})^a \ (LT^{-1})^b \ (L)^c \ (ML^{-1}T^{-1})^d$$
$$ML^{-1}T^{-2} = M^{\,a}L^{-3a} \ L^b \ T^{-b} \ L^c \ M^{\,d}L^{-d}T^{-d}$$
$$ML^{-1}T^{-2} = M^{\,a+b}L^{-3a+b+c-d} \ T^{\,-b-d}$$

Now comparing the power of L: $\quad -1 = -3a + b + c - d$ \hfill ...(2)

Comparing the power of M: $\qquad 1 = a + d$ \hfill ...(3)

Comparing the power of T: $\qquad -2 = -b - d$ \hfill ...(4)

There are four unknowns but only three equations. Let us therefore express three unknowns in terms of the other unknown.

From Eq. (3) $\qquad \boxed{a = 1 - d}$

From Eq. (4) $\qquad \boxed{b = 2 - d}$

From Eq. (2) $\qquad c = 3a - b + d - 1$
$$c = 3(1-d) - 2 + d + d - 1$$
$$c = 3 - 3d - 2 + d + d - 1$$

$$\boxed{c = -d}$$

Substituting the value of a, b and c in Eq. (1), we get

$$F = K\rho^{1-d}V^{2-d}d^{-d}\mu^d$$

$$F = K\rho\rho^{-d}V^2V^{-d}d^{-d}\mu^d$$

$$F = K\rho V^2 \frac{\mu^d}{\rho^d V^d d^d}$$

$$F = K\rho V^2 \left(\frac{\mu}{\rho V d}\right)^d$$

The above equation is written as

$$\frac{F}{\rho V^2} = \phi\left[\frac{\mu}{\rho V d}\right] \qquad\qquad \left| \because \mathrm{Re} = \frac{\rho V d}{\mu}\right.$$

$$\frac{F}{\rho V^2} = \phi[\mathrm{Re}]$$

or $\qquad\qquad F = \rho V^2 \phi[Re]$

Problem 7.6: A partially submerged body is towed in water. The resistance R to its motion depends on the density ρ, the viscosity μ of water, length l of the body, velocity v of the body and the acceleration due to gravity g. Show that the resistance to the motion can be expressed in the form

$$R = \rho l^2 v^2 \phi\left[\left(\frac{\mu}{\rho v l}\right), \left(\frac{lg}{v^2}\right)\right]$$

Solution: The resistance R is function of density ρ, the viscosity μ, length of the body l, velocity v and the acceleration due to gravity g.

i.e., $$R = f(\rho, \mu, l, v, g)$$

According to Rayleigh's method,

$$R = K(\rho^a, \mu^b, l^c, v^d, g^e) \qquad \qquad ...(1)$$

where K is dimensionless constant; dimensions of the variables involved in given problem:

Variables	Dimension
R	MLT^{-2}
ρ	ML^{-3}
μ	$ML^{-1}T^{-1}$
l	L
v	LT^{-1}
g	LT^{-2}

Performing a dimensional analysis:

Substituting the dimension on both sides.

$$MLT^{-2} = (ML^{-3})^a \ (ML^{-1}T^{-1})^b (L)^c (LT^{-1})^d (LT^{-2})^e$$

$$MLT^{-2} = M^a \ L^{-3a} \ M^b \ L^{-b} \ T^{-b} L^c L^d \ T^{-d} \ L^c \ T^{-2e}$$

Comparing the power of M, $1 = a + b$

Comparing the power of L, $1 = -3a - b + c + d + e$

Comparing the power of T, $-2 = -b - d - 2e$

There are five unknowns and equations are only three. Expressing the three unknowns in terms of two unknowns (μ and g).

Hence express a, c and d in terms of b and e

Solving, we get

$$\boxed{a = 1 - b}$$

$$-2 = -b - d - 2e$$

$$\boxed{d = 2 - b - 2e}$$

$$1 = -3 (1 - b) - b + c + 2 - b - 2e + e$$

$$1 = -3 + 3b - b + c + 2 - b - e$$

$$2 = b + c - e$$

or

$$\boxed{c = 2 - b + e}$$

Substituting the values of a, d and c in Eq. (1), we get

$$R = K\rho^{1-b} \mu^b l^{2-b+e} v^{2-b-2e} g^e$$

$$= K \frac{\rho}{\rho^b} \cdot \mu^b \frac{l^2}{l^b} \cdot l^e \frac{v^2}{v^b . v^{2e}} \cdot g^e$$

$$= K\rho l^2 v^2 \left[\frac{\mu}{\rho l v}\right]^b \left[\frac{lg}{v^2}\right]^e$$

Instead of this form where K, b and e are unknown quantities, the above equation is written in the functional form

$$\frac{R}{\rho l^2 v^2} = \phi\left[\left(\frac{\mu}{\rho l v}\right), \left(\frac{lg}{v^2}\right)\right]$$

or

$$R = \rho l^2 v^2 \phi\left[\left(\frac{\mu}{\rho l v}\right), \left(\frac{lg}{v^2}\right)\right]$$

7.4.2 Buckingham's π-Theorem

If the number of variables in an equation is less than or equal to four, Rayleigh's method for solving the equation is quite suitable but if the number of variables are more than four, the Rayleigh's method of dimensional analysis becomes more laborious. This difficulty is being overcome by using Buckingham's π-theorem. It provides an excellent tool by which these variables can be organised into the smallest number of significant dimensionless grouping from which an equation can be evaluated. The dimensionless groupings are called π-terms.

Buckingham's π-theorem states that if n number of variables (independent and dependent) are present in a physical process and these variables contain m numbers of fundamental dimension [M, L and T] then these variables are arranged into (n − m) dimensionless terms. Each term is called π-term.

Let X_1, X_2, X_3, X_4, ... X_n are the variables involved in a physical process.

Let X_1 be the dependent variable and X_2, X_3, X_4.......X_n are independent variables on which X_1 depends

$$X_1 = f(X_2, X_3, X_4.......X_n)$$

Above equation is also written as

$$f_1(X_1, X_2, X_3, X_4.......X_n) = 0 \qquad ...(1)$$

This equation as dimensionally homogeneous. It contains n number of variables. If there are m number of fundamental dimensions then according to Buckingham's π-theorem, the above Eq. (1) can be written as $(n - m)$ dimensionless groups of π-terms. [see examples in the Table 7.2].

Hence Equation (1) becomes

$$f(\pi_1, \pi_2, \pi_3, ...\pi_{n-m}) = 0$$

Each of these π-terms is dimensionless and is independent of the system. Division or multiplication by a constant does not change the character of the π-term. Each π-term contains $(m+1)$ variables, where m is the number of fundamental dimensions.

For example: Let X_1, X_2, X_3, X_4 and X_5 are variables involved in physical process m [*i.e., M, L* and *T*] = 3 number of fundamental dimensions are involved in process. Number of repeating variables is equal to number of fundamental dimension (m).

Choosing three repeating variables: X_2, X_3, X_4

Number of π-term $= n - m = 5 - 3$

$= 2$, each π-term contains $m + 1$ variables.

$$f(\pi_1, \pi_2) = 0$$

$$\pi_1 = X_2^a . X_3^b . X_4^c . X_1 \qquad \left[\begin{array}{l} \pi_1\text{-term contain } (m+1) = 4 \\ \text{variable } i.e. X_2, X_3, X_4 \text{ and } X_1 \end{array} \right]$$

and
$$\pi_2 = X_2^p . X_3^q . X_4^r . X_5$$

The above equations are solved by the principle of dimensional homogeneity and the value of a, b, c and p, q, r etc. are obtained. The value of π_1 and π_2 is obtained by substituting the value of a, b, c and p, q, r respectively. The final equation for the process is obtained by expressing any one of the π-terms as a function of other as

$$\pi_1 = \phi(\pi_2)$$
or
$$\pi_2 = \phi(\pi_1)$$

Write the final general equation for the phenomenon in terms of the π-terms.

In order to obtain the final expression in the desired manner, the following useful suggestion may be noted.

(*i*) Any π-term may be replaced by any power of that term, including negative as well as functional power. For example, π_1 may be replaced by π_1^{-1}, π_2 may be replaced by π_2^3, π_3 may be replaced by $\dfrac{1}{\pi_3^{\frac{1}{2}}}$ etc.

(*ii*) Any π-term may be replaced by multiplying it by a numerical constant. For example, π_1 may be replaced by $2\pi_1$ or and so on.

(*iii*) Any π-term may be replaced by multiplying or dividing it by adding or substracting an absolute numerical format.

(*iv*) Any π-term may be replaced by multiplying or dividing it by another π-term. For example: π may be replaced by $(\pi_1 \times \pi_2)$ or (π_1 / π_2).

Table 7.2

S. Nol.	Given Set of Variable	Number of Variables n	Number of Fundamental Dimension m	Number of Dimensionless Constant n − m	Number Repeating Variables = Number of Fundamental Dimension Involved in Process = m	π-term Contains m +1 Number of Variables
1	l, g, t	3	2 (L, T)	3 − 2 = 1	2	3
2	l, v, g	3	2 (L,T)	3 − 2 = 1	2	3
3	p, D, ρ, Q	4	3 (L, M, T)	4 − 3 = 1	3	4
4	F, D, v, ρ, μ	5	3 (L, M, T)	5 − 3 = 2	3	4
5	Q, H, g, l	4	2 (L, T)	4 − 2 = 2	2	3
6	D, N, μ, ρ, R	5	3 (L, M, T)	5 − 3 = 2	3	4
7	l, v, ρ, μ, g, R	6	3 (L, M, T)	6 − 3 = 3	· 3	4
8	Δp, D, l, ρ, μ, v, t	7	3 (L, M, T)	7 − 3 = 4	3	4
9	l, v, ρ, μ, E, R	6	3 (L, M, T)	6 − 3 = 3	3	4

7.5 METHOD OF SELECTING REPEATING VARIABLES

As discussed earlier, the number of repeating variables are equal to the number of fundamental dimensions of the problem. The choice of repeating variables is governed by the following consideration:

1. As far as possible, the dependent variable should not be selected as repeating variable.

2. The repeating variables should be chosen in such a way that one variable contains geometric property, other variable contains flow property and third variable contains fluid property. The repeating variables must not among themselves reduce to dimensionless parameter.

   Variables with geometric property are:

      (*i*) Length: *l*   (*ii*) Diameter: *D*   (*iii*) Height: *H* etc.

   Variables with flow property are:

      (*i*) Velocity: *V*   (*ii*) Rotational speed: *N* (*iii*) Acceleration: *a* etc.

   Variables with fluid property are:

      (*i*) Dynamic viscosity: μ (*ii*) Density: ρ

   If both of the fluid properties μ, ρ are involved in given problem, we select one of them which is more important, for example, for viscous fluid *i.e.*, lubricating oil we select μ and for working fluid like air, water etc. which are less viscous, we select ρ.

3. The repeating variables selected should not form a dimensionless group.

4. The repeating variables are equal to number of fundamental dimension involved in process.

5. No two repeating variables should have the same dimensions.

7.6 PROCEDURE FOR SOLVING PROBLEM BY BUCKINGHAM'S π-THEOREM

The procedure for solving problems by Buckingham's π-theorem is explained step by step.

Example: Assuming that the resistance force F exerted by a fluid on a sphere of diameter D depends on the viscosity μ, mass density of the fluid ρ and the velocity of the sphere V, obtain an expression for the resistance force.

Solution:

Step 1. The resistance force F is function of diameter D, viscosity μ, mass density of the fluid ρ and velocity of the sphere V.

   i.e., $F = f(D, \mu, \rho, V)$

   or it can written as $f_1(F, D, \mu, \rho, V) = 0$ ...(1)

   $\therefore$ Total number of variables: $n = 5$

Dimension of variables involved in given problem.

Variables	Dimension
F	MLT^{-2}
D	L
μ	$ML^{-1}T^{-1}$
ρ	ML^{-3}
V	LT^{-1}

Number of fundamental dimension: $m = 3$

Number of dimensionless π-terms $= n - m$

$$= 5 - 3 = 2$$

Thus two π-terms say π_1 and π_2 are formed. Hence Eq. (1) is written as

$$f_1(\pi_1, \pi_2) = 0 \qquad \qquad ...(2)$$

Step 2. Each π-term has $m + 1 = 3 + 1 = 4$ variables. Out of five variables F, D, μ, ρ and V, three variables are to be selected as repeating variable. F is a dependent variable and should not be selected as a repeating variable. Out of the four remaining variables, one variable should have geometric property, the second variable should have flow property and third one should have fluid property.

Variables	Repeating Variables
D	√
μ	
ρ	√
V	√

Choosing the repeating variables: D, ρ, V. They do not form dimensionless group.

Step-3. Each π-term is written as

$$\pi_1 = D^a \rho^b V^c F \qquad \qquad ...(3)$$

$$\pi_2 = D^p \rho^q V^r \mu \qquad \qquad ...(4)$$

Step-4. Each π-term is solved by the principle of dimensional homogeneity.

$$\pi_1 = D^a \rho^b V^c F$$

Substituting the dimensions on both sides, we get

$$M^0 L^0 T^0 = L^a (ML^{-3})^b (LT^{-1})^c MLT^{-2}$$

$$M^0 L^0 T^0 = L^a M^b L^{-3b} L^c T^{-c} MLT^{-2}$$

Comparing the power of M, we get

$$0 = b + 1$$

$$\boxed{b = -1}$$

Comparing the power of L, we get

$$0 = a - 3b + c + 1$$

Comparing the power of T, we get

$$0 = -c - 2$$

$$\boxed{c = -2}$$

$$0 = a - 3 \times (-1) - 2 + 1$$

$$\boxed{a = -2}$$

Substituting the values of a, b and c in Eq. (3), we get

$$\pi_1 = D^{-2}\rho^{-1}V^{-2}F = \frac{F}{\rho V^2 D^2}$$

Similarly for 2nd π-term, we get

$$\pi_2 = D^p \rho^q V^r \mu$$

Substituting the dimensions on both sides, we get

$$M^0 L^0 T^0 = L^p (ML^{-3})^q (LT^{-1})^r ML^{-1}T^{-1}$$
$$M^0 L^0 T^0 = L^p M^q L^{-3q}L^r T^{-r} ML^{-1}T^{-1}$$

Comparing the power of M, we get

$$0 = q + 1$$

$$\boxed{q = -1}$$

Comparing the power of L, we get

$$0 = p - 3q + r - 1$$

Comparing the power of T, we get

$$0 = -r - 1$$

$$\boxed{r = -1}$$

$$0 = p - 3(-1) - 1 - 1$$
$$0 = p + 3 - 2$$

$$\boxed{p = -1}$$

Substituting the values of p, q and r in Eq. (4), we get

$$\pi_2 = D^{-1}\rho^{-1}V^{-1}\mu = \frac{\mu}{D\rho V}$$

Step-5. Substituting the values of π_1 and π_2 in Eq. (2), we get

$$f_1\left(\frac{F}{\rho V^2 D^2}, \frac{\mu}{D\rho V}\right) = 0$$

or
$$\frac{F}{\rho V^2 D^2} = \phi\left[\frac{\mu}{D\rho V}\right]$$

$$F = \rho V^2 D^2 \phi \left[\frac{\mu}{D\rho V} \right]$$

$$F = \rho V^2 D^2 \phi [Re] \qquad \qquad \because \quad Re = \frac{D\rho V}{\mu}$$

Problem 7.7: The frictional torque T of a disc of diameter D rotating at a speed N in a fluid of viscosity μ and density ρ in a turbulent flow is given by

$$T = \rho N^2 D^5 \phi \left[\frac{\mu}{D^2 N \rho} \right]$$

Prove this by using Buckingham's π-theorem.

Solution: Frictional torque:

$$T = \text{function of } (D, N, \mu, \rho)$$

or $\qquad T = f(D, N, \mu, \rho)$

or $\quad f_1(T, D, N, \mu, \rho) = 0$

Dimension of variables involved in given problem.

Variables	Dimension
T	ML^2T^{-2}
D	L
N	T^{-1}
μ	$ML^{-1}T^{-1}$
ρ	ML^{-3}

Total numbers of variables involved in process: $\qquad n = 5$

Number of fundamental dimensions involved in process: $m = 3$

Number of dimensionless numbers or π-terms

$$= n - m$$
$$= 5 - 3 = 2$$

Number of repeating variables = number of fundamental dimensions = 3

Variables	Repeating Variables
D	$\sqrt{}$
N	$\sqrt{}$
μ	
ρ	$\sqrt{}$

Choosing repeating variables: D, N and ρ.

According to Buckingham's Pi-theorem, the number of π-terms involved is two,

$$f_1\,(\pi_1,\ \pi_2) = 0 \qquad\qquad ...(1)$$

$$\pi_1 \ = \ D^a N^b \rho^c T \qquad\qquad ...(2)$$

and $\qquad\qquad \pi_2 \ = \ D^p N^q \rho^r \mu \qquad\qquad ...(3)$

Now we take 1st π-term,

$$\pi_1 \ = \ D^a N^b \rho^c T$$

Substituting the dimensions on both sides, we get

$$L^0 M^0 T^0 = L^a T^{-b}\ (ML^{-3})^c\ ML^2 T^{-2}$$
$$L^0 M^0 T^0 = L^a T^{-b}\ M^c L^{-3c}\ ML^2 T^{-2}$$

Comparing the power of M, we get

$$0 = c + 1$$

$$\boxed{c = -1}$$

Comparing the power of L, we get

$$0 = a - 3c + 2$$
$$0 = a - 3(-1) + 2$$
$$0 = a + 3 + 2$$

$$\boxed{a = -5}$$

Comparing the power of T, we get

$$0 = -b - 2$$

$$\boxed{b = -2}$$

Substituting the values of a, b and c in Eq. (2), we get

$$\pi_1 = D^{-5} N^{-2} \rho^{-1} T \ = \ \frac{T}{\rho N^2 D^5}$$

Similarly for 2nd π-term, $\pi_2 = D^p N^q \rho^r \mu$

Substituting the dimensions on both sides, we get

$$L^0 M^0 T^0 = L^p T^{-q}\ (ML^{-3})^r\ ML^{-1} T^{-1}$$
$$L^0 M^0 T^0 = L^p T^{-q}\ M^r L^{-3r}\ ML^{-1} T^{-1}$$

Comparing the power of M, we get

$$0 = r + 1$$

$$\boxed{r = -1}$$

Comparing the power of L,

$$0 = p - 3r - 1$$
$$0 = p - 3(-1) - 1$$
$$0 = p + 3 - 1$$

$$\boxed{p = -2}$$

Substituting the values of p, q and r in Eq. (3), we get

$$\pi_2 = D^{-2}N^{-1}\rho^{-1}\mu = \frac{\mu}{D^2 N\rho}$$

Comparing the power of T, we get

$$0 = -q - 1$$

$$\boxed{q = -1}$$

Substituting the values of π_1 and π_2 in Eq. (1), we get

$$f_1\left(\frac{T}{\rho N^2 D^5}, \frac{\mu}{N\rho D^2}\right) = 0$$

$$\frac{T}{\rho N^2 D^5} = \phi\left[\frac{\mu}{N\rho D^2}\right]$$

$$\boldsymbol{T = \rho N^2 D^5 \phi\left[\frac{\mu}{N\rho D^2}\right]}$$

Problem 7.8: The resisting torque T against the motion of a shaft in a lubricated bearing depends on the viscosity μ, the rotational speed N, the diameter D and the bearing pressure intensity p. Show that

$$T = \mu N D^3 \phi\left(\frac{p}{\mu N}\right)$$

Solution: The resisting torque T is function of viscosity μ, the rotational speed N, the diameter D and pressure intensity p

i.e., $T = f(\mu, N, D, p)$

or $f_1(T, \mu, N, D, p) = 0$

Dimension of variables involved in given problem

Variables	Dimension
T	ML^2T^{-2}
μ	$ML^{-1}T^{-1}$
N	T^{-1}
D	L
p	$ML^{-1}T^{-2}$

Total numbers of variables (depended and independent) involved in process:

$$n = 5$$

Number of fundamental dimensions involved in problem: $m = 3$

Number of dimensionless constant or π-terms

$$= n - m = 5 - 3 = 2$$

Number of repeating variables = numbers of fundamental dimensions = 3

Variables	Repeating Variables
μ	√
N	√
D	√
p	

Choosing repeating variables: $\mu,\,N$ and D.

$$f_1(\pi_1, \pi_2) = 0 \qquad \qquad \text{...(1)}$$

$$\pi_1 = \mu^a N^b D^c T$$

and

$$\pi_2 = \mu^p N^q D^r p$$

Now we take Ist π-term,

$$\pi_1 = \mu^a N^b D^c T \qquad \qquad \text{...(2)}$$

Substituting the dimensions on both sides,

$$M^0 L^0 T^0 = (ML^{-1} T^{-1})^a (T^{-1})^b (L)^c \, ML^2 T^{-2}$$

$$M^0 L^0 T^0 = M^a \, L^{-a} T^{-a} \, T^{-b} L^c ML^2 \, T^{-2}$$

Comparing the power of M, we get

$$0 = a + 1$$

$$\boxed{a = -1}$$

Comparing the power of L, we get

$$0 = -a + c + 2$$

$$0 = -(-1) + c + 2$$

$$0 = 1 + c + 2$$

$$\boxed{c = -3}$$

Comparing the power of T, we get

$$0 = -a - b - 2$$

$$0 = 1 - b - 2$$

$$\boxed{b = -1}$$

Substituting the values of a. b and c in Eq. (2), we get

$$\pi_1 = \mu^{-1} N^{-1} D^{-3} T = \frac{T}{\mu N D^3}$$

Similarly for 2nd π-term,

$$\pi_2 = \mu^p N^q D^r p \qquad \qquad ...(3)$$

Substituting the dimensions on both sides,

$$M^0 L^0 T^0 = (ML^{-1}T^{-1})^p \ (T^{-1})^q \ L^r \ ML^{-1}T^{-2}$$
$$M^0 \ L^0 T^0 = M^p L^{-p} T^{-p} \ T^{-q} \ L^r \ ML^{-1}T^{-2}$$

Comparing the power of M, we get

$$0 = p + 1$$

$$\boxed{p = -1}$$

Comparing the power of L, we get

$$0 = -p + r - 1$$
$$0 = -(-1) + r - 1$$
$$0 = 1 + r - 1$$

$$\boxed{r = 0}$$

Comparing the power of T, we get

$$0 = -p - q - 2$$
$$0 = -(-1) - q - 2$$

$$\boxed{q = -1}$$

Substituting the values of p, q and r in Eq. (3), we get

$$\pi_2 = \mu^{-1} N^{-1} D^0 p = \frac{p}{\mu N}$$

Substituting the values of π_1 and π_2 in Eq. (1), we get

$$f_1\left(\frac{T}{\mu N D^3}, \frac{p}{\mu N}\right) = 0$$

$$\frac{T}{\mu N D^3} = \phi\left[\frac{p}{\mu N}\right]$$

$$\boldsymbol{T = \mu N D^3 \phi\left[\frac{p}{\mu N}\right]}$$

Problem 7.9: The force exerted by a flowing fluid on a stationary solid body depends upon the length (L) of the body, velocity (V) of fluid, the density (ρ) of fluid, viscosity (μ) of fluid and acceleration due to gravity (g). Find an expression for the force using dimensional analysis.

Solution: Force: F = function of [length (L), velocity (V), density (ρ), viscosity (μ), acceleration due to gravity (g)].

or $F = f(L, V, \rho, \mu, g)$

or $f_1(F, L, V, \rho, \mu, g) = 0$

Dimension of variables involved in given problem.

Variables	Dimension
F	MLT^{-2}
L	L
V	LT^{-1}
ρ	ML^{-3}
μ	$ML^{-1}T^{-1}$
g	LT^{-2}

Total number of variables involved in process: $n = 6$

Number of fundamental dimensions involved in process: $m = 3$

Number of dimensionless number or π-term involved

$$= n - m$$
$$= 6 - 3 = 3$$

Number of repeating variables $=$ number of fundamental dimension $= m$
$$= 3$$

Variables	Repeating Variables
L	√
V	√
ρ	√
μ	
g	

Choosing repeating variables: L, V and ρ.

According to Buckingham's π-theorem, the number of π-terms involved in present problem = 3

i.e., $f_1(\pi_1, \pi_2, \pi_3) = 0$ ...(1)

Each π-term contains $m + 1$ variables i.e., $3 + 1 = 4$ variables. Out of four variables three are repeating variables.

$$\pi_1 = L^a V^b \rho^c F \qquad \qquad ...(2)$$
$$\pi_2 = L^p V^q \rho^r \mu \qquad \qquad ...(3)$$
and $$\pi_3 = L^\alpha V^\beta \rho^\gamma g \qquad \qquad ...(4)$$

Now we take 1st π-term,

$$\pi_1 = L^a V^b \rho^c F$$

Substituting the dimensions on both sides, we have

$$L^0M^0T^0 = L^a(LT^{-1})^b (ML^{-3})^c MLT^{-2}$$
$$L^0M^0T^0 = L^a L^b T^{-b}M^c L^{-3c} MLT^{-2}$$
$$L^0M^0T^0 = L^{a+b-3c+1}M^{c+1}T^{-b-2}$$

Now,

Comparing the power of M, we have

$$0 = c + 1$$

or

$$\boxed{c = -1}$$

Comparing the power of T, we have

$$0 = -b - 2$$

$$\boxed{b = -2}$$

Comparing the power of L, we have

$$0 = a + b - 3c + 1$$

or

$$a = 3c - b - 1$$

Substituting the values of c and b in above equation, we have

$$a = 3(-1) - (-2) - 1$$
$$a = -3 + 2 - 1$$

$$\boxed{a = -2}$$

Now, substituting the values of a, b and c in Eq. (2), we get

$$\pi_1 = L^{-2}V^{-2}\rho^{-1}F = \frac{F}{\rho L^2 V^2}$$

Similarly for 2nd π-term, we have

$$\pi_2 = L^p V^q \rho^r \mu$$

Substituting the dimensions on both sides, we have

$$L^0M^0T^0 = L^P (LT^{-1})^q (ML^{-3})^r ML^{-1}T^{-1}$$
$$L^0M^0T^0 = L^P L^q T^{-q}M^r L^{-3r}ML^{-1} T^{-1}$$
$$L^0M^0T^0 = L^{p+q-3r-1}M^{r+1}T^{-q-1}$$

Now ;

Comparing the power of M, we have

$$0 = r + 1$$

or

$$\boxed{r = -1}$$

Comparing the power of T, we have

$$0 = -q - 1$$

or

$$\boxed{q = -1}$$

Comparing the power of L, we have

$$0 = p + q - 3r + 1$$

Substituting the values of r and q in above equation, we have

or $\qquad\qquad 0 = p - 1 - 3(-1) - 1$

or $\qquad\qquad 0 = p - 1 + 3 - 1$

or $\qquad\qquad \boxed{p = -1}$

Substituting the values of p, q and r in Eq. (3), we have

$$\pi_2 = L^{-1}V^{-1}\rho^{-1}\mu = \frac{\mu}{\rho VL}$$

and similarly for third π-term,

$$\pi_3 = L^{\alpha}V^{\beta}\rho^{\gamma}g$$

Substituting the dimensions on both sides, we get

$$L^0M^0T^0 = L^{\alpha}(LT^{-1})^{\beta}(ML^{-3})^{\gamma}(LT^{-2})$$
$$L^0M^0T^0 = L^{\alpha}L^{\beta}T^{-\beta}M^{\gamma}L^{-3\gamma}LT^{-2}$$
$$L^0M^0T^0 = L^{\alpha+\beta-3\gamma+1}M^{\gamma}T^{-\beta-2}$$

Now

Comparing the power of M, we get

$$0 = \gamma$$

or $\qquad\qquad \boxed{\gamma = 0}$

and comparing the power of T, we get

$$0 = -\beta - 2$$

or $\qquad\qquad \boxed{\beta = -2}$

Comparing the power of L, we get

$$0 = \alpha + \beta - 3\gamma + 1$$

Substituting the values of β and γ is above equation, we get

$$0 = \alpha - \beta - 3\gamma + 1$$
$$0 = \alpha - 2 - 3(0) + 1$$
$$0 = \alpha - 2 + 1$$

or $\qquad\qquad \boxed{\alpha = 1}$

Substituting the values of α, β and γ in Eq. (4), we get

$$\pi_3 = L^1V^{-2}\rho^0 g = \frac{Lg}{V^2}$$

Substituting the values of π_1, π_2 and π_3 in Eq. (1), we get

$$f_1\left(\frac{F}{\rho L^2 V^2}, \frac{\mu}{\rho VL}, \frac{Lg}{V^2}\right) = 0$$

$$\frac{F}{\rho L^2 V^2} = \phi\left[\frac{\mu}{\rho VL}, \frac{Lg}{V^2}\right]$$

$$F = \rho L^2 V^2 \phi\left[\frac{\mu}{\rho VL}, \frac{Lg}{V^2}\right]$$

Problem 7.10: The variables controlling the motion of a floating vessel through water are the drag force F, the speed V, the length L, the density ρ and dynamic viscosity μ of water and acceleration due to gravity g. Derive an expression for F by dimensional analysis. *(GGSIP. University, Delhi. Dec. 2006)*

Solution: Same as problem 7.9.

Problem 7.11: Given that pressure drop per unit length $\left(\dfrac{\Delta p}{l}\right)$ in a pipe is a function of the density ρ, viscosity μ, velocity V of the fluid and the diameter d of the pipe, show that:

$$\left(\frac{\Delta p}{l}\right) = \frac{\rho V^2}{d}\phi\left(\frac{dV\rho}{\mu}\right)$$

Solution: Pressure drop per unit length: $\dfrac{\Delta p}{l}$ = function of [density (ρ), viscosity (μ), velocity (V), diameter (d)]

or
$$\frac{\Delta p}{l} = f(\rho, \mu, V, d)$$

or
$$f_1\left(\frac{\Delta p}{l}, \rho, \mu, V, d\right) = 0$$

Dimension of variables involved in given problem.

Variables	Dimension
$\dfrac{\Delta p}{l}$	$ML^{-2}T^{-2}$
ρ	ML^{-3}
μ	$ML^{-1}T^{-1}$
V	LT^{-1}
d	L

Total number of variables involved in given problem: $n = 5$

Number of fundamental dimensions involved in problem: $m = 3$

Number of dimensionless number or π-terms involved $= n - m = 5 - 3 = 2$

Number of repeating variables $=$ number of fundamental dimensions $= m$
$$= 3$$

Inedependent Variables	Repeating Variables
ρ	√
μ	
V	√
d	√

Choosing repeating variables: ρ, V and d.

According to Buckingham's π-theorem, the number of π-terms involved in the problem = 2

i.e., $f_1(\pi_1, \pi_2) = 0$ ...(1)

Each π-term contains $m + 1$ variables *i.e.,* $3 + 1 = 4$ variables. Out of four variables, three are repeating variables.

$$\pi_1 = \rho^a V^b d^c \left(\frac{\Delta p}{l}\right)$$ ...(2)

and $$\pi_2 = \rho^p V^q d^r \mu$$ ...(3)

Now we take 1st π-term

Substituting the dimensions on both sides, we get

$$M^0 \; L^0 \; T^0 = (ML^{-3})^a \; (LT^{-1})^b L^c ML^{-2} T^{-2}$$
$$M^0 \; L^0 \; T^0 = M^a L^{-3a} \; L^b T^{-b} L^c \; ML^{-2} T^{-2}$$
$$M^0 \; L^0 \; T^0 = M^{a+1} L^{-3a+b+c-2} \; T^{-b-2}$$

Now, comparing the power of M, we get

$$0 = a + 1$$

or $\boxed{a = -1}$

Comparing the power of T, we get

$$0 = -b - 2$$

or $\boxed{b = -2}$

Comparing the power of L, we get

$$0 = -3a + b + c - 2$$

Substituting the values of a and b in above equation, we get

$$0 = -3 \times (-1) + (-2) + c - 2$$
$$0 = 3 - 2 + c - 2$$

or $\boxed{c = 1}$

Substituting the values of a, b and c in Eq. (2), we get

$$\pi_1 = \rho^{-1} V^{-2} d^1 \left(\frac{\Delta p}{l}\right) = \frac{d}{\rho V^2}\left(\frac{\Delta p}{l}\right)$$

Similarly for 2nd π-term,

$$\pi_2 = \rho^p V^q d^r \mu$$

Substituting the dimensions on both sides, we get

$$M^0 \ L^0 \ T^0 = (ML^{-3})^p \ (LT^{-1})^q L^r ML^{-1}T^{-1}$$

$$M^0 \ L^0 \ T^0 = M^p L^{-3p} \ L^q T^{-q} L^r \ ML^{-1}T^{-1}$$

$$M^0 \ L^0 \ T^0 = M^{p+1}L^{-3p+q+r-1}T^{-q-1}$$

Comparing the power of T, we get

$$0 = -q - 1$$

or

$$\boxed{q = -1}$$

Comparing the power of M, we get

$$0 = p + 1$$

$$\boxed{p = -1}$$

Comparing the power of L, we get

$$0 = -3p + q + r - 1$$

Substituting the values of p and q in above Eq., we get

$$0 = -3(-1) - 1 + r - 1$$
$$0 = 3 - 1 + r - 1$$

or

$$\boxed{r = -1}$$

Substituting the values of p, q and r in Eq. (3), we get

$$\pi_2 = \rho^{-1}V^{-1}d^{-1}\mu = \frac{\mu}{\rho V d}$$

Now, substituting the values of π_1 and π_2 in Eq. (1), we get

$$f_1\left[\frac{d}{\rho V^2}\left(\frac{\Delta p}{l}\right), \frac{\mu}{\rho V d}\right] = 0$$

$$\frac{d}{\rho V^2}\left(\frac{dp}{l}\right) = \phi\left[\frac{\mu}{\rho V d}\right]$$

or

$$\frac{\Delta p}{l} = \frac{\rho V^2}{d}\phi\left[\frac{\mu}{\rho V d}\right]$$

or

$$\frac{\Delta p}{l} = \frac{\rho V^2}{d}\phi\left[\frac{\rho V d}{\mu}\right]$$

Problem 7.12: The pressure difference (Δp) for viscous flow through a circular pipe depends upon the diameter (D) of the pipe, length (L) of pipe, velocity of fluid (V), viscosity (μ) and density (ρ) of the fluid. Find an expression for Δp using Buckingham's π-theorem.

Solution: Pressure difference: Δp = function of [diameter (D), length (L), velocity (V), viscosity (μ), density (ρ)]

or $\qquad\qquad\qquad \Delta p = f(D, L, V, \mu, \rho)$

or $\quad f_1(\Delta p, D, L, V, \mu, \rho) = 0$

Dimension of variables involved in given problem.

Variables	Dimension
Δp	$ML^{-1}T^{-2}$
D	L
L	L
V	LT^{-1}
μ	$ML^{-1}T^{-1}$
ρ	ML^{-3}

Total number of variables involved in a flowing viscous fluid: $n = 6$

Number of fundamental dimensions involved in process: $m = 3$

$\therefore$ Number of dimensionless numbers or π-terms involved $= n - m = 6 - 3 = 3$

Variables	Repeating Variables
D	$\surd$
L	
V	$\surd$
μ	$\surd$
ρ	

Choosing repeating variable: D, V and μ ($\because$ The fluid property μ is to be chosen instead of ρ because Δp is to be calculated for viscous flow).

According to Buckingham's π-theorem, the number of π-terms involved in the problem $n - m = 3$

i.e., $\qquad\qquad f_1(\pi_1, \pi_2, \pi_3) = 0$ $\qquad\qquad\qquad\qquad$ ...(1)

Each π-term contains $m + 1$ variables i.e., $3 + 1 = 4$ variables. Out of four variables, three are repeating variables.

$$\pi_1 = D^a V^b \mu^c \Delta p \qquad\qquad\qquad\qquad ...(2)$$

$$\pi_2 = D^p V^q \mu^r L \qquad\qquad\qquad\qquad ...(3)$$

and $\qquad\qquad \pi_3 = D^\alpha V^\beta \mu^\gamma \rho \qquad\qquad\qquad\qquad ...(4)$

Now we take 1st π_1-term,

$$\pi_1 = D^a V^b \mu^c \Delta p$$

Substituting the dimensions on both sides, we get

$$L^0 \, M^0 \, T^0 = L^a (LT^{-1})^b \ (ML^{-1}T^{-1})^c (ML^{-1}T^{-2})$$

$$L^0 \ M^0 \ T^0 = L^a L^b T^{-b} \ M^c L^{-c} T^{-c} M L^{-1} T^{-2}$$
$$L^0 \ M^0 \ T^0 = L^{a+b-c-1} M^{\ c+1} T^{\ -b-c-2}$$

Now comparing the power of M, we get

$$0 = c + 1$$

or $\boxed{c = -1}$

Comparing the power of T, we get

$$0 = -b - c - 2$$
$$0 = -b - (-1) - 2 \qquad\qquad (\because C = -1)$$
$$0 = -b + 1 - 2$$

or $\boxed{b = -1}$

and comparing the power of L, we get

$$0 = a + b - c - 1$$

Substituting the values of b and c in above equation, we get

$$0 = a - 1 - (-1) - 1$$
$$0 = a - 1 + 1 - 1$$

or $\boxed{a = 1}$

Substituting the values of a, b and c in Eq. (2), we get

$$\pi_1 = D^1 V^{-1} \mu^{-1} \Delta p = \frac{D \Delta p}{\mu V}$$

Similarly for 2nd π-term,

$$\pi_2 = D^p V^q \mu^r L$$

Substituting the dimensions on both sides, we get

$$L^0 \ M^0 \ T^0 = L^p (LT^{-1})^q \ (ML^{-1} T^{-1})^r L$$
$$L^0 \ M^0 \ T^0 = L^p L^q T^{-q} \ M^{\ r} L^{-r} T^{-r} L$$
$$L^0 \ M^0 \ T^0 = L^{p+q-r+1} M^r T^{\ -q-r}$$

Now comparing the power of M, we get

$$0 = r$$

or $\boxed{r = 0}$

Comparing the power of T, we get

$$0 = -q - r$$

or $\boxed{q = 0}$ $\qquad\qquad [\because r = 0]$

Comparing the power of L, we get

$$0 = p + q - r + 1$$

Substituting the values of r and q in above equation, we get

$$0 = p + 0 - 0 + 1$$

or $\boxed{p = -1}$

Substituting the values of p, q and r in Eq. (3), we get

$$\pi_2 = D^{-1}V^0\mu^0 L = \frac{L}{D}$$

and similarly for IIIrd π-term,

$$\pi_3 = D^\alpha V^\beta \mu^\gamma \rho$$

Substituting the dimensions on both sides, we get

$$L^0 M^0 T^0 = L^\alpha (LT^{-1})^\beta (ML^{-1}T^{-1})^\gamma (ML^{-3})$$

$$L^0 M^0 T^0 = L^\alpha L^\beta T^{-\beta} M^\gamma L^{-\gamma} T^{-\gamma})ML^{-3}$$

$$L^0 M^0 T^0 = L^{\alpha+\beta-\gamma-3} M^{\gamma+1} T^{-\beta-\gamma}$$

Now comparing the power of M, we get

$$0 = \gamma+1$$

or $\boxed{\gamma = -1}$

Comparing the power of T, we get

$$0 = -\beta - \gamma$$

or $\beta = -\gamma$ $[\because \gamma = -1]$

$$\beta = -(-1)$$

$\boxed{\beta = 1}$

Comparing the power of L, we get

$$0 = \alpha + \beta - \gamma - 3$$

Substituting the values of β and γ in above equation, we get

$$0 = \alpha + 1 - (-1) - 3$$

$$0 = \alpha + 1 + 1 - 3$$

or $\boxed{\alpha = 1}$

Substituting the values of α, β and γ in Eq. (4), we get

$$\pi_3 = D^1 V^1 \mu^{-1}\rho = \frac{DV\rho}{\mu}$$

Now, substituting the values of π_1, π_2 and π_3, in equation, we get

$$f_1\left(\frac{D\Delta p}{\mu V}, \frac{L}{D}, \frac{DV\rho}{\mu}\right) = 0$$

$$\frac{D\Delta p}{\mu V} = \phi\left[\frac{L}{D}, \frac{DV\rho}{\mu}\right]$$

or $$\Delta p = \frac{\mu V}{D}\phi\left[\frac{L}{D}, \frac{DV\rho}{\mu}\right]$$

$$\Delta p = \frac{\mu V}{D} \times \frac{L}{D}\phi\left[\frac{DV\rho}{\mu}\right]$$

[∵ Pressure difference Δp is a linear function $\dfrac{L}{D}$. Hence $\dfrac{L}{D}$ can be taken out of the functional.]

Dividing by ρg on both sides, we get

$$\frac{\Delta p}{\rho g} = \frac{\mu V}{\rho g D}\cdot\frac{L}{D}\phi\left[\frac{DV\rho}{\mu}\right]$$

where $\qquad\dfrac{\Delta p}{\rho g} = h_f$, head loss in pipe

and $\qquad\dfrac{DV\rho}{\mu}$ = Re, Reynolds number hence, the above is reduced.

$$h_f = \frac{\mu VL}{\rho g D^2}\,\phi[\text{Re}]$$

Problem 7.13: Using Buckingham's π-theorem show that velocity through a circular orifice is given by

$$V = \sqrt{2gH}\,\phi\left[\frac{D}{H},\frac{\mu}{\rho VH}\right]$$

where H is the head causing flow, D is the diameter of the orifice, μ is the coefficient of viscosity, ρ is the mass density and g is the acceleration due to gravity.

Solution: $\qquad\qquad V$ = function of (H, D, μ, ρ, g)

$\qquad\qquad\qquad V = f(H, D, \mu, \rho, g)$

or $\quad f_1(V, H, D, \mu, \rho, g) = 0$

Dimension of variables involved in given problem.

Variables	Dimension
V	LT^{-1}
H	L
D	L
μ	$ML^{-1}T^{-1}$
ρ	ML^{-3}
g	LT^{-2}

Total number of variables involved in process: $n = 6$

Number of fundamental dimensions involved in process: $m = 3$

Number of dimensionless number or π-terms

$$= n - m$$
$$= 6 - 3 = 3$$

Number of repeating variables (R.V.) = number of fundamental dimension = 3

Variables	Repeating Variables
H	√
D	
μ	
ρ	√
g	√

Choosing repeating variables: H, ρ and g

According Buckingham's Pi-theorem, the number of π-terms involved is 3

$$f_1(\pi_1, \pi_2, \pi_3) = 0 \qquad \qquad ...(1)$$

We get three π-terms as

$$\pi_1 = H^a \rho^b g^c V \qquad \qquad ...(2)$$

$$\pi_2 = H^p \rho^q g^r D \qquad \qquad ...(3)$$

and $\qquad \qquad \pi_3 = H^\alpha \rho^\beta g^\gamma \mu \qquad \qquad ...(4)$

Now we take 1st π-term,

$$\pi_1 = H^a \rho^b g^c V$$

Substituting the dimensions on both sides, we get

$$L^0 \, M^0 \, T^0 = L^a (ML^{-3})^b (LT^{-2})^c \, LT^{-1}$$

$$L^0 \, M^0 \, T^0 = L^a M^b L^{-3b} L^c T^{-2c} LT^{-1}$$

Comparing the power of L, we get

$$0 = a - 3b + c + 1$$

Comparing the power of M, we get

$$0 = b$$

or $\qquad \qquad \boxed{b = 0}$

Comparing the power of T, we get

$$0 = -2c - 1$$

$$\boxed{c = -\frac{1}{2}}$$

$$0 = a - 0 - \frac{1}{2} + 1$$

$$\boxed{a = -\frac{1}{2}}$$

Substituting the values of a, b and c in Eq. (2), we get

$$\pi_1 = H^{-\frac{1}{2}} \rho^0 g^{-\frac{1}{2}} V = \frac{V}{\sqrt{gH}}$$

2nd π-term, $\qquad \qquad \pi_2 = H^p \rho^q g^r D$

Substituting the dimensions on both sides, we get

$$L^0 \ M^0 \ T^0 = L^p (ML^{-3})^q (LT^{-2})^r L$$

$$L^0 \ M^0 \ T^0 = L^p M^q L^{-3q} L^r T^{-2r} L$$

Comparing the power of L, we get

$$0 = p - 3q + r + 1$$

Comparing the power of M, we get

$$0 = q$$

or

$$\boxed{q = 0}$$

Comparing the power of T, we get

$$0 = -2r$$

$$\boxed{r = 0}$$

$$0 = p - 0 + 0 + 1$$

$$\boxed{p = -1}$$

Substituting the values of p, q and r in Eq. (3), we get

$$\pi_2 = H^{-1} \rho^0 g^0 D = \frac{D}{H}$$

Similarly for 3rd π-term,

$$\pi_3 = H^\alpha \rho^\beta g^\gamma \mu$$

Substituting the dimensions on both sides, we get

$$L^0 \ M^0 \ T^0 = L^\alpha (ML^{-3})^\beta (LT^{-2})^\gamma ML^{-1} T^{-1}$$

$$L^0 \ M^0 \ T^0 = L^\alpha M^\beta L^{-3\beta} L^\gamma T^{-2\gamma} ML^{-1} T^{-1}$$

Comparing the power of L, we get

$$0 = \alpha - 3\beta + \gamma - 1$$

Comparing the power of M, we get

$$0 = \beta + 1$$

$$\boxed{\beta = -1}$$

Comparing the power of T, we get

$$0 = -2\gamma - 1$$

$$\boxed{\gamma = -\frac{1}{2}}$$

$$0 = \alpha - 3(-1) - \frac{1}{2} - 1$$

$$\boxed{\alpha = -\frac{3}{2}}$$

Substituting the values of α, β and γ in Eq. (4), we get

$$\pi_3 = H^{-3/2} \rho^{-1} g^{-1/2} \mu = \frac{\mu}{\rho \sqrt{g} \ H^{3/2}}$$

$$\pi_3 = \frac{\mu}{\rho\sqrt{g}\, H^{\frac{1}{2}}.H} = \frac{\mu}{\rho H\sqrt{gH}} \times \frac{V}{V} \qquad \left[\pi_1 = \frac{V}{\sqrt{gH}} \right]$$

$$= \pi_1 \frac{\mu}{\rho HV} = \frac{\mu}{\rho HV}$$

[**Note:** π-terms are not unique, any product of the π-terms is also a π. Multiplying, squaring, dividing by one π-term is another π-term or by any constant does not change the character of π-term].

Substituting the values of π_1, π_2 and π_3 in Eq. (1), we get

$$f_1 \left[\frac{V}{\sqrt{gH}}, \frac{D}{H}, \frac{\mu}{\rho HV} \right] = 0$$

$$\frac{V}{\sqrt{gH}} = \phi \left[\frac{D}{H}, \frac{\mu}{\rho HV} \right]$$

$$V = \sqrt{gH}\,\phi \left[\frac{D}{H}, \frac{\mu}{\rho HV} \right]$$

$$V = \sqrt{2gH}\,\phi \left[\frac{D}{H}, \frac{\mu}{\rho HV} \right]$$

Problem 7.14: The drag force F on a partially submerged body depends on the relative velocity V between the body and the fluid, characteristic linear dimension l, height of surface roughness k, fluid density ρ (rho), the viscosity μ (mu), and the acceleration due to gravity g. Obtain an expression for the drag force, using the method of Buckingham, s π– theorem.

Solution: Drag force: F = function of [relative velocity (V), linear
                    dimension (l), height of surface roughness
                    (k), fluid density (ρ), viscosity (μ),
                    acceleration due to gravity (g)]

or $F = f\,[V,\ l,\ k,\ \rho,\ \mu,\ g]$

or $f_1\,(F,\ V,\ l,\ k,\ \rho,\ \mu,\ g) = 0$

Dimension of variables involved in given problem

Variables	Dimension
F	MLT^{-2}
V	LT^{-1}
l	L
k	L
ρ	ML^{-3}
μ	$ML^{-1}T^{-1}$
g	LT^{-2}

Total number of variables: $n = 7$

Number of fundamental dimensions involved: $m = 3$

Number of dimensionless number or π-terms involved $= n - m$

$$= 7 - 3 = 4$$

Number of repeating variables = number of fundamental dimensions = 3

Variables	Repeating Variables
V	$\sqrt{}$
l	$\sqrt{}$
k	
ρ	$\sqrt{}$
μ	
g	

Choosing repeating variables: V, l and ρ

According to Buckingham's π-theorem, the number of π-terms are involved in given problem $= 4$.

i.e., $\quad f_1 (\pi_1, \pi_2, \pi_3, \pi_4) = 0$ ...(1)

where $\qquad \pi_1 = V^a \, l^b \, \rho^c \, F$ ...(2)

$\qquad\qquad\quad \pi_2 = V^p \, l^q \, \rho^r \, k$ ...(3)

$\qquad\qquad\quad \pi_3 = V^x \, l^y \, \rho^z \, \mu$ ...(4)

$\qquad\qquad\quad \pi_4 = V^\alpha \, l^\beta \, \rho^\gamma \, g$ ...(5)

Now we take Ist π-term,

$$\pi_1 = V^a \, l^b \, \rho^c \, F$$

Substituting the dimensions on both sides, we get

$$L^0 M^0 T^0 = (LT^{-1})^a \, (L)^b \, (ML^{-3})^c \, MLT^{-2}$$

$$L^0 M^0 T^0 = L^a T^{-a} L^b \, M^c \, L^{-3c} \, MLT^{-2}$$

$$L^0 M^0 T^0 = L^{a+b-3c+1} \, M^{c+1} \, T^{-a-2}$$

Comparing the power of M, we get

$$0 = c + 1$$

or $\qquad\qquad\qquad c = -1$

Comparing the power of T, we get

$$0 = -a - 2$$

or $\qquad\qquad\qquad a = -2$

Comparing the power of L, we get

$$0 = a + b - 3c + 1$$

$$0 = -2 + b - 3 (-1) + 1$$

$$0 = -2 + b + 3 + 1$$

or $\qquad\qquad\qquad b = -2$

Substituting the value of a, b, c in Eq. (2), we have

$$\pi_1 = V^{-2} l^{-1} \rho^{-1} F = \frac{F}{\rho V^2 l^2}$$

Similarly for 2nd π-term; we have

$$\pi_2 = V^p l^q \rho^r k$$

Substituting the dimensions on both sides, we have

$$L^0 M^0 T^0 = (LT^{-1})^p (L)^q (ML^{-3})^r L$$
$$L^0 M^0 T^0 = L^p T^{-p} L^q M^r L^{-3r} L$$
$$L^0 M^0 T^0 = L^{p+q-3r+1} M^r T^{-p}$$

Comparing the power of M, we have

$$0 = r$$

or $r = 0$

Comparing the power of T, we have

$$0 = -p$$

or $p = 0$

Comparing the power of L, we have

$$0 = p + q - 3r + 1$$
$$0 = 0 + q - 0 + 1$$

or $q = -1$

Substituting the values of p, q and r in Eq. (3), we have

$$\pi_2 = V^0 l^{-1} \rho^0 k = \frac{k}{l}$$

for third π-term, we have

$$\pi_3 = V^x l^y \rho^z \mu$$

Substituting the dimensions on both sides, we have

$$L^0 M^0 T^0 = (LT^{-1})^x (L)^y (ML^{-3})^z ML^{-1} T^{-1}$$
$$L^0 M^0 T^0 = L^x T^{-x} L^y M^z L^{-3z} ML^{-1} T^{-1}$$
$$L^0 M^0 T^0 = L^{x+y-3z-1} M^{z+1} T^{-x-1}$$

Comparing the power of M, we have

$$0 = z + 1$$

or $z = -1$

Comparing the power of T, we get

$$0 = -x - 1$$

or $x = -1$

Comparing the power of L, we get

$$0 = x + y - 3z - 1$$
$$0 = -1 + y - 3(-1) - 1$$
$$0 = -1 + y + 3 - 1$$

or $y = -1$

Substituting the values of x, y and z in Eq. (4), we have

$$\pi_3 = V^{-1} l^{-1} \rho^{-1} \mu = \frac{\mu}{\rho l V}$$

and similarly for 4th π-term, we have

$$\pi_4 = V^{\alpha} l^{\beta} \rho^{\gamma} g$$

Substituting the dimensions on both sides, we have

$$L^0 M^0 T^0 = (LT^{-1})^{\alpha} (L)^{\beta} (ML^{-3})^{\gamma} LT^{-2}$$
$$L^0 M^0 T^0 = L^{\alpha} T^{-\alpha} L^{\beta} M^{\gamma} L^{-3\gamma} LT^{-2}$$
$$L^0 M^0 T^0 = L^{\alpha+\beta-3\gamma+1} M^{\gamma} T^{-\alpha-2}$$

Comparing the power of M, we have

$$0 = \gamma$$

or

$$\gamma = 0$$

Comparing the power of T, we have

$$0 = -\alpha - 2$$

or

$$\alpha = -2$$

Comparing the power of L, we have

$$0 = \alpha + \beta - 3\gamma + 1$$
$$0 = -2 + \beta - 0 + 1$$

or

$$\beta = 1$$

Substituting the values of α, β and γ in Eq. (5), we have

$$\pi_4 = V^{-2} l^1 \rho^0 g = \frac{gl}{V^2}$$

Substituting the values of π_1, π_2, π_3 and π_4 in Eq. (1), we get

$$f_1\left(\frac{F}{\rho V^2 l^2}, \frac{k}{l}, \frac{\mu}{\rho l V}, \frac{gl}{V^2}\right) = 0$$

or

$$\frac{F}{\rho V^2 l^2} = \phi\left[\frac{k}{l}, \frac{\mu}{\rho l V}, \frac{gl}{V^2}\right]$$

or

$$F = \rho V^2 l^2 \phi\left[\frac{k}{l}, \frac{\mu}{\rho l V}, \frac{gl}{V^2}\right]$$

Problem 7.15: Develop an expression for the thrust developed by a propeller which depends upon the angular velocity ω, velocity V, dynamic viscosity μ, density ρ, propeller diameter D, and compressibility for the medium measured by the local velocity of round C.

Solution: Thrust:

F = function of [angular velocity (ω), velocity

(V), viscosity (μ), density (ρ), diameter

(D), velocity of sound (C)]

or

$$F = f[\omega, V, \mu, \rho, D, C]$$

$$f_1[F, \omega, V, \mu, \rho, D, C] = 0$$

Dimension of variables involved in given problem

Variables	Dimension
F	MLT^{-2}
ω	T^{-1}
V	LT^{-1}
μ	$ML^{-1}T^{-1}$
ρ	ML^{-3}
D	L
C	LT^{-1}

Total number of variables: $\qquad n = 7$

Number of fundamental dimensions involved: $\qquad m = 3$

Number of dimensionless number or π-terms involved $= n - m = 7 - 3 = 4$

Number of repeating variables = number of fundamental dimensions = 3

Variables	Repeating Variables
ω	
V	$\checkmark$
μ	
ρ	$\checkmark$
D	$\checkmark$
C	

Choosing repeating variables: V, ρ and D

According to Buckingham's π-theorem, the number of π-terms are involved in given problem = 4.

i.e., $\quad f_1(\pi_1, \pi_2, \pi_3, \pi_4) = 0$ $\qquad$ ...(1)

where $\qquad \pi_1 = V^a \rho^b D^c F$ $\qquad$ ...(2)

$\pi_2 = V^p \rho^q D^r \omega$ $\qquad$ ...(3)

$\pi_3 = V^l \rho^m D^n \mu$ $\qquad$ ...(4)

$\pi_4 = V^\alpha \rho^\beta D^\gamma F$ $\qquad$ ...(5)

Now we take Ist π-term,

$$\pi_1 = V^a \rho^b D^c F$$

Substituting the dimensions on both sides, we get

$$L^0 M^0 T^0 = (LT^{-1})^a (ML^{-3})^b L^c MLT^{-2}$$
$$L^0 M^0 T^0 = L^a T^{-a} M^b L^{-3b} L^c MLT^{-2}$$
$$L^0 M^0 T^0 = L^{a-3b+c+1} M^{b+1} T^{-a-2}$$

Comparing the power of M, we get

$$0 = b + 1$$

or $\qquad b = -1$

Comparing the power of T, we get
$$0 = -a - 2$$
or
$$a = -2$$

Comparing the power of L, we get
$$0 = a - 3b + c + 1$$
$$0 = -2 - 3(-1) + c + 1$$
$$0 = -2 + 3 + c + 1$$
or
$$c = -2$$

Substituting the values of a, b, c in Eq. (2), we get

$$\pi_1 = V^{-2} \rho^{-1} D^{-2} F = \frac{F}{\rho D^2 V^2}$$

Similarly for 2nd π-term, we get
$$\pi_2 = V^p \rho^q D^r \omega$$

Substituting the dimensions on both sides, we get
$$L^0 M^0 T^0 = (LT^{-1})^p (ML^{-3})^q L^r T^{-1}$$
$$L^0 M^0 T^0 = L^p T^{-p} M^q L^{-3q} L^r T^{-1}$$
$$L^0 M^0 T^0 = L^{p-3q+r} M^q T^{-p-1}$$

Comparing the power of M, we get
$$0 = q$$
or
$$q = 0$$

Comparing the power of T, we get
$$0 = -p - 1, \ p = -1$$

Comparing the power of L, we get
$$0 = p - 3q + r$$
$$0 = -1 - 0 + r$$
or
$$r = 1$$

Substituting the values of p, q and r in Eq. (3), we get

$$\pi_2 = V^{-1} \rho^0 D^1 \omega = \frac{D\omega}{V}$$

$$\pi_2 = \frac{D\omega}{V}$$

and similarly for third π-term, we get
$$\pi_3 = V^l \rho^m D^n \mu$$

Substituting the dimensions on both sides, we get
$$L^0 M^0 T^0 = (LT^{-1})^l (ML^{-3})^m L^n ML^{-1}T^{-1}$$
$$L^0 M^0 T^0 = L^l T^{-l} M^m L^{-3m} L^n ML^{-1}T^{-1}$$
$$L^0 M^0 T^0 = L^{l-3m+n-1} M^{m+1} T^{-l-1}$$

Comparing the power of M, we get
$$0 = m + 1$$
or
$$m = -1$$

Comparing the power of T, we get
$$0 = -l - 1$$
or
$$l = -1$$
Comparing the power of L, we get
$$0 = 1 - 3m + n - 1$$
$$0 = 1 - 3(-1) + n - 1$$
$$0 = -1 + 3 + n - 1$$
or
$$n = -1$$
Substituting the value of l, m and n in Eq. (4), we get

$$\pi_3 = V^{-1}\, \rho^{-1}\, D^{-1}\, \mu = \frac{\mu}{V\rho D}$$

$$\pi_3 = \frac{1}{Re} \qquad\qquad \because \text{ Reynolds number: } Re = \frac{V\rho D}{\mu}$$

or
$$\pi_3 = Re$$
Similarly for fourth π-term, we get
$$\pi_4 = V^\alpha\, \rho^\beta\, D^\gamma\, C$$
Substituting the dimensions on both sides, we get
$$L^0 M^0 T^0 = (LT^{-1})^\alpha\, (ML^{-3})^\beta\, L^\gamma\, LT^{-1}$$
$$L^0 M^0 T^0 = L^\alpha T^{-\alpha}\, M^\beta L^{-3\beta}\, L^\gamma\, LT^{-1}$$
$$L^0 M^0 T^0 = L^{\alpha-3\beta+\gamma+1}\, M^\beta\, T^{-\alpha-1}$$

Comparing the power of T, we get
$$0 = -\alpha - 1$$
or
$$\alpha = -1$$
Comparing the power of M, we get
$$0 = \beta$$
or
$$\beta = 0$$
Comparing the power of L, we get
$$0 = \alpha - 3\beta + \gamma + 1$$
$$0 = -1 - 0 + \gamma + 1$$
or
$$\gamma = 0$$
Substituting the values of α, β and γ in Eq. (5), we get
$$\pi_4 = V^{-1}\, \rho^0\, D^0\, C$$

$$\pi_4 = \frac{C}{V}$$

$$\pi_4 = \frac{1}{M} \qquad\qquad \because \text{ Mach number: } M = \frac{V}{C}$$

or
$$\pi_4 = M$$
Substituting the values of π_1, π_2, π_3 and π_4 in Eq. (1), we get

$$f_1\left(\frac{F}{\rho D^2 V^2}, \frac{D\omega}{V}, Re, M\right) = 0$$

or
$$\frac{F}{\rho D^2 V^2} = \phi \left[\frac{D\omega}{V}, \text{Re}, M \right]$$

or
$$F = \rho D^2 V^2 \phi \left[\frac{D\omega}{V}, \text{Re}, M \right]$$

Problem 7.16: The resisting force F and a supersonic plane using flight can be considered as dependent upon the length of aircraft l, velocity V, air viscosity μ, air density ρ, and bulk modulus of air K. Express the functional relationship between these variables and the resisting force by Buckingham's π-theorem.

(*GGSIP University, Delhi, Dec. 2008*)

Solution: Resisting force: F = function of [length (l), velocity (V), viscosity (m), density (ρ), bulk modulus (K)]

or $\qquad F = f[l, V, \mu, \rho, K]$

or $\quad f_1[F, l, V, \mu, \rho, K] = 0$

Dimension of variables involved in given problem

Variables	Dimension
F	MLT^{-2}
l	L
V	LT^{-1}
μ	$ML^{-1}T^{-1}$
ρ	ML^{-3}
K	$ML^{-1}T^{-2}$

Total number of variables: $n = 6$

Number of fundamental dimensions involved: $m = 3$

Number of dimensionless number or π-terms involved $= n - m = 6 - 3 = 3$

Number of repeating variables = number of fundamental dimensions = 3

Variables	Repeating Variables
l	$\checkmark$
V	$\checkmark$
μ	
ρ	$\checkmark$
K	

Choosing repeating variables: l, V and ρ

According to Buckingham's π-theorem, the number of π-terms involved in given problem = 3.

i.e., $\qquad f_1(\pi_1, \pi_2, \pi_3) = 0$ $\qquad$ ...(1)

where $\qquad \pi_1 = l^a V^b \rho^c F$ $\qquad$ ...(2)

$$\pi_2 = l^P V^q \rho^r \mu \qquad \qquad ...(3)$$
$$\pi_3 = l^\alpha V^\beta \rho^\gamma K \qquad \qquad ...(4)$$

Now we take Ist π-term,

$$\pi_1 = l^a V^b \rho^c F$$

Substituting the dimension on both sides, we have

$$L^0 M^0 T^0 = L^a (LT^{-1})^b (ML^{-3})^c MLT^{-2}$$
$$L^0 M^0 T^0 = L^a L^b T^{-b} M^c L^{-3c} MLT^{-2}$$
$$L^0 M^0 T^0 = L^{a+b-3c-1} M^{c+1} T^{-b-2}$$

Comparing the power of M, we get

$$0 = c + 1$$

or
$$c = -1$$

Comparing the power of T, we get

$$0 = -b - 2$$

or
$$b = -2$$

Comparing the power of L, we get

$$a + b - 3c + 1 = 0$$
$$a = 3c - b - 1$$
$$a = 3 \times (-1) - (-2) - 1 = -3 + 2 - 1 = -2$$

Substituting the values of a, b, c in Eq. (2), we have

$$\pi_1 = l^{-2} V^{-2} \rho^{-1} F = \frac{F}{\rho l^2 V^2}$$

Similarly for 2nd π-term, we have

$$\pi_2 = l^P V^q \rho^{-r} \mu$$

Substituting the dimensions on both sides, we have

$$L^0 M^0 T^0 = L^P (LT^{-1})^q (ML^{-3})^r ML^{-1} T^{-1}$$
$$L^0 M^0 T^0 = L^P L^q T^{-q} M^r L^{-3r} ML^{-1} T^{-1}$$
$$L^0 M^0 T^0 = L^{p+q-3r-1} M^{r+1} T^{-q-1}$$

Comparing the power of M, we get

$$0 = r + 1$$

or
$$r = -1$$

Comparing the power of T, we get

$$0 = -q - 1$$

or
$$q = -1$$

Comparing the power of L, we get

$$0 = p + q - 3r - 1$$
$$0 = p - 1 - 3(-1) - 1$$

or
$$0 = p - 1 + 3 - 1$$

or
$$p = -1$$

Substituting the values of p, q and r in Eq. (3), we get

$$\pi_2 = l^{-1} V^{-1} \rho^{-1} \mu$$

$$\pi_2 = \frac{\mu}{\rho l V}$$

and similarly for third π-term, we have

$$\pi_3 = l^\alpha \, V^\beta \, \rho^\gamma \, K$$

Substituting the dimensions on both sides, we get

$$L^0 M^0 T^0 = L^\alpha (LT^{-1})^\beta \, (ML^{-3})^\gamma \, ML^{-1} T^{-2}$$
$$L^0 M^0 T^0 = L^\alpha L^\beta T^{-\beta} \, M^\gamma L^{-3\gamma} \, ML^{-1} T^{-2}$$
$$L^0 M^0 T^0 = L^{\alpha+\beta-3\gamma-1} \, M^{\gamma+1} \, T^{-\beta-2}$$

Comparing the power of M, we get

$$0 = \gamma + 1$$

or $\qquad\qquad \gamma = -1$

Comparing the power of T, we have

$$0 = -\beta - 2$$

or $\qquad\qquad \beta = -2$

Comparing the power of L, we get

$$0 = \alpha + \beta - 3\gamma - 1$$
$$0 = \alpha - 2 - 3\,(-1) - 1$$
$$0 = \alpha - 2 + 3 - 1$$

or $\qquad\qquad \alpha = 0$

Substituting the value of α, β and γ in Eq. (1), we get

$$\pi_3 = l^0 \, V^{-2} \, \rho^{-1} \, K$$

$$\pi_3 = \frac{K}{\rho V^2}$$

Substituting the values of π_1, π_2 and π_3 in Eq. (1), we get

$$f_1\!\left(\frac{F}{\rho l^2 V^2}, \frac{\mu}{\rho V l}, \frac{K}{\rho V^2} \right) = 0$$

or $\qquad\qquad \dfrac{F}{\rho l^2 V^2} = \phi\!\left[\dfrac{\mu}{\rho V l}, \dfrac{K}{\rho V^2} \right]$

or $\qquad\qquad F = \rho l^2 V^2 \phi\!\left[\dfrac{\mu}{\rho V l}, \dfrac{K}{\rho V^2} \right]$

7.7 MODEL ANALYSIS

In different fields like engineering, medical science *etc.* when work on searching, modification or designing is done, a new structure or machine is developed. But before constructing or manufacturing new structure or machine there is need to check the performance or feasibility of the new structure or machine [prototype]. But the task of checking the feasibility of new structure being developed is difficult because it may either be too small, too large, too fast or too slow or it may be of very high cost. To tackle this problem, we make a small scale model that is proportional to the actual structure being developed and exhibit similar behaviour to the prototype (*i.e.,* actual system).

The model may be large, small or even of the same size that of prototype normally in case of fluid system models smaller than prototype and prototype is the full size machine or structure aircraft, ship or any other device, which is being studied by model tests. In other words, we can define model as small scale replica of the actual structure or the machine while the actual structure or m/c being developed is known as prototype. The following Fig. 7.1 shows the prototype of overflow spillway and its model.

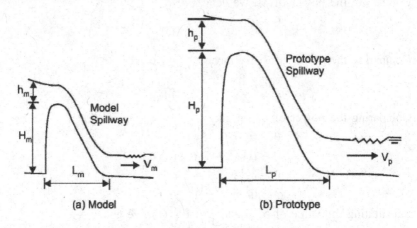

(a) Model   (b) Prototype

Fig. 7.1: Model spillway and its prototype.

Mostly models are much smaller than its corresponding prototype but in some cases it may be larger than prototype.

Now in case of fluids, the complete similarity exist between the flow in the prototype and in its model. Every dimensionless parameter (like Reynolds number, Froude's number etc.) has the same numerical value for both prototype and its corresponding model. And also all geometric, kinematic and dynamic π-terms are taken same in prototype and its model. It is evident that both must be similar in geometric, kinematic and dynamic aspects.

Model and its prototype, Fluid flow through pipe.

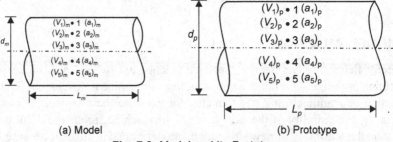

(a) Model   (b) Prototype

Fig. 7.2: Model and its Prototype

7.8 SIMILITUDE

Types of Similarities: Similitude is defined as the similarity between the model and its prototype which means that the model and prototype have similar properties or model and prototype are completely similar. The following three types of similarities must exist between the model and prototype:

1. Geometric similarity,
2. Kinematic similarity, and
3. Dynamic similarity.

7.8.1 Geometric Similarity

This is similarity of shape. This similarity is said to exist between a model and prototype if the ratio of corresponding linear dimension of the model and prototype is the same.

If L_m, b_m, D_m etc. refer to length, breadth, diameter of model respectively, and L_p, b_p, D_p etc. refer to length, breadth, diameter of prototype.

Then the condition of geometric similarity is

$$\frac{l_p}{l_m} = \frac{b_p}{b_m} = \frac{D_p}{D_m} = L_r$$

where L_r is called the scale ratio or length scale factor.

For area's ratio and volume's ratio the relation should be given below:

$$\frac{\Delta A_p}{\Delta A_m} = \frac{L_p \times b_p}{L_m \times b_m} = L_r \times L_r = L_r^2 = A_r$$

where A_r = area's ratio

and $$\frac{V_p}{V_m} = \left(\frac{L_p}{L_m}\right)^3 = \left(\frac{b_p}{b_m}\right)^3 = \left(\frac{D_p}{D_m}\right)^3 = L_r^3 = V_r$$

where V_r = volume's ratio.

7.8.2 Kinematic Similarity

This is similarity of motion. This similarity is said to exist between the model and the prototype if the ratio of velocities and acceleration at the corresponding points are same.

If V_{m1}, V_{m2} are velocities of fluid at points '1' and '2' in model and a_{m1}, a_{m2} are accelerations of fluid at points '1' and '2' in model as shown in Fig. 17.2.

V_{P1}, V_{p2} are velocities of fluid at points '1' and '2' in prototype.

a_{P1}, a_{P2} are accelerations of fluid at points '1' and '2' in prototype.

For kinematic similarity

$$\frac{V_{p1}}{V_{m1}} = \frac{V_{p2}}{V_{m2}} = V_r \text{ where } V_r \text{ is the velocity ratio or velocity scale factor.}$$

and $\dfrac{a_{p_1}}{a_{m_1}} = \dfrac{a_{p_2}}{a_{m_2}} = a_r$ where a_r is the acceleration ratio or acceleration scale factor.

Also the directions of the velocities and accelerations in the model and prototype should be same because both velocity and acceleration are vector quantities.

7.8.3 Dynamic Similarity

Dynamic Similarity is similarity of forces between the model and prototype. Thus, dynamic similarity is said to exist between the model and prototype if the ratios of the corresponding forces acting at the corresponding points of model and prototype are equal. A fluid mass, in general, may be subjected to the following forces:

Inertia force: $\qquad F_i$
Gravitational force: $\qquad F_g$
Viscous force: $\qquad F_v$
Pressure force: $\qquad F_P$
Surface tension force: $\qquad F_t$
Elastic force: $\qquad F_e$

The inertia force is the resultant force acting on a fluid mass and produces the acceleration.

$\therefore \qquad F_i = ma = F_g + F_v + F_P + F_l + F_e$

Hence the condition for dynamic similarity is

$$\frac{(F_i)_p}{(F_i)_m} = \frac{(F_g)_p}{(F_g)_m} = \frac{(F_v)_p}{(F_v)_m} = \frac{(F_P)_p}{(F_P)_m} = \frac{(F_t)_p}{(F_t)_m} = \frac{(F_e)_p}{(F_e)_m} = F_r$$

where F_r is the force ratio or force scale factor.

The above conditions can be expressed as follows:

$$\left[\frac{F_i}{F_g}\right]_m = \left[\frac{F_i}{F_g}\right]_p$$

$$\left[\frac{F_i}{F_P}\right]_m = \left[\frac{F_i}{F_P}\right]_p$$

$$\left[\frac{F_i}{F_v}\right]_m = \left[\frac{F_i}{F_v}\right]_p$$

$$\left[\frac{F_i}{F_t}\right]_m = \left[\frac{F_i}{F_t}\right]_p$$

$$\left[\frac{F_i}{F_e}\right]_m = \left[\frac{F_i}{F_e}\right]_p$$

Also the direction of the corresponding forces at the corresponding points in the model and prototype should be same. Dynamic similarity also includes geometric and kinematic similarities.

[**Note:** Angle scale factor or angle ratio must be unity *i.e.,* $\dfrac{\theta_p}{\theta_m} = 1$]

7.9 TYPES OF FORCES ACTING IN MOVING FLUID

For the fluid flow problems, the forces acting on a fluid mass may be any one or a combination of the several of the following forces:

1. Inertia force: F_i
2. Viscous force: F_v
3. Gravity force: F_g
4. Pressure force: F_P
5. Surface tension force: F_t
6. Elastic force: F_e

7.9.1 Inertia Force (F_i)

It is equal to the product of mass and acceleration of the flowing fluid and acts in the direction opposite to the direction of flow. It always exists in the fluid flow problem.

Mathematically,

Intertia force: F_i = mass × acceleration

= density × volume × acceleration

$$= \rho L^3 \times \frac{dV}{dt} = \rho L^3 \frac{V}{L/V} = \rho L^2 V^2$$

7.9.2 Viscous Force (F_v)

It is equal to the product of shear stress (t) due to viscosity and surface area of the flow. It is present in fluid flow problem where viscosity is having an important role to play.

Mathematically,

Viscous force: F_v = shear stress × area

$= \tau \times A$ | According to Newton's law of viscosity:

$= \mu \dfrac{du}{dy}.L^2$ | $\tau = \mu \dfrac{du}{dy}$

$= \mu \dfrac{V}{L}.L^2 = \mu LV$

7.9.3 Gravity Force (F_g)

It is equal to the product of mass and acceleration due to gravity of the flowing fluid. It is present in case of open surface flow.

Mathematically,

Gravity force: F_g = mass × acceleration due to gravity

= density × volume × $g = \rho.L^3.g = \rho g L^3$

7.9.4 Pressure Force (F_P)

It is equal to the product of pressure intensity and cross-sectional area of the flowing fluid. It is present in case of pipe-flow.

Mathematically,

Pressure force: F_P = pressure intensity × area

$= p.\ A = pL^2$

7.9.5 Surface Tension Force (F_t)

It is equal to the product of surface tension and length of the surface of the flowing fluid.

Mathematically,

Surface tension force: F_t = surface tension × length of the surface

$$= \sigma L$$

where σ = surface tension.

7.9.6 Elastic Force (F_e)

It is equal to the product of elastic stress and area of the flowing fluid.

Mathematically,

Elastic force: F_e = elastic stress × area = $K. A = K L^2$

where K = elastic stress.

7.10   DIMENSIONLESS NUMBERS

Dimensionless numbers are those numbers which are obtained by dividing the intertia force by any one of the remaining forces. It is a ratio of one force to the other force, it will be a dimensionless number. The following five dimensionless constants are important in fluid mechanics.

1. Reynolds Number: Re
2. Froude's Number: Fr
3. Euler's Number: Eu
4. Weber's Number: We
5. Mach's Number: M

7.10.1 Reynolds Number (Re)

It is defined as the ratio of the inertia force to the viscous force. Mathematically,

$$\text{Reynolds Number:} \quad Re = \frac{\text{Inertia force} : F_i}{\text{Viscous force} : F_v}$$

$$= \frac{F_i}{F_v} = \frac{\rho V^2 L^2}{\mu L V} = \frac{\rho V L}{\mu}$$

$$= \frac{LV}{v} \qquad \left| \; \because \text{Kinematic viscosity:} \; v = \frac{\mu}{\rho} \right.$$

In case of pipe flow, the linear dimension L is taken as diameter d.

Hence Reynolds number for pipe flow: $Re = \dfrac{\rho V d}{\mu} = \dfrac{V d}{v}$

Significance:

(a) Reynolds number must have the same value for model and prototype to achieve dynamic similarity.

(b) The value of the Reynolds number in a flow gives an idea about its nature. At higher Reynolds number, the magnitude of viscous force is small as compared to the inertia force. At low Reynolds number, the magnitude of viscous force is large as compared to the inertia force.

7.10.2 Froude's Number (Fr)

The Froude's number is defined as the square root of the ratio of inertia force of a flowing fluid to the gravity force.

Mathematically,

$$\text{Froude's Number: } Fr = \sqrt{\frac{\text{Inertia force}: F_i}{\text{Gravity force}: F_g}} = \sqrt{\frac{F_i}{F_g}}$$

$$= \sqrt{\frac{\rho L^2 V^2}{\rho g L^3}} = \sqrt{\frac{V^2}{gL}} = \frac{V}{\sqrt{gL}}$$

Significance:

(a) Froude's number must have the same value for model and prototype to achieve dynamic similarity.

(b) Froude's number is important in the flow situation; where gravitational force is predominant and all other forces are comparatively negligible.

For Example:

(i) Flow over notches and weirs.

(ii) Flow over the spillway of a dam.

(iii) Flow through open channels, considering waves and jumps.

(iv) Surface wave.

7.10.3 Euler's Number (Eu)

It is defined as the square root of the ratio of the inertia force of a flowing fluid to the pressure force.

Mathematically,

$$\text{Euler's Number: } Eu = \sqrt{\frac{\text{Inertia force}: F_i}{\text{Pressure force}: F_p}} = \sqrt{\frac{F_i}{F_p}}$$

$$= \sqrt{\frac{\rho L^2 V^2}{p L^2}} = \sqrt{\frac{\rho V^2}{p}} = \frac{V}{\sqrt{p/\rho}}$$

Significance:

(a) Euler's number must have the same value for model and prototype to achieve dynamic similarity.

(b) Euler's number is important in the flow situation in which a pressure gradient exists.

For example:

(*i*) Discharge through orifices, mouth pieces.

(*ii*) Pressure rise due to sudden closure of valves.

(*iii*) Flow through pipes and

(*iv*) Water hammer created in penstocks.

7.10.4 Weber's Number (*We*)

It is defined as the square root of the ratio of the inertia force of a flowing fluid to the surface tension force.

Mathematically,

$$\text{Weber's Number: } We = \sqrt{\frac{\text{Inertia force}: F_i}{\text{Surface tension force}: F_t}}$$

$$= \sqrt{\frac{F_i}{F_t}} = \sqrt{\frac{\rho L^2 V^2}{\sigma . L}} = \sqrt{\frac{\rho L V^2}{\sigma}} = \frac{V}{\sqrt{\sigma / \rho L}}$$

Significance:

(*a*) Weber's number must have the same value for model and prototype to achieve dynamic similarity.

(*b*) Weber's number is used in following flow situations:

     (*i*) Capillary tube flow, (*ii*) Human blood flow studies, and

     (*iii*) Liquid atomisation.

7.10.5 Mach's Number (*M*)

It is defined as the square root of the ratio of the inertia force of a flowing fluid to the elastic force.

Mathematically,

$$\text{Mach's Number: } M = \sqrt{\frac{\text{Inertia force}: F_i}{\text{Elastic force}: F_e}} = \sqrt{\frac{F_i}{F_e}} = \sqrt{\frac{\rho L^2 V^2}{K L^2}}$$

$$= \sqrt{\frac{\rho V^2}{K}} = \frac{V}{\sqrt{K / \rho}} = \frac{V}{a}$$

[where *a* is called velocity of sound in the fluid $= \sqrt{K / \rho}$]

$$\boxed{M = \frac{V}{a}}$$

Hence Mach's number is also defined as the ratio of the velocity of fluid to the velocity of sound in the fluid.

Significance:

(*a*) Mach's number must have the same value for model and prototype to achieve dynamic similarity.

(b) Mach's number is important in compressible fluid flow problem at high velocities, such as high velocity flow in pipes.

(c) Fluid motion may be classified according to Mach number as follows:

    (i) $M < 0.2$; for incompressible motion e.g. low speed fluid flow.

    (ii) M lies between $0.9 - 1.1$, for transonic motion.

    (iii) When $1 < M < 7$, super sonic motion.

    (iv) When $M > 7$, hypersonic motion.

    (v) $M = 1$, sonic motion.

    (vi) When $M < 1$, sub-sonic motion.

7.11 MODEL LAWS OR SIMILARITY LAWS

For dynamic similarity between the model and prototype, the dimensionless numbers should be same for model and the prototype. But it is quite difficult to satisfy the condition that all the dimensionless numbers (i.e., Reynolds Number, Froude's Number, Weber's Number, Euler's Number and Mach Number) are the same for the model and prototype. Hence, models are designed on the basis of ratio of the forces. The laws on which the models are designed for dynamic similarity are called model laws or laws of similarity. The following main model laws are used in fluid mechanics.

    1. Reynolds Law   2. Froude's Law

    3. Euler's Law   4. Weber's Law

    5. Mach's Law.

7.11.1 Reynolds Law

The model design according to this law, the Reynolds number for the model must be equal to the Reynolds number for the prototype.

Mathematically,

Reynolds number for the model = Reynolds number for the prototype

$$(Re)_m = (Re)_P$$

$$\left(\frac{\rho V L}{\mu}\right)_m = \left(\frac{\rho V L}{\mu}\right)_p \qquad \left| \because Re = \frac{\rho V L}{\mu} \right.$$

$$\frac{\rho_m V_m L_m}{\mu_m} = \frac{\rho_p V_p L_p}{\mu_p}$$

where

        ρ_m = density of fluid in model

        V_m = velocity of fluid in model

        L_m = length or linear dimension of the model

        μ_m = viscosity of fluid in model and

        ρ_p = density of fluid in prototype

        V_p = velocity of fluid in prototype

        L_p = length or linear dimension of the model

        μ_p = viscosity of fluid in prototype.

$$\frac{\rho_p}{\rho_m} \times \frac{V_p}{V_m} \times \frac{L_p}{L_m} \times \frac{1}{\mu_p / \mu_m} = 1$$

$$\frac{\rho_r \times V_r \times L_r}{\mu_r} = 1$$

where $\qquad \rho_r = \dfrac{\rho_p}{\rho_m}$ is density scale ratio

$$V_r = \frac{V_p}{V_m} \text{ is velocity scale ratio}$$

$$L_r = \frac{L_p}{L_m} \text{ is linear dimension scale ratio}$$

$$\mu_r = \frac{\mu_p}{\mu_m} \text{ is viscosity scale ratio.}$$

Reynolds law has application only in those problems where the viscous forces alone are predominant. The following are some of the examples where Reynolds law has applications:

(*i*) Incompressible fluid flow in closed pipes.

(*ii*) Motion of a body fully submerged in a fluid.

7.11.2 Froude's Law

The model design according to this law, states the Froude's number for the model must be equal to the Froude's number for the prototype.

Mathematically,

Froude's number for model = Froude's number for the prototype.

$$(Fr)_m = (Fr)_p$$

$$\left(\frac{V}{\sqrt{gL}}\right)_m = \left(\frac{V}{\sqrt{gL}}\right)_p$$

$$\frac{V_m}{\sqrt{g_m L_m}} = \frac{V_p}{\sqrt{g_p L_p}}$$

If the tests on the model are performed on the same place where prototype is to operate, then $g_m = g_p$ and above equation becomes as

$$\frac{V_m}{\sqrt{L_m}} = \frac{V_p}{\sqrt{L_p}}$$

$$\frac{V_p}{V_m} = \sqrt{\frac{L_p}{L_m}}$$

$$\frac{V_p}{V_m} = \sqrt{\frac{L_p}{L_m}} = \sqrt{L_r}$$

where scale ratio for length:

$$L_r = \frac{L_p}{L_m}$$

$$\frac{V_p}{V_m} = V_r$$

where V_r is scale ratio for velocity.

$$\therefore \qquad \frac{V_p}{V_m} = V_r = \sqrt{L_r}$$

Scale ratios for various quantities based on Froude's law are:

(*i*) **Scale ratio for time:**

Time: $\qquad T = \dfrac{\text{Length} : L}{\text{Velocity} : V}$

Then ratio of time for prototype and model is

$$T_r = \frac{T_p}{T_m} = \frac{\left(\dfrac{L}{V}\right)_p}{\left(\dfrac{L}{V}\right)_m} = \frac{\dfrac{L_p}{V_p}}{\dfrac{L_m}{V_m}} = \frac{L_p}{L_m} \cdot \frac{V_m}{V_p}$$

$$= L_r \cdot \frac{1}{\sqrt{L_r}} = \sqrt{L_r}$$

(*ii*) **Scale ratio for acceleration:**

Acceleration: $\qquad a = \dfrac{\text{Velocity} : V}{\text{Time} : T}$

$\therefore$ Ratio for acceleration: $a_r = \dfrac{a_p}{a_m} = \dfrac{(V/T)_p}{(V/T)_m} = \dfrac{V_p}{T_p} \cdot \dfrac{T_m}{V_m}$

$$= \frac{V_p}{V_m} \cdot \frac{T_m}{T_p}$$

$$= \sqrt{L_r} \cdot \frac{1}{\sqrt{L_r}} \qquad \left| \; \because \; \frac{V_p}{V_m} = \sqrt{L_r}, \frac{T_p}{T_m} = \sqrt{L_r} \right.$$

$$= 1$$

(*iii*) **Scale ratio for discharge:**

Discharge: $\qquad Q = AV = L^2 \dfrac{L}{T} = \dfrac{L^3}{T}$

$\therefore$ Ratio for discharge: $Q_r = \dfrac{Q_p}{Q_m} = \dfrac{\left(\dfrac{L^3}{T}\right)_p}{\left(\dfrac{L^3}{T}\right)_m} = \left(\dfrac{L_p}{L_m}\right)^3 \cdot \left(\dfrac{T_m}{T_p}\right)$

$$= L_r^3 \times \frac{1}{\sqrt{L_r}} = L_r^{2.5}$$

(*iv*) **Scale ratio for force:**

Force: $\quad\quad\quad F = $ mass $\times$ acceleration $=$ density $\times$ volume $\times$ acceleration

$$= \rho L^3 . \frac{V}{T} = \rho L^2 . \frac{L}{T} . V = \rho L^2 V^2$$

$\therefore$ Ratio for force: $F_r = \dfrac{F_p}{F_m} = \dfrac{\rho_p L_p^2 V_p^2}{\rho_m L_m^2 V_m^2} = \dfrac{\rho_p}{\rho_m} \times \left(\dfrac{L_p}{L_m}\right)^2 \times \left(\dfrac{V_p}{V_m}\right)^2$

If the working fluid in model and prototype is same, then

$$\frac{\rho_p}{\rho_m} = 1 \quad i.e., \quad \rho_p = \rho_m$$

Hence $\quad\quad\quad F_r = \left(\dfrac{L_p}{L_m}\right)^2 . \left(\dfrac{V_p}{V_m}\right)^2 = L_r^2 . \left(\sqrt{L_r}\right)^2 = L_r^2 . L_r = L_r^3$

(*v*) **Scale ratio for pressure intensity:**

Pressure intensity: $\quad p = \dfrac{\text{Force} : F}{\text{Area} : A} = \dfrac{\rho L^2 V^2}{L^2} = \rho V^2$

$\therefore$ Pressure ratio: $p_r = \dfrac{p_p}{p_m} = \dfrac{\rho_p V_p^2}{\rho_m V_m^2}$

If the working fluid in model and prototype is same, then $\rho_p = \rho_m$

(*vi*) (*a*) **Scale ratio for energy or work:**

Energy or work $=$ force $\times$ distance

$$E \text{ or } W = F. L$$

$\therefore$ Energy or work ratio $= \dfrac{E_p}{E_m}$

$$E_r = W_r = \frac{F_p L_p}{F_m L_m} = \frac{F_p}{F_m} . \frac{L_p}{L_m}$$

$$= L_r^3 . L_r = L_r^4$$

(*b*) **Scale ratio for torque or moment:**

Torque or moment $=$ force $\times$ distance

$$T^* \text{ or } M = F. L$$

$\therefore$ Torque or moment ratio $= \dfrac{T_p^*}{T_m^*}$

$$T_r^* = M_r = \frac{F_p L_p}{F_m L_m} = \frac{F_p}{F_m} . \frac{L_p}{L_m} = L_r^3 . L_r = L_r^4$$

Hence scale ratios for Energy, Work, Torque and moment are same:

i.e., $\quad\quad\quad \boxed{E_r = W_r = T_r^* = M_r = L_r^4}$

(*vii*) (*a*) **Scale ratio for power:**

Power: $P = $ Work per unit time.

$$= \frac{\text{Force} \times \text{distance}}{\text{time}} = \frac{F.L}{T}$$

∴ Power ratio: $P_r = \dfrac{P_p}{P_m} = \dfrac{\left(\dfrac{FL}{T}\right)_p}{\left(\dfrac{FL}{T}\right)_m} = \dfrac{F_p}{F_m}.\dfrac{L_p}{L_m}.\dfrac{1}{\dfrac{T_p}{T_m}}$

$$= F_r.L_r.\frac{1}{T_r} = L_r^3.L_r.\frac{1}{\sqrt{L_r}} = L_r^3\sqrt{L_r}$$

$$= L_r^{7/2} = L_r^{3.5}$$

(*b*) **Scale ratio for momentum or impulse:**

Momentum = Impulse = mV

$$M^* = I = \rho L^3 V$$

Mass = density × volume

$$m = \rho L^3$$

∴ Momentum or impulse ratio $= \dfrac{M_p^*}{M_m^*}$

$$M_r^* = I_r = \frac{\rho_p L_p^3 V_p}{\rho_m L_m^3 V_m}$$

If working fluid in model and prototype is same, then $\dfrac{\rho_p}{\rho_m} = 1$ *i.e.,* $\rho_p = \rho_m$

Hence $\quad M_r^* = I_{fr} = \dfrac{L_p^3}{L_m^3}.\dfrac{V_p}{V_m} = L_r^3\sqrt{L_r} = L_r^{7/2} = L_r^{3.5}$

Hence scale ratios for power, momentum and impulse are same

$$\boxed{P_r = M_r^* = I_r = L_r^{3.5}}$$

Froude's law has application only in those problems where the gravity force alone is predominent. The following are some of the examples where Froude's law has applications.

(*i*) Free surface flow such as flow over spillways, weirs, sluices, channels etc.

(*ii*) Flow of jet from an orifice or nozzle.

(*iii*) Where fluids of different densities flow over one another.

(*iv*) Where fluids are likely to be formed on surface.

7.12.3 Euler's Law

The model design according to this law states the Euler's number for model must be equal to the Euler's number for the prototype.

Mathematically,

Euler's number for the model = Euler's number for the prototype.

$$(Eu)_m = (Eu)_p$$

$$\left(\frac{V}{\sqrt{p/\rho}}\right)_m = \left(\frac{V}{\sqrt{p/\rho}}\right)_p$$

$$\frac{V_m}{\sqrt{P_m/\rho_m}} = \frac{V_p}{\sqrt{P_p/\rho_p}}$$

where V_m = velocity of fluid in model.
                         P_m = pressure of fluid in model.
                         ρ_m = density of fluid in model.
and V_p = velocity of fluid in prototype.
                         P_p = pressure of fluid in prototype.
                         ρ_p = density of fluid in prototype.

If the working is same in model and prototype, then $\rho_m = \rho_p$, the above equation becomes as

$$\frac{V_m}{\sqrt{P_m}} = \frac{V_p}{\sqrt{P_p}}$$

Euler's law has application only in those problems, where the pressure forces alone are predominant. The following are some of the examples where Euler's law has applications:

(*i*) Pressure rise due to sudden closure of valves,

(*ii*) Flow through pipes, and

(*iii*) Water hammer created in Penstocks.

7.12.4 Weber's Law

The model design according to this Law states the Weber's number for the model must be equal to Weber's number for the prototype.

Mathematically,

Weber's number for the model = Weber's number for the prototype.

$$(We)_m = (We)_p$$

$$\left(\frac{V}{\sqrt{\sigma/\rho L}}\right)_m = \left(\frac{V}{\sqrt{\sigma/\rho L}}\right)_p$$

$$\frac{V_m}{\sqrt{\sigma_m/\rho_m L_m}} = \frac{V_p}{\sqrt{\sigma_p/\rho_p L_p}}$$

where V_m = velocity of fluid in model.

$$\sigma_m = \text{surface tension of fluid in model.}$$
$$\rho_m = \text{density of fluid in model.}$$
$$L_p = \text{length of surface in prototype.}$$
and $\qquad V_p = \text{velocity of fluid in prototype.}$
$$\sigma_p = \text{surface tension of fluid in prototype.}$$
$$\rho_p = \text{density of fluid in prototype.}$$
$$L_p = \text{length of surface in prototype.}$$

If working fluid is same in model and prototype, then $\rho_m = \rho_p$, the above equation becomes as

$$\frac{V_m}{\sqrt{\sigma_m / L_m}} = \frac{V_p}{\sqrt{\sigma_p / L_p}}$$

Weber's law has application only in such problems, where the surface tension forces alone are predominant. The following are some of the examples where Weber's law has application:

(*i*) Capillary movement of water in soils.

(*ii*) Flow of a liquid at a very small depth over a surface.

(*iii*) Flow over weirs at very small heads.

(*iv*) Spray of liquid from exit of a discharging tube resulting in the formation of drop of liquids.

7.12.5 Mach's Law

The model design according to this law states the Mach number for the model must be equal to the Mach number for the prototype.

Mathematically,

Mach number for the model = Mach number for the prototype.

$$(M)_m = (M)_p$$

$$\left(\frac{V}{\sqrt{K/\rho}} \right)_m = \left(\frac{V}{\sqrt{K/\rho}} \right)_p$$

$$\frac{V_m}{\sqrt{K_m / \rho_m}} = \frac{V_p}{\sqrt{K_p / \rho_p}}$$

where $\qquad V_m = \text{velocity of fluid in model.}$

$$K_m = \text{elastic stress for model.}$$

$$\rho_m = \text{density of working fluid in model.}$$

and $\qquad V_p = \text{velocity of fluid in prototype.}$

$$K_p = \text{elastic stress for prototype.}$$

$$\rho_p = \text{density of working fluid in prototype.}$$

Mach's law has application only in those problems, where the compressibility forces alone are predominant.

The following are some of examples where Mach's law has application:

(i) Flow of aeroplane and projectile through air at supersonic speed [i.e., velocity of fluid more than the velocity of sound].

(ii) Aerodynamic testing.

(iii) Under water testing of torpedoes.

(iv) Water-hammer problems.

Problem 7.17: A 200 mm diameter pipe conveys water at a velocity of 3.50 metre per second. For the condition of dynamic similarity, what is the velocity of oil flowing in a 80 mm diameter pipe? Take kinematic viscosity of water and oil equal to 0.01 stoke and 0.03 stoke respectively.

Solution: This is pipe flow problem, the dynamic similarity will be obtained if the Reynolds numbers in the model and prototype are equal.

$$[Re]_{model} = [Re]_{prototype}$$

$$\frac{V_m D_m}{v_m} = \frac{V_p D_p}{v_p} \quad ...(1)$$

$$\because Re = \frac{\rho VD}{\mu}, \ Re = \frac{VD}{v}$$

where kinematic viscosity:

$$v = \frac{\text{Dynamic viscosity}}{\text{Density}} = \frac{\mu}{\rho}$$

Particulars of the model and prototype are given below:

Model	Prototype
Working fluid is oil	Working fluid is water
D_m = 80 mm = 0.080 m	D_p = 200 mm = 0.20 m
V_m = ?	V_p = 3.50 m/s
v_m = 0.03 stoke	v_p = 0.01 stoke

Substituting the values of D_m, v_m and D_p, V_p, v_p in Eq. (1), we get

$$\frac{V_m \times 0.08}{0.03} = \frac{3.50 \times 0.20}{0.01}$$

$$V_m = 26.25 \text{ m/s}$$

Problem 7.18: A pipe of diameter 1.5 m is required to transport an oil of specific gravity 0.90 and viscosity 3×10^{-2} poise at the rate of 3000 litre/s. Tests were conducted on a 15 cm diameter pipe using water at 20°C. Find the velocity and rate of flow in the model. Viscosity of water at 20°C = 0.01 poise.

Solution: This is pipe flow problem, the dynamic similarity will be obtained if the Reynolds numbers in the model and prototype are equal.

$$\frac{\rho_m V_m D_m}{\mu_m} = \frac{\rho_p V_p D_p}{\mu_p} \qquad (1)$$

Particulars of the model and prototype are given below:

Model	Prototype
Working fluid is water	Working fluid is oil
D_m = 150 mm = 0.15 m	D_p = 1.5 m
μ_m = 0.01 poise	Specific gravity of oil: S_p = 0.9
V_m = ?	μ_p = 3×10^{-2} poise
Q_m = ?	Q_p = 3000 litre/s = 3.0 m³/s

$$\rho_m = 1000 \text{ kg/m}^3 \qquad V_p = \frac{Q_p}{A_p} = \frac{3.0}{\frac{\pi}{4}D_p^2} = \frac{3.0}{\frac{\pi}{4}(1.5)^2} = 1.698 \text{ m /s}$$

$$\rho_p = S_p \times \rho_{\text{water}} = 0.9 \times 1000 = 900 \text{ kg/m}^3$$

Substituting the values of ρ_m, D_m, μ_m and ρ_p, V_p, D_p, μ_p in Eq. (1), we get

$$\frac{1000 \times V_m \times 0.15}{0.01} = \frac{900 \times 1.698 \times 1.5}{3 \times 10^{-2}}$$

$$V_m = \textbf{5.094 m/s}$$

Rate of flow through model: $Q_m = A_m \cdot V_m$

$$= \frac{\pi}{4}(D_m)^2 \times 5.094 = \frac{\pi}{4}(0.15)^2 \times 5.094$$

$$= 0.08997 \text{ m}^3/\text{s} = 0.8997 \times 1000 \text{ litre/s} = \textbf{89.97 litre/s}$$

Problem 7.19: The ratio of length of a sub-marine and its model is 30: 1. The speed of submarine (prototype) is 10 m/s. The model is to be tested in wind tunnel. Find the speed of air in wind tunnel. Also determine the ratio of the drag (resistance) between the model and its prototype. Take the value of kinematic viscosities for sea water and air as 0.012 stoke and 0.016 stoke respectively. The density for sea-water and air is given as 1030 kg/m³ and 1.24 kg/m³ respectively.

Solution: This is submarine problem, the viscous resistance is to be overcome and hence for fully submerged submarine, the dynamic similarity will be obtained if the Reynolds numbers in the model and prototype are equal.

$$(Re)_m = (Re)_p$$

$$\frac{\rho_m V_m D_m}{\mu_m} = \frac{\rho_p V_p D_p}{\mu_p} \qquad \left[\because \text{ Kinematic viscosity: } v = \frac{\mu}{\rho} \right]$$

$$\frac{V_m D_m}{v_m} = \frac{V_p D_p}{v_p} \qquad \ldots(1)$$

Particulars of the model and prototype are given below:

Scale ratio: $L_r = \dfrac{L_p}{L_m} = 30$

<table>
<tr><td align="center">Model</td><td align="center">Prototype</td></tr>
<tr><td>Working fluid is air</td><td>Working fluid is sea water</td></tr>
<tr><td>$v_m = 0.016$ stoke $= 0.016$ m^2 /s</td><td>$V_p = 10$ m/s</td></tr>
<tr><td>$= 0.016 \times 10^{-4}$ m^2/s</td><td></td></tr>
<tr><td>$\rho_m = 1.24$ kg/m^3</td><td>$v_p = 0.012$ stoke $= 0.012$ cm^2/s</td></tr>
<tr><td></td><td>$= 0.012 \times 10^{-4}$ m^2/s</td></tr>
<tr><td>$V_m = ?$</td><td>$\rho_p = 1030$ kg/m^3</td></tr>
</table>

Substituting the values of v_m and V_p, v_p in Eq. (1), we get

$$\frac{V_m \times D_m}{0.016 \times 10^{-4}} = \frac{10 \times D_p}{0.012 \times 10^{-4}}$$

$$V_m = \frac{10 \times 0.016 \times 10^{-4}}{0.012 \times 10^{-4}} \times \frac{D_p}{D_m} \qquad \left[\because \frac{D_p}{D_m} = \frac{L_p}{L_m} = L_r = 30 \right]$$

$$V_m = \frac{10 \times 0.016 \times 10^{-4}}{0.012 \times 10^{-4}} \times 30 = 400 \text{ m/s}$$

Drag force: $F = $ mass $\times$ acceleration

$\qquad\qquad\qquad\quad = $ density $\times$ volume $\times$ acceleration

$$= \rho L^3 \frac{V}{t} = \rho L^3 \frac{V}{L/V} = \rho L^2 V^2 \qquad \left[\because \text{Time: } t = \frac{L}{V} \right]$$

The ratio of drag force (resistance):

$$\frac{\text{Drag force for the prototype } (F_p)}{\text{Drag force for the model } (F_m)} = \frac{\rho_p L_p^2 V_p^2}{\rho_m L_m^2 V_m^2}$$

$$\frac{F_p}{F_m} = \frac{\rho_p}{\rho_m} \left(\frac{L_p}{L_m} \right)^2 \frac{V_p^2}{V_m^2} = \frac{1030}{1.24} \times (30)^2 \times \left(\frac{10}{400} \right)^2 = \textbf{467.238}$$

Problem 7.20: A model of spillway is made to test the flow. The discharge and the velocity of flow over the model were measured as 3 m^3/s and 1.5 m/s respectively. Determine the discharge and the velocity over the prototype which is 40 times larger than its model.

Solution: Given data:

<table>
<tr><td align="center">Prototype</td><td align="center">Model</td></tr>
<tr><td>Discharge: $Q_p = ?$</td><td>Discharge: $Q_m = 3$ m^3/s</td></tr>
<tr><td></td><td>Velocity: $V_m = 1.5$ m/s</td></tr>
</table>

Scale ratio: $L_r = 40$

According to Froude's law,

$$(Fr)_p = (Fr)_m$$

$$\frac{V_p}{\sqrt{L_p g_p}} = \frac{V_m}{\sqrt{L_m g_m}}$$

Assuming: $g_p = g_m$

$$\frac{V_p}{\sqrt{L_p}} = \frac{V_m}{\sqrt{L_m}}$$

$$V_p = V_m \sqrt{\frac{L_p}{L_m}} = 1.5 \times \sqrt{40} = 9.48 \text{ m/s}$$

Discharge: $Q = AV$

$$\therefore \quad \frac{Q_p}{Q_m} = \frac{A_p V_p}{A_m V_m}$$

$$\therefore \quad Q_p = \frac{A_p}{A_m} \cdot \frac{V_p}{V_m} \cdot Q_m \qquad \left| \text{Area scale ratio}: A_r = L_r^2 \right.$$

$$= L_r^2 \frac{V_p}{V_m} \cdot Q_m = 40^2 \times \frac{9.48}{1.5} \times 3 = \textbf{30336 m}^3\textbf{/s}$$

Problem 7.21: In 1:20 model of stilling basin, the height of hydraulic jump in the model is observed to be 20 cm. What would be the corresponding height of jump in the prototype? If energy dissipation in the model is 0.1 kW, what would be the corresponding value of the prototype?

Solution: Given data:

Prototype	**Model**

Linear scale ratio: $L_r = \dfrac{L_p}{L_m} = 20$

$h_p = ?$ $h_m = 20 \text{ cm} = 0.2 \text{ m}$

$P_p = ?$ $P_m = 0.1 \text{ kW} = 0.1 \times 10^3 \text{ W}$

We know, $L_r = \dfrac{h_p}{h_m}$

or $h_p = L_r \cdot h_m$

$$= 20 \times 0.2 = \textbf{4 m}$$

We know, power scale ratio: $P_r = \dfrac{P_p}{P_m} = L_r^{3.5}$

or $\dfrac{P_p}{P_m} = L_r^{3.5}$

or
$$P_p = L_r^{3.5} \times P_m$$
$$= (20)^{3.5} \times 0.1 \times 10^3 \text{ W} = 3577708.76 \text{ W}$$
$$= \mathbf{3577.708 \text{ kW}}$$

Problem 7.22: A ship model 1m long with negligible friction is tested in a towing tank at a speed of 0.7 m/s. To what ship velocity does this correspond if the ship is 50 m long. A force of 5 N is required to tow the model. What propulsive force does this represent in the prototype ?

Solution: Given data:

Prototype	Model
$L_p = 50$ mm	$L_m = 1$ m
$V_p = ?$	$V_m = 0.7$ m/s
$F_p = ?$	$F_m = 5$ N

According to Froude's law,

$$\frac{V_p}{\sqrt{L_p g_p}} = \frac{V_m}{\sqrt{L_m g_m}}$$

$$\frac{V_p}{\sqrt{L_p}} = \frac{V_m}{\sqrt{L_m}} \qquad \qquad \Big| \text{Assuming} : g_p = g_m$$

$$V_P = \frac{\sqrt{L_p}}{\sqrt{L_m}} V_m = \sqrt{\frac{L_p}{L_m}} . V_m$$

$$V_P = \sqrt{\frac{50}{1}} \times 0.7 = \mathbf{4.95 \text{ m/s}}$$

We know, Force ratio: $F_r = L_r^3$

$$\frac{F_p}{F_m} = L_r^3 \qquad \qquad \Big| L_r = \frac{L_p}{L_m} = \frac{50}{1} = 50$$

$$F_p = (50)^3 \times F_m = (50)^3 \times 5 = 625000 \text{ N} = \mathbf{625 \text{ kN}}$$

Problem 7.23: A model of a harbour has a length scale of 1: 200. Storm waves of 20 m amplitude and 10 m/s velocity strike against the break water of the prototype harbour. Determine for the model:

(*i*) the size and velocity of the waves, neglecting viscous effects,

(*ii*) if the time between the tides in the prototype is 12 hrs., what should be the tidal period in the model?

Solution: Given data:

Prototype	Model
$L_r = 200$	
Amplitude of wave: $A_p = 20$ m	Size (amplitude of wave): $A_m = ?$
Velocity: $V_p = 10$ m/s	Velocity: $V_m = ?$
Time: $T_p = 12$ hrs.	Time: $T_m = ?$

According to Froude's law,

$$\frac{V_p}{\sqrt{L_p g_p}} = \frac{V_m}{\sqrt{L_m g_m}} \qquad \left| g_p = g_m \text{ (assuming)} \right.$$

$$\frac{V_p}{\sqrt{L_p}} = \frac{V_m}{\sqrt{L_m}}$$

or $$V_m = \frac{V_p}{\sqrt{L_p / L_m}} = \frac{V_p}{\sqrt{L_r}}$$

(i) $$V_m = \frac{10}{\sqrt{200}} = 0.707 \text{ m/s}$$

The size (amplitude of wave in model): A_m

$$\frac{A_p}{A_m} = L_r$$

or $$A_m = \frac{A_p}{A_r} = \frac{20}{200} = 0.1 \text{ m}$$

(ii) As we know that the time scale ratio:

$$T_r = \frac{T_p}{T_m} = \sqrt{L_r}$$

$$T_m = \frac{T_p}{\sqrt{L_r}} = \frac{12}{\sqrt{200}} = 0.8450 \text{ hr} = 50.91 \text{ min.}$$

Problem 7.24: The force required to propel a 1: 40 scale model of a motor boat in a lake at a speed of 2 m/s is 0.5 N. Assuming that the viscous resistance due to water and air is negligible in comparison with the wave resistance, determine the corresponding velocity of the prototype of 45 m for dynamically similar conditions. What would be the force required to propel the prototype at that velocity in the same lake ?

Solution: Given data:

Prototype	Model
	Scale ratio: $L_r = 40$
$V_p = ?$	$V_m = 2$ m/s
$F_p = ?$	$F_m = 0.5$ N
$L_p = 45$ m	

According to Froude's law,

$$\frac{V_p}{\sqrt{L_p g_p}} = \frac{V_m}{\sqrt{L_m g_m}} \qquad \left| g_p = g_m \text{ (assuming)} \right.$$

$$V_p = V_m \sqrt{\frac{L_p}{L_m}} = 2 \times \sqrt{L_r} = 2 \times \sqrt{40} = 12.64 \text{ m/s}$$

We know, force scale: $F_r = \dfrac{F_p}{F_m} = L_r^3$

or $$F_p = L_r^3 \times F_m = (40)^3 \times 0.5 = 32000 \text{ N}$$

Problem 7.25: The wave motion inside a harbour is to be studied by means of a geometrically similar model constructed to a scale of 1: 50. Neglecting viscous and surface tension effects, obtain the prototype to model scale ratios for:

           (i) Velocity   *(ii) Time*

          (iii) Acceleration and   *(iv) Force.*

In the harbour, a 1.5 m high wave travels a certain distance in 25 second. What will be the corresponding wave height in the model and what time it will take to travel the corresponding distance ?

Solution. Given data:

              Prototype   **Model**

           Scale ratio: $L_r = 50$

           Wave height in prototype: Wave height in model:

                    $H_p = 1.5$ m $H_m = ?$

          Time: $T_p = 25$ s   Time: $T_m = ?$

    (i) Velocity ratio: $V_r = \sqrt{L_r} = \sqrt{50} = \mathbf{7.07}$

    (ii) Time ratio: $T_r = \sqrt{L_r} = \sqrt{50} = \mathbf{7.07}$

    (iii) Acceleration ratio: $a_r = \mathbf{1}$

    (iv) Force ratio: $F_r = L_r^3 = (50)^3 = \mathbf{125000}$

    Now, $\dfrac{\text{Wave height in the prototype: } H_p}{\text{Wave height in the model: } H_m} = L_r$, scale ratio

  or $H_m = \dfrac{H_p}{L_r} = \dfrac{1.5}{50} = \mathbf{0.03}$ **m**

  and Time ratio: $T_r = \dfrac{T_p}{T_m}$

  or $T_m = \dfrac{T_p}{T_r} = \dfrac{25}{7.07} = \mathbf{3.53}$ **s**

7.12 TYPES OF MODELS

The hydraulic models are classified as:

    1. Undistorted models, and   2. Distorted models.

7.12.1 Undistorted Models

Undistorted models are those models which are geometrically similar to their prototypes or in other words if the scale ratio for the linear dimensions of the model and its prototype is same, the model is called undistorted model. The behaviour of the prototype can be easily predicted from the results of undistorted model.

7.12.2 Distorted Models

Distorted models are those models which are not geometrically similar to their prototypes. For a distorted model, different scale ratios for the linear dimensions are adopted. For example:

Models of rivers and harbours are usually constructed following different scales for horizontal and vertical dimensions. It is not feasible to provide geometrical similarity in these problems. Because, if geometric similarity is strictly followed the depths of flow will become too small to be measured accurately. Moreover, due to very small depths of flow, surface tensile effects may alter the flow condition. Further it is necessary to maintain a turbulent flow condition as it exists in the prototype. Hence the vertical scale is exaggerated resulting in a distorted model. As emphasized earlier the spirit behind the design of a model is that is that it is not enough if it looks like the prototype but should behave like the prototype. In other words, it is hydraulic similitude and not geometric similitude which is the main governing factor in model designs. Hence in order to achieve hydraulic similitude it may become necessary to adopt distorted models.

In general, the distortion introduced in a model may be of the following types:

- (*i*) **Geometric Distortion:** As mentioned earlier this is a distortion introduced by adopting different scales for horizontal and vertical dimensions.
- (*ii*) **Configuration Distortion:** In this case the bed slope of the model is increased by keeping it in a tilted position compared to the position of the prototype.
- (*iii*) **Hydraulic Distortion:** This distortion occurs due to a change in some hydraulic quantity such as velocity or discharge.
- (*iv*) **Material Distortion:** This type of distortion involves the use of different materials for the model and prototype, *i.e.,* surface materials, surface roughness or the medium in which the model works, may be changed.

Merits of Distorted Models

- (*i*) The necessary hydraulic similitude is obtained.
- (*ii*) Depth of flow is increased affording precise measurements.
- (*iii*) Height of waves is increased affording precise measurements.
- (*iv*) Viscous effects which are practically absent in the prototype can be practically eliminated in the model, for instance, by increasing the bed slope in an otherwise geometrically similar model, the velocity can be increased, thus decreasing viscous effects.
- (*v*) Movement of silt and sand can be satisfactorily brought about to match the corresponding behaviour of the prototype.
- (*vi*) By adopting a distorted model the size of the model can be reduced, thus simplifying the operation of the model.

Demerits of Distorted Models

(i) Due to unequal horizontal and vertical scales the pressure and velocity distribution are not truly reproduced in the model.

(ii) The wave pattern in the model will be different from that in the prototype due to depth distortion.

(iii) Slope's bends and earth cuts are not truly reproduced.

Scale Ratios for comparison of quantities in geometrically distorted models:

Let the ratio of linear horizontal dimension of prototype to corresponding linear horizontal dimension of the model be $(L_r)_H$

Let $\quad (L_r)_H$ = scale ratio for horizontal dimension.

i.e., $\qquad (L_r)_H = \dfrac{\text{Length of fluid in prototype}}{\text{Length of fluid in model}}$

$$= \dfrac{\text{Breadth of fluid in prototype}}{\text{Breadth of fluid in model}}$$

$$(L_r)_H = \frac{L_p}{L_m} = \frac{B_p}{B_m}$$

Let the ratio of linear vertical dimension of prototype to corresponding linear vertical dimension of model be (L_r)

Let, $\qquad (L_r)_v$ = scale ratio for vertical dimension.

i.e., $\qquad (L_r)_v = \dfrac{\text{Height of fluid in prototype}}{\text{Height of fluid in models}}$

(i) Scale factor for area of flow:

$$\dfrac{\text{Area of fluid in prototype}}{\text{Area of flow in model}} = \dfrac{\text{Breadth} \times \text{height of fluid in prototype}}{\text{Breadth} \times \text{height of fluid in model}}$$

$$\frac{A_p}{A_m} = \frac{B_p h_p}{B_m h_m} = (L_r)_H (L_r)_v$$

$$A_p = (L_r)_H (L_r)_v \times A_m$$

(ii) Scale factor for velocity:

$$\dfrac{\text{Velocity of fluid in prototype}}{\text{Velocity of fluid in model}} = \text{Scale factor for velocity}$$

$$\frac{V_p}{V_m} = \frac{\left(\sqrt{2gh}\right)_p}{\left(\sqrt{2gh}\right)_m} = \text{Scale factor for velocity}$$

$$\frac{V_p}{V_m} = \sqrt{\frac{h_p}{h_m}} = \left(\frac{h_p}{h_m}\right)^{1/2} = (L_r)_v^{0.5}$$

$$\boxed{V_p = (L_r)_v^{0.5}\, V_m}$$

(*iii*) Scale factor for time:

$$\frac{T_p}{T_m} = \frac{L_p / V_p}{L_m / V_m}$$

$$\frac{T_p}{T_m} = \frac{(L_r)_H}{(L_r)_V^{0.5}}$$

$$T_p = \frac{(V_r)_H}{(L_r)_V^{0.5}} T_m$$

(*iv*) Scale factor for discharge:

$$\frac{Q_p}{Q_m} = \frac{A_p V_p}{A_m V_m} = \frac{A_p}{A_m} \times \frac{V_p}{V_m} = (L_r)_H (L_r)_V \times (L_r)_V^{0.5}$$

$$\frac{Q_p}{Q_m} = (L_r)_H (L_r)_V^{0.5}$$

(*v*) Scale factor for bed slope:

$$\frac{i_p}{i_m} = \frac{h_p / L_p}{h_m / L_m} = \frac{h_p / h_m}{L_p / L_m} = \frac{n}{m}$$

$$\frac{i_p}{i_m} = \frac{n}{m}$$

$$i_p = \left(\frac{n}{m}\right) i_m$$

SUMMARY

1. Dimensional analysis is a mathematical technique that helps us to set up the empirical relation about a fluid flow or several engineering problems.
2. The fundamental quantities are described by symbols such as M for mass, L for length, T for time and θ for the temperature. The symbols L, M, T and θ are known as primary or fundamental dimensions.
3. The following two methods are used for dimensional analysis:
   (*i*) Rayleigh's method and
   (*ii*) Buckingham's π (*Pi*)-theorem.

Contd...

4. The number of variables in an equation are less than ro equal to four. Rayleigh's method for solving the equation is quite suitable but if the number of variables are more than four, the Rayleigh's method of dimensional analysis becomes more laborious. This difficulty is being overcome by using Buckingham's π-theorem. There is no restriction on the number of variables for Buckingham's π-theorem.

5. Buckingham's π-theorem states that if n numbers of variables (independent and dependent) are present in a physical process and these variables contain m numbers of fundamental dimensions $[M, L$ and $T]$ then the variables are arranged into $(n - m)$ dimensionless terms. Each term is called π-term.

6. The actual machine or structure is called prototype. A model is a small scale replica of the actual machine.

7. The following three types of similarities must exist between the model and prototype:

   (*i*) Geometric similarity

   (*ii*) Kinematic similarity and

   (*iii*) Dynamic similarity.

8. Geometric similarity is said to exist between a model and prototype if the ratio of corresponding linear dimension of the model and prototype is the same.

9. Kinematic similarity is said to exist between a model and prototype if the ratio of velocities and accelerations at the corresponding points are same.

10. Dynamic similarity is said to exist between the model and prototype if the ratios of the corresponding forces acting at the corresponding points are equal.

11. **Types of force acting in moving fluid:**

   (*i*) **Inertia force:** F_i = mass × acceleration

   $$= \rho L^2 \, V^2$$

   (*ii*) **Viscous force:** F_v = shear stress × area

   $$= \mu LV$$

   (*iii*) **Gravity force:** F_g = mass × acceleration due to gravity

   $$= \rho g L^3$$

   (*iv*) **Pressure force:** F = pressure intensity × area

   $$= p.A$$
   $$= pL^2$$

   (*v*) **Surface tension force:** F_t = surface tension × length of the surface

   $$= \sigma L$$

   (*vi*) **Elastic force:** F_e = elastic stress × area

   $$= KL^2$$

Contd...

12. Dimensionless number:

(i) **Reynolds Number:** $Re = \dfrac{\text{Inertia force} : F_i}{\text{Viscous force} : F_v}$

$$= \frac{\rho V^2 L^2}{\mu LV} = \frac{\rho VL}{\mu} = \frac{LV}{v}$$

$$= \frac{\rho VD}{\mu} = \frac{DV}{v} \text{ for pipe flow.}$$

(ii) **Froude's Number:** $Fr = \sqrt{\dfrac{\text{Inertia force} : F_i}{\text{Gravity force} : F_g}}$

$$= \sqrt{\frac{\rho V^2 L^2}{\rho g L^3}} = \sqrt{\frac{V^2}{gL}} = \frac{V}{\sqrt{gL}}$$

(iii) **Euler's Number:** $Eu = \sqrt{\dfrac{\text{Inertia force} : F_i}{\text{Pressure force} : F_p}}$

$$= \sqrt{\frac{\rho L^2 V^2}{pL^2}} = \frac{V}{\sqrt{p/\rho}}$$

(iv) **Weber's Number:** $We = \sqrt{\dfrac{\text{Inertia force} : F_i}{\text{Surface tension force} : F_t}}$

$$= \sqrt{\frac{\rho L^2 V^2}{\sigma.L}} = \frac{V}{\sqrt{\sigma/\rho L}}$$

(v) **Mach's Number:** $M = \sqrt{\dfrac{\text{Inertia force} : F_i}{\text{Elastic force} : F_e}}$

$$= \sqrt{\frac{\rho L^2 V^2}{KL^2}} = \frac{V}{\sqrt{K/\rho}} = \frac{V}{a}$$

where a is called velocity of sound in the fluid $= \sqrt{K/\rho}$

Hence Mach's number is also defined as the ratio of the velocity of fluid to velocity of sound in the fluid.

13. Model laws or similarity laws:

(i) **Reynolds Law:** According to this law; the Reynolds number for the model must be equal to the Reynolds number for the prototype.

Mathematically, $(Re)_m = (Re)_p$

$$\left(\frac{\rho VL}{\mu}\right)_m = \left(\frac{\rho VL}{\mu}\right)_p$$

$$\frac{\rho_m V_m L_m}{\mu_m} = \frac{\rho_p V_p L_p}{\mu_p}$$

Contd...

(*ii*) **Froude's Law:** The model design according to this law, the Froude's number for the model must be equal to the Froude's number for the prototype.

Mathematically, $(Fr)_m = (Fr)_p$

$$\frac{V_m}{\sqrt{g_m L_m}} = \frac{V_p}{\sqrt{g_p L_p}}$$

$$\frac{V_m}{\sqrt{L_m}} = \frac{V_p}{\sqrt{L_p}}$$

$$\frac{V_p}{V_m} = \sqrt{\frac{L_p}{L_m}} = \sqrt{L_r}$$

$$V_r = \frac{V_p}{V_m} = \sqrt{L_r}$$

where V_r = scale ratio for velocity,

L_r = scale ratio for length.

(*a*)   **Scale ratio for time:** $T_r = \dfrac{T_p}{T_m} = \sqrt{L_r}$

(*b*)   **Scale ratio for acceleration:** $a_r = \dfrac{a_p}{a_m} = 1$

(*c*)   **Scale ratio for discharge:** $Q_r = \dfrac{Q_p}{Q_m} = L_r^{2.5}$

(*d*)   **Scale ratio for force:** $F_r = \dfrac{F_p}{F_m} = L_r^3$

(*e*)   **Scale ratio for pressure intensity:** $P_r = \dfrac{p_p}{p_m} = L_r$

(*f*)   (*i*) **Scale ratio for energy or work:** $E_r = W_r = L_r^4$

(*ii*) **Scale ratio for torque or moment:** $T_r^* = M_r = L_r^4$

Scale ratios for energy, work, torque and moment are same:

i.e., $\boxed{E_r = W_r = T_r^* = M_r = L_r^4}$

(*g*)   (*i*) **Scale ratio for power:** $P_r = \dfrac{P_p}{P_m} = L_r^{3.5}$

(*ii*) **Scale ratio momentum or impulse:**

$$M_r^* = I_r = L_r^{3.5}$$

Hence scale ratios for power, momentum and impulse are same

i.e., $\boxed{P_r = M_r^* = I_r = L_r^{3.5}}$

Contd...

(*iii*) **Euler's Law:** The model design according to this law, the Euler's number for model must be equal to the Euler's number for the prototype.

Mathematically, $(Eu)_m = (Eu)_p$

$$\left(\frac{V}{p/\rho}\right)_m = \left(\frac{V}{p/\rho}\right)_p$$

$$\frac{V_m}{p_m/\rho_m} = \frac{V_p}{p_p/\rho_p}$$

(*iv*) **Weber's Law:** The model design according to this law, the Weber's number for the model must be equal to Weber's number for the prototype.

Mathematically, $(We)_m = (We)_p$

$$\left(\frac{V}{\sigma/\rho L}\right)_m = \left(\frac{V}{\sigma/\rho L}\right)_p$$

$$\frac{V_m}{\sigma_m/\rho_m L_m} = \frac{V_p}{\sigma_p/\rho_p L_p}$$

(*v*) **Mach's Law:** The model design according to this law, the Mach's number for the model must be equal to Mach's number for the prototype.

Mathematically, $(M)_m = (M)_p$

$$\left(\frac{V_m}{\sqrt{K_m/\rho_m}}\right) = \left(\frac{V_p}{\sqrt{K_p/\rho_p}}\right)$$

14. **Types of Models:** The hydraulic models are classified as:
    (*i*) Undistorted models, and
    (*ii*) Distorted models.

(*i*) **Undistorted Models:** Undistorted models are those models which are geometrically similar to their prototypes.

(*ii*) **Distorted Models:** Distroted models are those models which are not geometrically similar to their prototypes.

ASSIGNMENT - 1

1. What do you mean by dimensional analysis?
2. Differentiate between fundamental quantities and derived quantities.
3. What is a dimensionally homogeneous equation? Give examples.
4. Write down the dimensions of:
   (*i*) Specific weight   (*ii*) Surface tension
   (*iii*) Momentum   (*iv*) Bulk-modulus.

5. What are the methods of dimensional analysis ? Describe the Rayleigh's method for dimensional analysis.

6. State Buckingham's pi-theorem. Why this theorem is considered superior over the Rayleigh's method for dimensional analysis.

7. What do you mean by repeating variables? How are the repeating variables selected for dimensional analysis?

8. What do you mean by model analysis?

9. Define and explain the terms: Geometric similarity, kinematic similarity and dynamic similarity applied to model analysis.

10. Define and explain the terms: Reynolds number, Froude's number and Mach number. Derive expressions for any two numbers.

11. Explain the significance of Reynolds number. What are the fields of application of Froude's law and Reynolds law?

10. What do you mean by distorted model? Why models of rivers and harbours are made as distorted models.

13. Can a geometrically distorted model be dynamically similar to the prototype?

ASSIGNMENT - 2

1. A pump develops a power P which is a function of the discharge Q, the head H and the specific weight w of the fluid. Show that $P = KQwH$ where K is a dimensionless constant.

2. Show by Rayleigh's method the resistance R to the motion of a sphere of diameter D as it moves at a velocity V in a fluid of density ρ and viscosity μ is given by

$$R = \rho D^2 V^2 \phi\left(\frac{DV\rho}{\mu}\right)$$

3. The discharge Q through a passage of circular selection is a function of the diameter D of the passage, the pressure intensity p of the moving fluid and its density ρ. Show that

$$Q = KD^2 \sqrt{\frac{p}{\rho}}$$

4. Show that the discharge over a triangular notch of given angle is given by

$$Q = H^{5/2} g^{1/2} \phi\left(\frac{H^{3/2} g^{1/2}}{\gamma}\right)$$

5. The power P developed by a turbine depends on the diameter D of the rotor, the head H the rotational speed N, the density of the fluid ρ and the acceleration due to gravity g. Show that

$$P = \rho D^5 N^3 \phi\left(\frac{D^2 N^2}{gH}\right)$$

6. The resistance R, to the motion of a completely submerged body upon the length of the body L, velocity of flow V, mass density of fluid ρ and kinematic viscosity of fluid v. By using Buckingham pi-theorem, prove that

$$R = \rho V^2 L^2 \phi\left(\frac{VL}{v}\right)$$

7. A pipe of diameter 1.5 m is required to transport on oil of specific gravity 0.92 and viscosity 0.04 poise at the rate of 2000 litre/sec. Tests were conducted on a 130 mm diameter pipe using water at 20°C. Find the velocity and rate of flow in the model. Viscosity of water at 20°C = 0.01 poise.

Ans. V_m = 3 m/s, Q_m = 39.79 litre/s

8. A 1.25 m diameter pipe has to be provided to convey oil of specific gravity 0.85 and kinematic viscosity 2.75 × 10^{-2} stoke at 1.25 m^3/s. In order to model the flow if a 120 mm diameter pipe is used to convey water of kinematic viscosity 0.01 stoke, what should be the velocity and the discharge in the model? **Ans.** V_m = 3.85 m/s, Q_m = 0.0435 m^3/s

9. Water flows at 20°C through a venturimeter at a mean velocity of 1.25 metre per second at the throat section. The diameters of the venturimeter at the main and throat sections are 1.50 mm and 0.75 m respectively. Find the velocity at the throat section of another venturimeter whose throat diameter is 100 mm and the water flows at 10°C. Kinematic viscosity of water at 20°C and 10°C may be taken as 1 × 10^{-2} stoke and 1.30 × 10^{-2} stoke respectively.

Ans. V_m = 12.19 m/s

10. Water flows at 3 metre per second in a 200 mm diameter pipe at a temperature 120°C. Find the corresponding velocity of oil at a temperature of 40°C through a 100 mm diameter pipe for the condition of dynamic similarity.
Kinematic viscosity of water at 20°C = 0.0101 stoke.
Kinematic viscosity of water at 40°C = 0.0130 stoke. **Ans.** 7.72 m/s

11. The model of a boat is made to a scale 1: 60. The model boat has a wave resistance of 0.025 N while operating in water at a velocity of 1 m/s. Determine the corresponding wave resistance of the prototype. Find also the power required for the prototype. What velocity does this test represent in the prototype? **Ans.** R_m = 5400 N, P_m = 41.82 kW

12. A hydroplane has to move at a velocity of 30 metre per second. The resistance of the $\frac{1}{20}$ scale model was found to be 2.22 N. Assuming that the resistance is entirely due to wave formation, calculate the resistance of the hydroplane. What is the corresponding speed of the model? **Ans.** 17.76 kN

13. A 1: 16 scale model was tested in fresh water at a corresponding velocity. The prototype flying boat has to move in sea water of specific weight 10105 N/m^3 at a velocity of 25 m/s. Find the corresponding speed of the

model. If the wave making resistance of the prototype is estimated to be 5490 N, what would be the corresponding wave making resistance of the model?

Ans. $V_m = 6.25$ m/s, $R_m = 1.30$ N

14. A 1: 20 scale model of a canal is made for studying wave motion. Find the scale ratios for:

   (*i*) Time   (*ii*) Velocity
   (*iii*) Acceleration   (*iv*) Force
   (*v*) Work

Ans. (*i*) 4.47, (*ii*) 4.47, (*iii*) 1, (*iv*) 8000, (*v*) 160000

□□□

References

1. Yunus A. Cengel and John M. Cimbala; *Fluid Mechanics*, Tata McGraw Hill Publishing Company Limited, New Delhi, 2006.

2. A.K. Jain; *Fluid Mechanics*, Khanna Publishers, Delhi, 2002.

3. Shiv Kumar; *Fluid Systems*, Satya Prakashan, New Delhi, 2008.

4. H. Schlichting; *Boundary Layer Theory*, McGraw Hill, New York, 1987.

5. F.M. White; *Fluid Mechanics*, McGraw - Hill, New York, 2003.

6. S.K. Som and C. Biswas; *Fluid Mechanics and Fluid Machines*, Tata McGraw Hill, New Delhi, 2006.

7. R.K. Bansal; *Fluid Mechanics and Hydraulic Machines*, Laxmi Publications (P) Ltd. New Delhi, 2002.

8. Waren L. McCabe, Julian C. Smith, Peter Harriott; *Unit Operations of Chemical Engineering*, McGraw-Hill, New York, 2005.

9. A.K. Mohanty; *Fluid Mechanics*, Prentice-Hall of India Private Limited, New Delhi, 1994.

10. R.W. Fox and A.T. McDonald; *Introduction of Fluid Mechanics*, Wiley, New York, 1999.

11. D. Rama Durgaiah; *Fluid Mechanics and Machinery*, New Age International (P) Limited, New Delhi, 2002.

12. C.T. Crowe, J.A. Roberson and D.F. *Engineering Fluid Mechanics*, Wiley, New York, 2001.

13. B.R. Munson, D.F. Young and T. Okiishi; *Fundamentals of Fluid Mechanics*, Wiley, New York, 2002.

© The Author(s) 2023
S. Kumar, *Fluid Mechanics (Vol. 1)*,
https://doi.org/10.1007/978-3-030-99762-5

Appendices

Standard Prefixes in SI Units

Prefix	Symbol	Multiple
exa	E	10^{18}
peta	P	10^{15}
tera	T	10^{12}
giga	G	10^{9}
mega	M	10^{6}
kilo	k	10^{3}
hecto	h	10^{2}
deka	da	10^{1}
deci	d	10^{-1}
centi	c	10^{-2}
milli	m	10^{-3}
micro	μ	10^{-6}
nano	n	10^{-9}
pico	p	10^{-12}
femto	f	10^{-15}
atto	a	10^{-18}

© The Author(s) 2023
S. Kumar, *Fluid Mechanics (Vol. 1)*,
https://doi.org/10.1007/978-3-030-99762-5

CONVERSION FACTORS

Length

$$1m = 1000 \text{ mm}$$
$$= 100 \text{ cm}$$
$$= 39.37 \text{ in}$$
$$= 3.281 \text{ ft}$$
$$= 1.0936 \text{ yd}$$
$$1 \text{ cm} = 10 \text{ mm}$$
$$1 \text{ in} = 2.54 \text{ cm}$$
$$1 \text{ ft} = 30.48 \text{ cm}$$
$$1 \text{ mile} = 1.609 \text{ km}$$
$$1 \text{ yd} = 3 \text{ ft}$$

Mass

$$1 \text{ kg} = 2.204 \text{ lb}$$
$$1 \text{ lb} = 0.4537 \text{ kg}$$
$$1 \text{ kg} = 1000 \text{ g}$$
$$1 \text{ tonne} = 1000 \text{ kg}$$
$$1 \text{ tonne} = 0.984 \text{ ton}$$
$$1 \text{ ton} = 1016.26 \text{ kg}$$
$$1 \text{ slug} = 14.59 \text{ kg}$$

Area

$$1 \text{ m}^2 = 10^4 \text{ cm}^2$$
$$1 \text{ km}^2 = 0.3862 \text{ mi}^2$$
$$1 \text{ ft}^2 = 0.09289 \text{ m}^2$$
$$1 \text{ in}^2 = 6.4516 \text{ cm}^2$$

Density

$$1 \text{ kg/m}^3 = 10^{-3} \text{ g/cm}^3$$
$$= 0.0624 \text{ lb/ft}^3$$
$$1 \text{ lb/ft}^3 = 16.025 \text{ kg/m}^3$$

Volume

$$1 \text{ litre} = 1000 \text{ cc or cm}^3$$
$$= 10^{-3} \text{ m}^3$$
$$= 0.0353 \text{ ft}^3$$
$$1 \text{ m}^3 = 35.32 \text{ ft}^3$$
$$1 \text{ m}^3 = 1000 \text{ litres}$$
$$= 10^6 \text{ cm}^3$$
$$1 \text{ gallon} = 4.546 \text{ litres}$$
$$= 4.546 \times 10^{-3} \text{ m}^3$$
$$= 8 \text{ pints}$$
$$1 \text{ pint} = 568.25 \text{ cc}$$
$$= 568.25 \times 10^{-6} \text{ m}^3$$

Force

$$1 \text{ N} = 1 \text{ kg m/s}^2$$
$$= 0.102 \text{ kgf}$$
$$= 10^5 \text{ dyne}$$
$$1 \text{ dyne} = 1 \text{ gm cm/s}^2$$
$$1 \text{ kgf} = 9.807 \text{ N}$$
$$= 2.204 \text{ lbf}$$

Discharge

$$1 \text{ litre/s} = 10^{-3} \text{m}^3\text{/s}$$
$$1\text{m}^3\text{/s} = 10^3 \text{ litre/s}$$
$$= 35.32 \text{ ft}^3\text{/s}$$
$$1 \text{ cusecs} = 0.02831 \text{ m}^3\text{/s}$$

Energy and Work

$1 \text{ Nm} = 1 \text{ J} = 10^7 \text{ erg}$

$1 \text{ erg} = 1 \text{ dyn. cm}$ $1 \text{ kWh} = 3600 \text{ kJ}$

    $= 10^{-5} \times 10^{-2} \text{ Nm}$ $= 860 \text{ kcal}$

    $= 10^{-7} \text{ J}$ $= 3.6 \text{ MJ}$

$1 \text{ kgfm} = 9.807 \text{ Nm}$

    $= 7.229 \text{ ft lbf}$

$1 \text{ ft lbf} = 0.1383 \text{ kgfm}$

    $= 1.356 \text{ Nm}$

    $= 1.356 \text{ J}$

Heat

$1 \text{ kJ} = 0.2388 \text{ kcal}$

    $= 0.9478 \text{ Btu}$ $\text{Btu} = \text{British thermal units}$

$1 \text{ Btu} = 1.055 \text{ kJ}$

    $= 0.252 \text{ kcal}$

    $= 778 \text{ ft lbf}$

$1 \text{ kcal} = 427 \text{ kgf m}$

    $= 4.187 \text{ kJ}$

    $= 3.968 \text{ Btu}$

$1 \text{ cal} = 4.187 \text{ J}$

Power

$1 \text{ Nm/s} = 1 \text{ J/s} = 1 \text{ W}$

$1 \text{ kW} = 1000 \text{ W}$

    $= 860 \text{ kcal/h}$

    $= 102 \text{ kgf m/s}$

    $= 737.5 \text{ ft lbf/s}$

    $= 1.359 \text{ hp (metric)}$

    $= 1.341 \text{ hp (FPS)}$

$1 \text{ hp (metric)} = 75 \text{ kgf m/s}^2 \text{ (MKS)}$

    $= 75g \text{ watt}$

    $= 735.75 \text{ W}$

$1 \text{ hp (FPS)} = 745.70 \text{ W}$

$1 \text{ hp (metric)} = 4500 \text{ kgf m/min}$

Pressure

$$1 \ N/m^2 = 1 \ Pa$$

$$1 \ bar = 10^5 \ N/m^2$$
$$= 10^2 \ kPa$$
$$= 0.1 \ M \ N/m^2$$
$$= 1.0197 \ kgf/cm^2$$

$$1 \ mbar = 10^{-3} \ bar$$
$$= 100 \ N/m^2$$
$$= 100 \ Pa$$
$$= 10.2 \ mm$$

$$1 \ atm = 101.325 \ kPa$$
$$= 1.01325 \ bar$$
$$= 760 \ mm \ of \ Hg$$
$$= 10.33 \ m \ of \ water$$
$$= 760 \ torr$$

$$1 \ torr = 1 \ mm \ of \ Hg$$
$$= 13.6 \ mm \ of \ water$$
$$= 1.334 \ mbar$$
$$= 133.38 \ Pa$$

$$1 \ mm \ of \ water = 1 \ kgf/m^2$$
$$= 9.807 \ N/m^2$$
$$= 0.0981 \ mbar$$

$$1 \ ata = 1 \ kgf/cm^2$$
$$= 9.807 \times 10^4 \ N/m^2$$
$$= 0.981 \times 10^5 \ N/m^2$$
$$= 0.981 \ bar$$

Temperature

$$T(K) = T \ °C + 273.15$$
$$1 \ °F = 1.8 \ °C + 32$$

Dynamic Viscosity

$$1 \ Ns/m^2 = 1 \ Pa.s$$
$$= 1 \ kg/ms$$
$$= 10 \ poise$$

$$1 \ poise = \frac{1}{10} \ Ns/m^2 = 0.1 \ Ns/m^2$$
$$= dyne\text{-}s/cm^2$$

$$1 \ centipoise = 10^{-2} \ poise$$
$$= 10^{-3} \ Ns/m^2$$
$$1 \ kgf \ s/m^2 = 98.1 \ dyn\text{-}s/cm^2$$
$$= 98.1 \ poise$$
$$1 \ lbf \ s/ft = 47.847 \ Ns/m^2$$

Kinematic Viscosity

$$1 \ m^2/s = 10^4 \ cm^2/s$$
$$= 10^4 \ stokes$$
$$1 \ stokes = 1 \ cm^2/s$$
$$1 \ centistokes = 10^{-2} \ stokes$$
$$= 10^{-6} \ m^2/s$$
$$1 \ ft^2/s = 0.0929 \ m^2/s$$

Mathematical Formulae

1. TRIGONOMETRICAL FORMULAE

$$\text{cosec A} = \frac{1}{\sin A} \qquad \sec A = \frac{1}{\cos A}$$

$$\tan A = \frac{\sin A}{\cos A} = \frac{1}{\cot A} \qquad \cot A = \frac{\cos A}{\sin A} = \frac{1}{\tan A}$$

$$\sin^2 A + \cos^2 A = 1$$

$$\sec^2 A = 1 + \tan^2 A \qquad \text{cosec}^2 A = 1 + \cot^2 A$$

$$\sin(90° + A) = \cos A \qquad \sin(90° - A) = \cos A$$

$$\cos(90° + A) = -\sin A \qquad \cos(90° - A) = \sin A$$

$$\tan(90° + A) = -\cot A \qquad \tan(90° - A) = \cot A$$

$$\sin(180° + A) = -\sin A \qquad \sin(180° - A) = \sin A$$

$$\cos(180° + A) = -\cos A \qquad \cos(180° - A) = -\cos A$$

$$\tan(180° + A) = \tan A \qquad \tan(180° - A) = -\tan A$$

$$\sin(360° + A) = \sin A \qquad \sin(360° - A) = -\sin A$$

$$\cos(360° + A) = \cos A \qquad \cos(360° - A) = \cos A$$

$$\cos(360° + A) = \tan A \qquad \tan(360° - A) = -\tan A$$

$$\sin(A + B) = \sin A \cos B + \cos A \sin B$$

$$\sin(A - B) = \sin A \cos B - \cos A \sin B$$

$$\cos(A + B) = \cos A \cos B - \sin A \sin B$$

$$\cos(A - B) = \cos A \cos B - \sin A \sin B$$

$$\tan(A + B) = \frac{\tan A + \tan B}{1 - \tan A \tan B}$$

$$\tan(A - B) = \frac{\tan A - \tan B}{1 + \tan A \tan B}$$

$$\sin 2A = 2 \sin A \cos A$$

$$\cos 2A = \cos^2 A - \sin^2 A = 2 \cos^2 A - 1 = 1 - 2 \sin^2 A$$

$$\tan 2A = \frac{2 \tan A}{1 - \tan^2 A}$$

$$\sin 2A = \frac{2\tan A}{1+\tan^2 A}$$

$$\cos 2A = \frac{1-\tan^2 A}{1+\tan^2 A}$$

$$\sin 3A = 3\sin A - 4\sin^3 A$$

$$\cos 3A = 4\cos^3 A - 3\cos A$$

$$\tan 3A = \frac{3\tan A - \tan^3 A}{1-3\tan^2 A}$$

$$\sin(A+B)\sin(A-B) = \sin^2 A - \sin^2 B = \cos^2 B - \cos^2 A$$

$$\cos(A+B)\cos(A-B) = \cos^2 A - \sin^2 B = \cos^2 B - \sin^2 A$$

$$\sin\frac{A}{2} = \pm\sqrt{\frac{1}{2}(1-\cos A)}$$

$$\cos\frac{A}{2} = \pm\sqrt{\frac{1}{2}(1+\cos A)}$$

$$\tan\frac{A}{2} = \frac{\sin A}{1+\cos A}$$

$$\sin\frac{A}{2} + \cos\frac{A}{2} = \pm\sqrt{1+\sin A}$$

$$\sin\frac{A}{2} - \cos\frac{A}{2} = \pm\sqrt{1-\sin A}$$

$$\sin(A+B) + \sin(A-B) = 2\sin A \cos B$$

$$\sin(A+B) - \sin(A-B) = 2\cos A \sin B$$

$$\cos(A+B) + \cos(A-B) = 2\cos A \cos B$$

$$\cos(A+B) - \cos(A-B) = -2\sin A \sin B$$

$$\sin A + \sin B = 2\sin\frac{A+B}{2}\cos\frac{A-B}{2}$$

$$\sin A - \sin B = 2\cos\frac{A+B}{2}\sin\frac{A-B}{2}$$

$$\cos A + \cos B = 2\cos\frac{A+B}{2}\cos\frac{A-B}{2}$$

$$\cos A - \cos B = -2\sin\frac{A+B}{2}\sin\frac{A-B}{2}$$

2. DIFFERENTIAL CALCULUS FORMULAE

(a) $\dfrac{d(x)^n}{dx} = nx^{n-1}; \quad \dfrac{dx^5}{dx} = 5x^4, \quad \dfrac{d(x)}{dx} = 1$

(b) $\dfrac{d}{dx}(ax + b)^n = n(ax + b)^{n-1} \times a$

(c) $\dfrac{d(C)}{dx} = 0$

(d) $\dfrac{d(u \times v)}{dx} = u\dfrac{dv}{dx} + v\dfrac{du}{dx}$

(e) $\dfrac{d}{dx}\left(\dfrac{u}{v}\right) = \dfrac{v\dfrac{du}{dx} + u\dfrac{dv}{dx}}{v^2}$

(f) $\dfrac{d(\sin x)}{dx} = \cos x, \quad \dfrac{d(\cos x)}{dx} = -\sin x.$

(g) $\dfrac{d(\tan x)}{dx} = \sec^2 x, \quad \dfrac{d(\cot x)}{dx} = -\operatorname{cosec}^2 x$

(h) $\dfrac{d(\sec x)}{dx} = \sec x \tan x, \quad \dfrac{d(\operatorname{cosec} x)}{dx} = -\operatorname{cosec} x \cot x$

3. INTEGRAL CALCULUS FORMULAE

(a) $\displaystyle\int x^n dx = \dfrac{x^{n+1}}{n+1}$

(b) $\displaystyle\int 7dx = 7x, \quad \int Cdx = Cx$

(c) $\displaystyle\int (ax+b)^n dx = \dfrac{(ax+b)^{n+1}}{(n+1)\times n}$

(d) $\displaystyle\int_0^l x^5 dx = \left[\dfrac{x^4}{4}\right]_0^l = \dfrac{l^4}{4} - \dfrac{0}{4} = \dfrac{l^4}{4}$

(e) $\displaystyle\int \log_e x dx = \dfrac{1}{x}$

❑❑❑

Index

A

Absolute pressure 63
Adiabatic process 23
Adhesion 5
Aneroid barometer 56
Angular (or shear) deformation 241
Archimedes Principle 161

B

Barometer 55
Bellows pressure gauge 89
Bernoulli's equation 295
Bingham Plastic 11
Bourdon Tube Pressure Gauge 87
Boiling 13
Buckingham's Theorem 372
Buoyancy 159
Bulk Modulus of elasticity 21

C

Capillarity 18
Cavitation 14
Centre of buoyancy 159
Centre of pressure 97
Circulation 269
Coefficient of compressibility 22
Coefficient of discharge 317
Coefficient of contraction 339
Coefficient of discharge 340
Cohesion 6

Compressibility 21

Compressibility 21
Compressible flow 204
Continuity equation 208
Centre of Buoyancy 161
Convective acceleration 220
Control mass system 289
Control volume system 290
Current meter 350

D

Dead-weight pressure gauge 88
Density 3
Derived dimensions 362
Determination of metacentric height 168
Differential manometers 78
Diaphragm pressure gauge 87
Dilatant Fluid 11
Doublet 270
Dynamic Similarity 406

E

Elastic force 288
Equilibrium of floating bodies 166
Equipotential line 261
Eulerian method 197
Euler's equation of motion 287
Euler's equation of motion 288
Euler's equation of motion 295
Evaporisation 13

Printed in the United States
by Baker & Taylor Publisher Services